Machine Learning and Internet of Things in Fire Ecology

Christian Kaunert
Dublin City University, Ireland

Sahil Lal
Sharda University, India

Rishabha Malviya
Galgotias University, India

Manmeet Kaur Arora
Sharda University, India

Bhupinder Singh
Sharda University, India

IGI Global
Scientific Publishing
Publishing Tomorrow's Research Today

Vice President of Editorial Melissa Wagner
Managing Editor of Acquisitions Mikaela Felty
Managing Editor of Book Development Jocelynn Hessler
Production Manager Mike Brehm
Cover Design Phillip Shickler

Published in the United States of America by
IGI Global Scientific Publishing
701 East Chocolate Avenue
Hershey, PA, 17033, USA
Tel: 717-533-8845
Fax: 717-533-8661
Website: https://www.igi-global.com E-mail: cust@igi-global.com

Library of Congress Cataloging-in-Publication Data

Names: Kaunert, Christian, editor.
Title: Machine learning and internet of things in fire ecology / edited by Christian Kaunert, Rishabha Malviya, Bhupinder Singh, Sahil Lal, Manmeet Arora.
Description: Hershey, PA : IGI Global Scientific Publishing, [2025] | Includes bibliographical references and index. | Summary: "This book elucidates and explores the interface of Fire Ecology with AI, Machine Learning & IoTs, as it has emerged as a pivotal domain with transformative potential"-- Provided by publisher.
Identifiers: LCCN 2024052926 (print) | LCCN 2024052927 (ebook) | ISBN 9798369375655 (h/c) | ISBN 9798369375662 (s/c) | ISBN 9798369375679 (eISBN)
Subjects: LCSH: Forest fires--Prevention and control--Data processing. | Fire ecology--Data processing. | Forest fires--Prevention and control--Technological innovations. | Forests and forestry--Fire management. | Forest management--Technological innovations.
Classification: LCC SD421.45.D37 M33 2025 (print) | LCC SD421.45.D37 (ebook) | DDC 634.9/6180285--dc23/eng/20250122
LC record available at https://lccn.loc.gov/2024052926
LC ebook record available at https://lccn.loc.gov/2024052927

British Cataloguing in Publication Data
A Cataloguing in Publication record for this book is available from the British Library.

Table of Contents

Preface xiii

Chapter 1
Addressing the Impacts of Annual Wildfires on Biodiversity and Ecosystem Services: A Review of Global Efforts 1
Saurabh Chandra, Bennett University, India
Bhumika Muchan, Delhi High Court, India

Chapter 2
Advancing Artificial Intelligence for Wildfire Prediction and Control: New Edges in Environmental Safety and Fire Ecology 25
Tarun Kumar Kaushik, Sharda University, India
Ravish, Meerut College, India
Anurag Singh, Meerut College, India
Anjali Raghav, Sharda University, India
Bhupinder Singh, Sharda University, India
Kittisak Jermsittiparsert, Shinawatra University, Thailand

Chapter 3
AI and ML Applications in Wildlife Conservation and Forest Management: A Comprehensive Review 49
Hind Hammouch, Faculty of Legal, Economic, and Social Sciences of Fez, Morocco
Anjali Raghav, Sharda University, India

Chapter 4
Artificial Intelligence and Machine Learning Technological Tools in Enhancing Efficiency in Wildlife and Forest Conservation 69
Tarun Kumar Kaushik, Sharda University, India
Ravish, Meerut College, India
Anurag Singh, Meerut College, India
Anjali Raghav, Sharda University, India
Bhupinder Singh, Sharda University, India

Chapter 5
Discovering Advances in Using Artificial Techniques (AI) Techniques to Understand and Manage Wildfires 93
Gunjan Srivastava, Babu Banarasi Das University, Lucknow, India
Bhupinder Singh, Sharda University, India

Chapter 6
Emerging Trends of IoT in Conservation of Biodiversity: A Futuristic Approach 113
Fatima Zahra Ouariach, Abdelmalek Essaadi University, Morocco
Anjali Raghav, Sharda University, India

Chapter 7
Integrating Machine Learning Into Wildlife and Forest Conservation: Ethical Considerations and Sustainable Practices 129
Souvik Dhar, Brainware University, India
Arya Dhar, Laromed GmbH, Germany

Chapter 8
Integrating UAV-IoT Networks for Proactive Wildfire Detection: Innovations and Future Directions 157
Rishabha Malviya, Galgotias University, India
Manmeet Kaur Arora, Sharda University, India
Sahil Lal, Sharda University, India

Chapter 9
IoT Networks and ML in Fire Ecology Innovations and Impacts 175
Manjunatha Badiger, NMAM Institute of Technology, Nitte, India
Mehnaz Fathima C., Sahyadri College of Engineering and Management, India
Sumiksha Shetty, Sahyadri College of Engineering and Management, India
Smitha A. B., Sahyadri College of Engineering and Management, India
Kripa T., Sahyadri College of Engineering and Management, India
Prajna U. R., Sahyadri College of Engineering and Management, India

Chapter 10
Modernizing Wildfire Management Through Deep Learning and IoT in Fire Ecology 203
V. Valarmathi, Sri Krishna Arts and Science College, India
J. Ramkumar, Sri Krishna Arts and Science College, India

Chapter 11
Prospects and Challenges of Artificial Intelligence in Forest Protection 231
Laeeq Razzak Janjua, WSB University, Poland
Saquib Ahmed, Sharda University, India
Bhupinder Singh, Sharda University, India

Chapter 12
Role of Machine Learning and Deep Learning in Forest Management Through Data Analytics 247
Ravish, Meerut College, India
Anurag Singh, Meerut College, India
Anjali Raghav, Sharda University, India
Bhupinder Singh, Sharda University, India
Christian Kaunert, Dublin City University, Ireland
Tarun Kumar Kaushik, Sharda University, India

Chapter 13
Role of UAV-IoT Networks in Future Wildfire Detection 273
Ujjwal Agrawal, UM-DAE Centre for Excellence in Basic Sciences, Mumbai, India

Chapter 14
Wildlife and Forest Resource Management With Artificial Intelligence 301
Rhishikesh Dave, Sharda University, India
Christian Kaunert, Dublin City University, Ireland & University of South Wales, UK
Bhupinder Singh, Sharda University, India

Compilation of References 325

About the Contributors 363

Index 371

Detailed Table of Contents

Preface.. xiii

Chapter 1
Addressing the Impacts of Annual Wildfires on Biodiversity and Ecosystem Services: A Review of Global Efforts.. 1
Saurabh Chandra, Bennett University, India
Bhumika Muchan, Delhi High Court, India

Annual disasters caused by wildfires affect millions of hectares of forests, woodlands, and other vegetation, which are destructive forces that affect species and human beings. These fires lead to loss of human and animal lives, cause severe economic losses and consequently affect the ecosystem. In addition to the direct impacts, wildfires also have wide extended impacts that affect the strategies towards the attainment of sustainable development goals (SDGs) and implementation of the Paris Agreement. The Food and Agricultural Organisation (FAO) of the United Nations has stepped up its activity as a coordinator of international actions regarding the effective prevention and management of wildfires. The FAO's approach to wildfire management is grounded in the concept of Integrated Fire Management (IFM), which is structured around the 5Rs; This framework is clearly and wisely organized with incontestable stress on the preparation and prevention coupled with the reaction when the wildfires do happen.

Chapter 2
Advancing Artificial Intelligence for Wildfire Prediction and Control: New Edges in Environmental Safety and Fire Ecology 25
Tarun Kumar Kaushik, Sharda University, India
Ravish, Meerut College, India
Anurag Singh, Meerut College, India
Anjali Raghav, Sharda University, India
Bhupinder Singh, Sharda University, India
Kittisak Jermsittiparsert, Shinawatra University, Thailand

The advancements in data science, along with the advancements in digital and satellite technologies, have increased the capabilities for using artificial intelligence (AI) in the fields of forestry and wildlife. However, the rapid expansion of construction projects, agricultural activities, and urban regions presents a substantial risk to biodiversity worldwide. Therefore, using cutting-edge technology like artificial intelligence (AI) in the domains of forests and biodiversity might enhance the effective monitoring, management, and conservation of biodiversity and forest resources. This chapter aims to provide a thorough overview of the global use of AI and machine learning (ML) algorithms in the forestry sector and biodiversity protection. Moreover, this study investigates the challenges faced during the use of AI technology in the domains of forestry and biodiversity. The results of this study would motivate forest officials, scientists, researchers, and conservationists to explore the possibilities of using AI technology for forest management and biodiversity protection.

Chapter 3
AI and ML Applications in Wildlife Conservation and Forest Management: A Comprehensive Review 49
Hind Hammouch, Faculty of Legal, Economic, and Social Sciences of Fez, Morocco
Anjali Raghav, Sharda University, India

AI and ML have opened new avenues into wildlife conservation, linking efforts worldwide to improve forest management. However, a comprehensive review highlights the manifold applications of these new technologies for biodiversity conservation and natural resource management. These technologies are automated species identification and population monitoring to predict habitat degradation and anti-poaching strategies, all contributing immensely to the preservation of Wildlife. This chapter outlines how AI-powered drones and sensor networks can quickly monitor environments, find illegal actions as they happen, and collect reams of data on wildlife behavior. It also explores the opportunities for ML algorithms to study intricate ecological inter-relationships, predict climate change effects & optimize forest restoration and management activities. AI and ML provide a roadmap of the great promise that emergent technologies hold in providing new configurations in Wildlife Conservation and Forest Management to promote a Sustainable future.

Chapter 4
Artificial Intelligence and Machine Learning Technological Tools in Enhancing Efficiency in Wildlife and Forest Conservation 69
Tarun Kumar Kaushik, Sharda University, India
Ravish, Meerut College, India
Anurag Singh, Meerut College, India
Anjali Raghav, Sharda University, India
Bhupinder Singh, Sharda University, India

General enthusiasm surrounding the possibilities of AI, there are persistent concerns regarding its negative impacts, including substantial energy consumption and issues of ethics. The evaluation has looked at several uses in fields including building, transportation, healthcare, manufacturing, agriculture, and water. Among the numerous techniques used in sustainability regression, DSS-based (Decision Support

System) AI models and RL (Reinforcement Learning) are the most often used ones. The assessment also provides direction on the industrial sectors using strategies to include ideas of sustainable development into their operations.

Chapter 5
Discovering Advances in Using Artificial Techniques (AI) Techniques to Understand and Manage Wildfires 93
Gunjan Srivastava, Babu Banarasi Das University, Lucknow, India
Bhupinder Singh, Sharda University, India

Wildfires are a naturally occurring event but recent progress and usage of AI to enhance their monitoring and mitigation is revolutionizing the ability to predict, monitor, understand as well as mitigate these natural disasters. This kind of fine-granular analysis of satellite images, weather parameters and historical fire data using AI techniques which includes machine learning and deep learning can help to accurately predict how a forest-fire will behave and spread. They strengthened early warning and monitoring systems, leading to more efficient placement of firefighting equipment and evacuation plans. AI-based models help evaluate environmental consequences and post-fire recovery in relation to vegetation regrowth and ecosystem status. This chapter focuses on AI continuing to progress as intended, more and more of its capabilities will be utilized to help enhance resilience against wildfires - helping protect human communities as well nature from extinction by fire.

Chapter 6
Emerging Trends of IoT in Conservation of Biodiversity: A Futuristic Approach.. 113
Fatima Zahra Ouariach, Abdelmalek Essaadi University, Morocco
Anjali Raghav, Sharda University, India

The Internet of Things (IoT) is changing biodiversity conservation. Also, the Internet of Things (IoT) can help conservationists tackle biodiversity loss by providing them with all three: real-time data collection, information processing, and interpretation to find patterns. This chapter reviews how an IoT-based system could contribute to biodiversity conservation, focusing on habitat and ecosystem monitoring, wildlife tracking and protection, environmental monitoring, and pollution control. However, it explores the role of IoT in accessing real-time information and rapid interventions to protect ecosystems using instruments like wildlife tracking collars, anti-poaching systems, and smart air and water quality sensors. Furthermore, utilizing the capability of IoT will allow conservationists to receive accurate data, have a complete understanding, and involve everyone in the battle against nature for flourishing our precious biodiversity on this planet. It underscores the kind of power IoT has to deliver a sustainable future for humanity.

Chapter 7
Integrating Machine Learning Into Wildlife and Forest Conservation: Ethical Considerations and Sustainable Practices 129
Souvik Dhar, Brainware University, India
Arya Dhar, Laromed GmbH, Germany

This chapter has argued for clear ethical standards and good practices in the use of Machine Learning in the management of wildfire. Advanced technologies, such as Machine Learning, holds a lot of promise in enhancing wildfire management through better forecasts and improved strategies for quick response. However, the use of Machine Learning in the domain also raises some interesting and significant ethical issues, like data privacy, accountability, environmental justice, and community engagement. Before delving into ethical considerations, the chapter discusses the legal frameworks governing wildland fire management in the US, Australia, Canada, European Union, India, and other regions and how they adapted wildfire strategies in response to climate change and disaster recovery. With ethical considerations absorbed into the deployment of such technologies, stakeholders may work together to ensure that Machine Learning is a tool for equitable and sustainable management of wildfires.

Chapter 8
Integrating UAV-IoT Networks for Proactive Wildfire Detection: Innovations and Future Directions 157
Rishabha Malviya, Galgotias University, India
Manmeet Kaur Arora, Sharda University, India
Sahil Lal, Sharda University, India

The most effective solutions brought to all industries, especially for environmental monitoring and disaster management with this integration is the Unmanned Aerial Vehicles (UAVs) combined with Internet of Things (IoT) technologies. The aim of this paper is to investigate the architecture, applications and challenges of UAV-IoT networks that have potential to support real-time data collection capabilities along with communication range. First, we develop a new concept of passive monitoring using UAVs as mobile platforms fitted with advanced sensors to sense data from ground-based IoT devices in various environments efficiently. Through teamwork between researchers, industry professionals and policy makers to address the current challenges of these technologies, they can be leveraged (if successfully applied) for sustainable development with more safety.

Chapter 9
IoT Networks and ML in Fire Ecology Innovations and Impacts 175
Manjunatha Badiger, NMAM Institute of Technology, Nitte, India
Mehnaz Fathima C., Sahyadri College of Engineering and Management, India
Sumiksha Shetty, Sahyadri College of Engineering and Management, India
Smitha A. B., Sahyadri College of Engineering and Management, India
Kripa T., Sahyadri College of Engineering and Management, India
Prajna U. R., Sahyadri College of Engineering and Management, India

Fire ecology has seen great progress due to the combination of machine learning (ML) and networks connected to the Internet of Things (IoT). It has significantly boosted our chances of monitoring, predicting, and even containing these fires more effectively. This chapter investigates the incorporation of the Internet of Things (IoT) into the framework of fire ecology studies using machine learning. It focuses on several technical advances, real-world applications, and the main impacts of wildfire suppression over the last few years. Through IoT sensors, drones, and sateIoT sensors, drones, and satellites, along with highly developed ML algorithms and big data utilization, immediately collect and analyze data, leading to early fire identification, and precise fire behaviour prediction. This chapter also discusses the difficulties and limitations of these technologies, as well as offering perspectives on future advancements and possible enhancements in the sector.

Chapter 10
Modernizing Wildfire Management Through Deep Learning and IoT in Fire Ecology 203
V. Valarmathi, Sri Krishna Arts and Science College, India
J. Ramkumar, Sri Krishna Arts and Science College, India

The increasing frequency and severity of wildfires present critical challenges to ecosystems, human safety, and property and underline the inefficiency of traditional methods of fire detection and management. This chapter will present how integration between DL and IoT could give way to a revolution in fire ecology by providing innovative tools for real-time fire prediction, detection, monitoring, and response. DL, in particular, through Convolutional and Recurrent Neural Networks, looks into terabytes of data ranging from historical fire data to weather patterns and topography to predict and assess wildfire risks. IoT aids this with real-time data that emanates from a network of sensors, drones, and cameras spread across susceptible areas. This synergy therefore offers DL and IoT more accurate, timely, and proactive fire management. Future technologies will focus on 5G, blockchain, and advanced robotics for more resilient fire management strategies.

Chapter 11
Prospects and Challenges of Artificial Intelligence in Forest Protection 231
Laeeq Razzak Janjua, WSB University, Poland
Saquib Ahmed, Sharda University, India
Bhupinder Singh, Sharda University, India

The protection of Earth's ecology and balancing ecology rests heavily on forest preservation. Issues like trafficking of wildlife, illegal logging, and deforestation are still existing. Conventional methods of monitoring techniques and safety precautions have drawbacks and are inefficient to address these ecological problems. In the present era, Artificial Intelligence is advancing, and it has given a fresh hope in preserving the forest. The chapter examines the best possible use of Artificial Intelligence in forest preservation concentrating on certain situations like wildlife conservation, illegal logging surveillance and prediction of fire forests. Artificial intelligence (AI) technology makes available greater accuracy and efficiency than conventional techniques, allowing for quicker detection and reaction to forest damage activities. AI will become more significant in the future because of additional technical developments and the growth of application areas, opening new avenues for the preservation and sustainable use of forests.

Chapter 12
Role of Machine Learning and Deep Learning in Forest Management Through Data Analytics 247
Ravish, Meerut College, India
Anurag Singh, Meerut College, India
Anjali Raghav, Sharda University, India
Bhupinder Singh, Sharda University, India
Christian Kaunert, Dublin City University, Ireland
Tarun Kumar Kaushik, Sharda University, India

Everyone must create a good framework to handle environmental, financial, and social issues if we are to move towards sustainability. Many researchers have used artificial intelligence (AI) and machine learning to forward sustainable development goals by building a highly efficient system that supports a circular economy and aligns the needs of the current generation while safeguarding the capacities of future generations. Artificial intelligence has made significant progress in recent years, leading to significant changes in various industries including healthcare, transportation, agriculture, energy, and media. This paper offers a comprehensive analysis of the junction between artificial intelligence (AI) and sustainability together with several possible directions for the next investigation.

Chapter 13
Role of UAV-IoT Networks in Future Wildfire Detection 273
Ujjwal Agrawal, UM-DAE Centre for Excellence in Basic Sciences, Mumbai, India

Wildfires have posed a significant threat to ecosystems and human settlements throughout history, and their frequency and intensity have only increased in modern times due to climate change and other anthropogenic factors. This chapter explores the evolution of wildfire management, from early historical approaches to modern-day challenges. It emphasizes the growing need for prevention over reactive firefighting efforts, as prevention strategies offer more sustainable and effective long-term solutions. The chapter introduces the role of Unmanned Aerial Vehicles (UAVs) integrated with Internet of Things (IoT) networks as a transformative approach to wildfire detection and monitoring. It examines how real-time data collection, sensor networks, and machine learning algorithms enable early detection and prediction of wildfires, allowing for rapid response and prevention. The potential and limitations of these technologies are discussed, along with the challenges of implementation, such as data processing, scalability, and resource allocation.

Chapter 14
Wildlife and Forest Resource Management With Artificial Intelligence 301
Rhishikesh Dave, Sharda University, India
Christian Kaunert, Dublin City University, Ireland & University of South Wales, UK
Bhupinder Singh, Sharda University, India

The incorporation if AI in the management of wildlife and forest assets can transform conservation challenges by providing new ways to protect biodiversity and ensure ecosystem sustainability. Machine learning, computer vision and other AI technologies are to monitor wildlife populations with high accuracy. These include tools that enable the analyses of vast datasets collected from remotely sensed technologies including drones, satellite imagery for real-time biodiversity and forest dynamics. Though AI provides many benefits, there are numerous issues that must be tackled around all things from data privacy to ethical considerations and the necessity of a transactional or multi-disciplinary attempt on it. This chapter comprehensively explores the AI can be used for sustainable wildlife and forest resources that humans thought they might not manage forever because it is important to keep them alive.

Compilation of References 325

About the Contributors 363

Index 371

Preface

The intersection of fire ecology with Artificial Intelligence (AI), Machine Learning (ML), and the Internet of Things (IoT) has emerged as a dynamic and transformative domain, holding the potential to drastically reshape our approach to wildfire management. With each passing year, the devastation of forest landscapes due to wildfires intensifies, raising a global call for innovative solutions. AI, with its capacity to predict, monitor, and even prevent fire outbreaks, has become instrumental in addressing these urgent ecological challenges. The promise of AI and IoT in anticipating fire risks through advanced sensor technology and UAV-IoT systems is invaluable in mitigating the extent and frequency of these natural disasters.

This book, Machine Learning and Internet of Things in Fire Ecology, brings together perspectives from leading academics, researchers, and industry professionals to explore the essential role of AI and IoT in wildfire management. The chapters within delve deeply into how IoT enables real-time data collection and monitoring, paving the way for effective forest fire forecasts. Devices like the ECO-Guard, an AI-powered sensor system that incorporates YOLO models and various ML sensors, exemplify how technology can advance forest monitoring and support the achievement of Sustainable Development Goal 12 (SDG 12). The contributors of this volume examine the synergistic potential of these technologies to reduce fire risks, safeguard biodiversity, and support sustainable ecosystem management.

We envision this book as a valuable resource for a wide-ranging audience, including environmental policymakers, industry leaders, local government officials, fire management authorities, researchers, and students. Each chapter offers insights into AI-driven strategies, IoT applications, and machine learning algorithms, all aimed at mitigating the impact of forest fires through predictive analysis and innovative fire management approaches. This compilation of research also addresses emerging trends in UAV-IoT technologies, federated learning, deep learning, and RLSTM-NN models, each contributing to a proactive approach to wildfire management.

The content has been structured to provide a broad yet detailed exploration of these technologies within fire ecology. From policy frameworks that support sustainable wildfire management to the applications of UAV-IoT in future wildfire detection, the book covers a spectrum of topics essential to advancing fire ecology. By capturing the latest advancements and case studies, this book offers not only theoretical foundations but also practical applications of AI, ML, and IoT within the environmental sector.

Chapter Overview

Chapter 1: This chapter examines the destructive impact of wildfires on ecosystems, economies, and human life. It explores the role of the Food and Agricultural Organization (FAO) in coordinating global wildfire prevention and management efforts, emphasizing its Integrated Fire Management (IFM) framework. The 5Rs approach—emphasizing prevention, preparation, and response—is a central strategy towards achieving sustainable development goals and aligning with the Paris Agreement.

Chapter 2: With the growth of AI and digital technologies, the forestry and biodiversity sectors are seeing enhanced capabilities in conservation and monitoring. This chapter reviews AI and machine learning applications in biodiversity protection, addressing how they assist in managing challenges from urbanization and agriculture. It aims to motivate further exploration of AI's potential in conservation by showcasing its ability to improve biodiversity and forest resource management.

Chapter 3: AI and ML technologies are revolutionizing wildlife conservation efforts. This chapter outlines applications like automated species identification, habitat monitoring, and anti-poaching. Additionally, it covers AI-driven drones and sensor networks that collect critical data on wildlife and ecological relationships, offering strategies for forest management, climate prediction, and habitat restoration.

Chapter 4: This chapter addresses the broad use of AI in sectors such as construction, agriculture, and transportation, focusing on sustainability. Techniques like DSS-based models and reinforcement learning (RL) integrate sustainable practices within these industries. However, it also raises concerns about the energy demands and ethical implications of AI, guiding on sustainability best practices.

Chapter 5: The use of AI in wildfire monitoring has improved prediction accuracy and early warning systems. By analyzing satellite images, weather, and historical data, AI supports targeted response strategies for firefighting and evacuation. This chapter highlights how AI's role is growing in bolstering resilience against wildfires, aiding in both human and environmental protection.

Chapter 6: IoT is transforming biodiversity conservation by enabling real-time data collection and rapid interventions. This chapter examines IoT's applications in ecosystem monitoring, wildlife tracking, pollution control, and anti-poaching systems. It emphasizes how IoT data enhances understanding and collaborative efforts in conservation, providing a promising pathway to a sustainable future.

Chapter 7: This chapter covers the ethical considerations in applying machine learning for wildfire management, addressing data privacy, accountability, and community involvement. By reviewing legal frameworks from various regions, it emphasizes the need for ethical guidelines to ensure equitable and sustainable use of AI technologies in disaster management.

Chapter 8: Reiterating the importance of ethics, this chapter delves into the moral responsibilities of using AI in wildlife and forest management. Highlighting issues such as transparency and environmental justice, it calls for a collaborative approach to integrate ethical practices within ML applications, promoting balanced and fair use of AI.

Chapter 9: IoT combined with ML has advanced fire ecology, enabling early detection and monitoring of wildfires. This chapter explores the real-world applications of IoT sensors and drones in data collection, as well as the limitations and future potential of these technologies in reducing wildfire damage.

Chapter 10: The integration of deep learning (DL) and IoT is transforming fire prediction and response. This chapter discusses how Convolutional and Recurrent Neural Networks analyze large data sets for real-time fire management. It envisions future technologies, including 5G and robotics, to further enhance fire ecology and disaster resilience.

Chapter 11: AI brings new solutions to forest preservation, including detecting illegal logging and monitoring wildlife trafficking. This chapter emphasizes AI's ability to enhance conservation accuracy, efficiency, and responsiveness, while exploring how its future development will continue to improve forest sustainability.

Chapter 12: This chapter provides an in-depth analysis of AI's role in promoting sustainable development. Highlighting sectors like healthcare, energy, and agriculture, it examines how AI-driven systems contribute to the circular economy and sustainable practices, proposing future directions for AI's role in achieving sustainability goals.

Chapter 13: Tracing the history of wildfire management, this chapter highlights the shift from reactive to preventive strategies. It introduces UAVs and IoT for real-time fire monitoring, exploring the benefits and challenges of this approach in addressing modern wildfire threats.

Chapter 14: AI's impact on wildlife and forest conservation is profound, offering high-precision monitoring tools. This chapter covers AI applications in data privacy, ethics, and multidisciplinary collaboration for sustainable resource management. It underscores AI's potential to secure the future of biodiversity and forest health.

This book has been curated to inspire academicians, environmentalists, technologists, and policymakers to integrate modern technological advancements into wildfire management strategies. Our hope is that this volume will serve as a springboard for further research and development, encouraging a forward-thinking approach to protecting our forests and wildlife ecosystems. Together, may we advance the dialogue around fire ecology and contribute to a sustainable future.

Christian Kaunert
Dublin City University, Ireland & University of South Wales, UK

Rishabha Malviya
Galgotias University, India

Bhupinder Singh
Sharda University, India

Sahil Lal
Sharda University, India

Manmeet Arora
Sharda University, India

Chapter 1
Addressing the Impacts of Annual Wildfires on Biodiversity and Ecosystem Services: A Review of Global Efforts

Saurabh Chandra
https://orcid.org/0000-0003-4172-9968
Bennett University, India

Bhumika Muchan
https://orcid.org/0009-0003-5294-998X
Delhi High Court, India

ABSTRACT

Annual disasters caused by wildfires affect millions of hectares of forests, woodlands, and other vegetation, which are destructive forces that affect species and human beings. These fires lead to loss of human and animal lives, cause severe economic losses and consequently affect the ecosystem. In addition to the direct impacts, wildfires also have wide extended impacts that affect the strategies towards the attainment of sustainable development goals (SDGs) and implementation of the Paris Agreement. The Food and Agricultural Organisation (FAO) of the United Nations has stepped up its activity as a coordinator of international actions regarding the effective prevention and management of wildfires. The FAO's approach to wildfire management is grounded in the concept of Integrated Fire Management (IFM), which is structured around the 5Rs; This framework is clearly and wisely organized with incontestable stress on the preparation and prevention coupled with the reaction

DOI: 10.4018/979-8-3693-7565-5.ch001

when the wildfires do happen.

LITERATURE REVIEW

Annual wildfires significantly impact biodiversity and ecosystem services in forest ecosystems worldwide. While some argue that fires can increase biodiversity through mosaic pattern formation, many biologists consider wildfires destructive to biota, leading to species loss and ecosystem function deterioration (Geraskina et al., 2022; Geraskina et al., 2021). Wildfires affect various ecosystem services, including forest provisioning, water regulation, and climate regulation (Pacheco & Claro, 2023; Schmerbeck & Fiener, 2015). In tropical dry forests of India, intentionally lit fires are linked to local livelihoods but have adverse long-term effects on forest ecosystems and soil degradation (Schmerbeck & Fiener, 2015). The loss of keystone large mammal species in modern forests has reduced natural gap formation, which is crucial for maintaining biodiversity (Geraskina et al., 2022). To conserve biodiversity and ecosystem functions, alternatives to wildfires, such as special felling techniques and supporting keystone animal populations, are recommended (Geraskina et al., 2021).

Wildfires are a significant global phenomenon, annually burning an area half the size of Australia and contributing nearly 40% of anthropogenic $CO2$ emissions (Peduzzi et al., 2004). While often perceived as destructive, wildfires play a crucial role in ecosystem dynamics, providing various benefits to humankind, including the evolution of shade-intolerant plants and animals (Pausas & Keeley, 2019). These fires offer provisioning, regulating, and cultural services, with prescribed fires and management techniques mimicking their ancestral role in increasingly populated areas (Pausas & Keeley, 2019). Mediterranean ecosystems are particularly affected by wildfires, which are becoming a critical environmental factor worldwide (Chuvieco, 2009). Despite their ecological importance, wildfires pose significant threats to life and property globally. Recent research focuses on developing intelligent, software-assisted methodologies for prevention and mitigation of wildfire impacts (Adavala, 2023). Notably, while some fires occur naturally, 90% are caused by human activities (Peduzzi et al., 2004).

Wildfires are a global phenomenon with significant impacts on ecosystems, biodiversity, and climate (Bowman et al., 2009). They affect millions of square kilometers annually, causing substantial economic losses and ecosystem damage (Chatenoux & Peduzzi, 2013). Climate change is expected to increase fire occurrence and area burned in many regions, with lengthening fire seasons in temperate and boreal areas (Flannigan et al., 2009). In tropical dry forests of India, wildfires are often intentionally lit for livelihood purposes, but they can have adverse long-term effects on forest ecosystems and soil quality (Schmerbeck & Fiener, 2015). While fire's impact on biodiversity varies depending on the species involved, regular burning generally leads to biodiversity loss (Schmerbeck & Fiener, 2015). Fire also affects ecosystem services, potentially reducing forests' ability to regulate hydrological cycles (Schmerbeck & Fiener, 2015). Improved fire data and research on human factors are needed to better understand and manage global fire activity (Flannigan et al., 2009).

Wildfire is a global phenomenon with significant implications for sustainable development and climate change mitigation. While not explicitly mentioned in the United Nations Sustainable Development Goals (SDGs), fire plays a crucial role in ecosystem services, air quality, and land degradation (Martin, 2019).

While some argue that fires can increase biodiversity through mosaic pattern formation, many biologists consider wildfires destructive to biota, leading to species loss and ecosystem function deterioration (Geraskina et al., 2022; Geraskina et al., 2021). Wildfires affect various ecosystem services, including forest provisioning, water regulation, and climate regulation (Pacheco & Claro, 2023; Schmerbeck & Fiener, 2015). In tropical dry forests of India, intentionally lit fires are linked to local livelihoods but have adverse long-term effects on forest ecosystems and soil degradation (Schmerbeck & Fiener, 2015). The loss of keystone large mammal species in modern forests has reduced natural gap formation, which is crucial for maintaining biodiversity (Geraskina et al., 2022). To conserve biodiversity and ecosystem functions, alternatives to wildfires, such as special felling techniques and supporting keystone animal populations, are recommended (Geraskina et al., 2021).

Forest fires in India pose significant threats to biodiversity, ecosystem services, and local economies. Annually, about 1.4% of India's tropical dry forests burn, impacting forest provisioning and water regulation services (Schmerbeck & Fiener, 2015). These fires, often intentionally lit, have short-term benefits for livelihoods but cause long-term ecosystem degradation (Schmerbeck & Fiener, 2015). In Uttarakhand, forest fires put approximately 700 species at high risk of extinction (Binjola et al., 2022). The central forest cluster, covering 28% of India's forests, accounts for 56% of the total burnt area, making it the most vulnerable region (Sewak et al., 2021). Forest fires in India have wide-ranging impacts on biodiversity, climate, forest species, and soil characteristics (Gupta & Singh, 2023; Sewak et al., 2021). The economic impact is estimated at over INR 1101 crore annually (Sewak et al., 2021). Promoting changes in demand-side behavior could help limit potential trade-offs between climate action and other SDG objectives (Moreno et al., 2023).

The implementation of the Paris Agreement and the Sustainable Development Goals (SDGs) are closely linked, with potential for co-benefits and synergies (Gómez-Echeverri, 2018; Brandi et al., 2017). While fire is not explicitly mentioned in the SDGs, it significantly impacts ecosystem services and contributes to climate change through emissions, affecting multiple goals (Martin, 2019). Some SDG efforts, such as infrastructure development for vulnerable communities, may increase greenhouse gas emissions, making it harder to limit global temperature rise to 1.5°C (Seroa da Motta, 2019). However, other targets, like sustainable agriculture and resource-efficient cities, can support climate mitigation efforts (Seroa da Motta, 2019). Education is crucial for empowering individuals towards climate action (Seroa da Motta, 2019). Challenges in implementation include knowledge gaps, finance, and governance issues (Gómez-Echeverri, 2018). Stronger linkages between climate and development initiatives are necessary to avoid harmful trade-offs and maximize benefits (Gómez-Echeverri, 2018).

India has made significant strides in aligning its development agenda with the Sustainable Development Goals (SDGs) and Paris Agreement commitments. The country has integrated SDGs into its national policy framework, focusing on five P's: People, Planet, Peace, Prosperity, and Partnership (Singh, 2021). India's climate action efforts include the National Action Plan for Climate Change and Intended Nationally Determined Contributions (Sharma & Parameswaran, 2023). As the world's third-largest greenhouse gas emitter and fastest-growing economy, India faces the challenge of balancing development needs with climate action (Chitre, 2016). The country has made progress in renewable energy installation and is working towards climate neutrality (Sharma & Parameswaran, 2023). However, implementation challenges remain, including varying national priorities and the need for inclusive development (Khalid, 2023). Creating synergies between SDGs and climate action offers opportunities for global climate action based on just energy transformation (Khalid, 2023).

Forest fires have significant impacts on biological diversity and ecosystems in India. While some fires can have positive effects on wildlife in boreal forests (Nasi et al., 2002), severe fires can devastate forest vertebrates and invertebrates, leading to declines in species dependent on fruit trees (Nasi et al., 2002). In India, which supports 16% of the global population with only 2.5% of the world's land area, forest fires cause extensive damage to ecosystems and economies (Gupta & Singh, 2023). The central cluster of India's forests is most vulnerable, accounting for 56% of burnt areas despite covering only 28% of forest cover (Sewak et al., 2021). Large fires can affect ecosystems for extended periods, with partial recovery taking over a decade in some cases (Barabde & Mangle, 2022). The Indian government has implemented various strategies to detect and control forest fires, recognizing their significant economic impact, estimated at over INR 1101 crore annually (Sewak et al., 2021).

Wildfires have significant impacts on ecosystems and wildlife. They can cause direct mortality and injury to animals, as well as alter habitats and vegetation structure (Tekalign, 2016; Garcês & Pires, 2023). While fires can improve habitats for some species through post-fire succession, they can also lead to habitat fragmentation and reduced biodiversity (Tekalign, 2016; Gajendiran et al., 2023).

INTRODUCTION

Many forest ecosystems depend on fire as a natural and essential component of their operations. For thousands of years, humans have employed fire as a tool for managing their land. One of the natural factors that has shaped plant communities over time is fire, which also plays a crucial role in preserving the well-being of some ecosystems because it is a natural process. But by the latter half of the 20th century, El Niño frequency increases coupled with changes in the human-fire dynamic had made fires a serious threat to many forests and their biodiversity. Wildfires in the 1980s and 1990s destroyed tropical rainforests and cloud forests, which normally do not burn extensively.

Wildfires are any unplanned or uncontrolled fire affecting natural, cultural, industrial and residential landscapes. Wildfires, also known as unplanned or uncontrollably spreading landscape fires, are primarily caused by human activity and primarily originate from burning live or dead vegetation in natural or anthropogenically altered ecosystems. They can also be started by natural causes, such as lightning, volcanic activity, or coal seams burning. These include protected areas such as wilderness and conservation sites, as well as cultivated areas such as plantations, agricultural and pasture lands, forests, grasslands, bush (shrub, scrub), and organic terrain (Tedim, F., & Leone, V. 2020).

Forest fires affect biological diversity in a number of ways. Firstly, they are a major source of carbon emissions that contribute to global warming, which may alter biodiversity. Secondly, smoke from fires can significantly reduce photosynthetic activity and can be harmful to human and animal health. Finally, forest fires affect biomass stocks and modify the hydrological cycle, which can have an impact on marine systems like coral reefs and affect the functioning of plant and animal species. One of the most significant ecological consequences of burning is the increased likelihood of more fires in the years to come, as fallen trees expose the forest to sunlight-induced drying and increase the fuel load by attracting more species that can catch fire, like pyrophytic grasses. Recurrent burning has negative effects because it contributes significantly to the decline in biodiversity in rainforest ecosystems. Insect colonisation and infestation can occur after fires, upsetting the natural equilibrium. One of the most detrimental ecological effects of fires in tropical rain forests is the replacement of large tracts of forest with pyrophytic grasslands. Parts of Amazonia and Indonesia have previously seen these processes.

A few fire-resistant tree species and weedy grasses cover the once-dense, evergreen forest, turning it into a poor forest (Bhatt, D. 2022).

Causes of Wildfires

The risk of a fire developing is driven by three main factors:

- Dry fuel such as leaves, grass, branches, and other organic materials
- Oxygen in the air
- Heat to ignite and burn

Lightning is the most common ignition source that causes the vast majority of wildfires. Most wildfires are started by lightning, which is also their most frequent ignition source. There are two types of lightning: cold and hot. Since cold lightning rarely lasts long, it is rarely the source of wildfires. It is not possible to say the same about hot lightning, as its currents have lower voltage but last longer. Most natural fires are caused by hot lightning because of the extreme heat it produces. Despite the full unpredictability of this natural phenomenon, proper planning for landscape fire control and land management can greatly reduce the intensity of wildfires, avoid needless fatalities, and stop the displacement of people and animals. Without a doubt, the other main cause of intense storms and lightning is climate change. The land surface gets warmer with longer and warmer summers. Together with rising carbon emissions, this results in greater updrafts, which raise the possibility of more frequent and intense lightning strikes. According to a 2014 study, for every degree Celsius that the temperature rises, the probability of lightning strikes increases by 12% (Touge, Y. 2022).

One of the three things a wildfire needs to start is fuel. This frequently manifests as dry vegetation. Wildfire-prone vegetation is frequently created by high temperatures and little precipitation throughout the winter. This was the situation in California in 2021, as dry vegetation increased by 65% in a matter of months. In a similar vein, wildfires are common in many parts of Australia, where the environment is hot and dry and there has been a consistent drop in rainfall since 1970. 2019 had 43 exceptionally warm days, making it the warmest year ever recorded. Coincidentally, the nation encountered a wildfire disaster that same year, which caused 11 million hectares of parks, forests, and scrub to be destroyed in the states of New South Wales and Victoria (Lowry, R. 2021).

Whether on purpose or accidentally, humans, to a very large extent, are responsible for wildfires. A variety of human-caused incidents can start fires, including campfires, equipment malfunctions, engine failure, debris burning, careless cigarette disposal on dry land, and other deliberate acts of arson. One of the most frequent causes of wildfires is the latter.

40% of wildfires that impact British Columbia on a yearly basis are caused by humans, according to official sources. According to data from the National Park Service, human activity is to blame for almost 85% of the roughly 100,000 wildland fires that occur in North America each year, more than twice as many in the US. Here, the duration of North America's fire seasons has increased from 46 to 154 days between 1992 and 2012 due to man-made fires. During the course of the 21-year study, arson and debris burning were the main causes of fires; campfires and fireworks accounted for "only" 5% of all fires. Additionally, a study of more recent California burns revealed that wildfires started by humans are more severe and devastating than those caused by nature because they spread more than twice as quickly as fast, spreading about 1.83 kilometres per day.

Impact of Wildfires on Ecosystem and Wildlife

There are many different effects of wildfires on the ecology. The destruction of vegetation and fauna is immense. Reduced variety leads to a simpler and poorer habitat. Forests are replaced by grass and shrubs. Higher soil temperatures brought on by a loss of forest cover can have an impact on animal behaviour and plant growth. Birds and small mammals that nest in cavities are affected by rising temperatures. Extreme temperatures caused by dead trees in nest cavities have an impact on the survival of young animals and birds that are heat-sensitive as well as the incubation of eggs.

Animal injuries and the destruction of breeding and nesting grounds, shelters, and food supplies are strongly linked to the negative impacts of forest fires. Even while certain animals can help prevent wildfires, they nevertheless experience stress. Studies on the immediate and long-term impacts of stress, however, are scarce. Slower-moving species, such as turtles, badgers, and elderly and young animals that are unable to flee, are usually the most affected. Furthermore, stress impedes the population's ability to recover and reproduce because wildfires frequently happen in the late spring or summer.78% of the Pantanal (Brazil) jaguar range was destroyed by fire in 2020. Jaguar populations suffered as a result of the year's high fire activity and those of previous years (Prado, P. I. 2022).

Temporary displacements that complicated the search for new, suitable habitats, an increase in territorial conflicts, and a decline in survival and reproductive success were the main outcomes noted. A lot of animals have also perished from malnutrition, thirst, and other causes, which has made the already endangered jaguar population in Pantanal even more vulnerable. Depending on how much the fire changed the habitat's structure and food source, the animals that fled the area may come back.

Birds, reptiles, and non-burrowing animals might come back in a few days or hours. When animals return to a burned area after a fire, they examine the surroundings to decide where they might dwell. Some creatures are unable to survive and migrate to other locations, while some can return to benefit from the new habitat and modify their diet and behaviour (Verling, E. 2024).

Not all animals are impacted by stress factors after a fire in the same way. The impact on animals will depend on a number of elements, including the size and duration of the fire, the kind of vegetation, the climate, smoke, and water bodies. The animal's species, physical state, age, physiological state, reaction to stress, access to food and water, and injuries (such as burns or smoke inhalation) are additional factors to take into account. As a result, it's important to take into consideration a variety of subtleties in both animals and fireplaces that affect how likely an animal is to become unwell following a fire.

The fire causes an alteration in critical soil biological processes because exposed soil experiences increased light, temperature, and wind. Examples include a decrease in humidity, the release of carbon and nitrogen into the atmosphere, the deposit of ash and charcoal, and other physicochemical changes (such as population shifts and bacterial and fungal activity) in soil. This leads to increasing rates of tree fall, increased canopy fracture, plant diversity differentiation, and shift of the vertical foliage density stratification downward. Thus, housing and food are decreased, which causes a change in the distribution of wildlife. Furthermore, the Insufficient coverage renders tiny species more vulnerable, including insects, lizards, mice, and amphibians exposed to the elements and readily pursued by predators (Hongtu, X. I. E. 2023).

Nonetheless, several ecosystems have learnt to tolerate and even benefit from small-scale forest fires. In certain places, ash acts as a natural soil fertiliser and promotes the germination of fresh seeds. By removing unhealthy and dead vegetation and preventing overgrown forests, fires promote the growth of healthier plants by letting more sunlight reach the ground. By forming an island with a variety of microhabitats, it permits a higher degree of plant diversity, which contributes to an improvement in ecosystem resilience. Fires help certain animal species with their nesting and foraging habits. For certain species, the deadwood on the ground can serve as a cavity nest, food source, and place to hide. In this downed timber, insects, small mammals, birds, amphibians, and reptiles seek cover and protection. Because their food is more plentiful or exposed in burnt areas than in unburned ones, predators and scavengers are frequently drawn to these areas (Merino, A. 2022).

A fire's abundance of carcasses can be advantageous to omnivorous animals like vultures. For instance, because dead wood contains a large number of beetles and other invertebrates, woodpeckers are drawn to places that are burning.

Along with the destruction and loss of forests caused by blazes, there are immediate and long-term environmental impacts that dramatically affect vital resources. Burning forests release significant quantities of smoke into the air. This smoke consists of a complicated mixture of gases, tiny particles, and water vapour. The smoke particles are usually very tiny, typically less than 2.5 micrometres in size, which is about 1/70th the width of a human hair. PM2.5 is the official term for particulate matter that is less than 2.5 micrometres in size. Our bodies struggle to filter out these tiny particles because they are very small. They reside in the lungs, impact our breathing, and strain our hearts. Furthermore, they have the ability to annoy our eyes and result in nasal congestion. Individuals with long-term respiratory or heart conditions face the greatest potential health risks when exposed to a wildfire. Smoke pollution from fires can come from nearby fires or be carried by wind from fires far away. Smaller fires in the area usually have less force propelling the smoke column upwards. In these instances, the smoke lingers in the nearby area of the burn. Enormous wildfires in regions such as Northern California and Oregon are highly energetic. The smoke from these fires is pushed up into the sky and carried by the prevailing winds. After cooling down, the plume will come to rest. The smoke is commonly found resting in the valleys of Northern Utah. In conclusion, carbon monoxide is the result of forests burning incompletely. It reaches its peak during the slow-burning phases of a fire and can pose several threats to human health (Ostad-Ali-Askari, K. 2022).

The loss of standing cavity trees and dead logs on the ground is detrimental to the majority of small mammal species (such as tarsiers, bats, and lemurs) as well as cavity-nesting birds. Fires can drive territorial birds and mammals away, upsetting the local balance and ultimately leading to the extinction of wildlife because displaced individuals have nowhere to go. The 1998 Russian Federation fires caused high water temperatures and carbon dioxide levels in lakes and waterways, which negatively impacted salmon spawning. In regions where burning occurs frequently, protecting a variety of microhabitats can significantly contribute to biodiversity conservation.

The decline in fruit trees leads to a decrease in bird and animal species that depend on fruits for food, especially in tropical forests. Following the 1982-1983 fires in Kutai National Park, East Kalimantan, there was a significant decrease in fruit-eating birds like hornbills, while insectivorous birds like woodpeckers became more prevalent due to the increased number of wood-eating insects. Forests that have been burned lose small mammals, birds, and reptiles, and carnivores typically steer clear of these areas. Decreases in small mammal densities, like rodents, can harm the food source for small carnivores. Fires also eradicate leaf litter and the arthropod community it supports, leading to less food for omnivores and carnivores.

Wildfires have the potential to impact the quality of streams, rivers, lakes, and reservoirs in terms of their physical, chemical, and biological characteristics. These changes remain visible for many years, even decades following a fire. The primary and most important consequence of wildfires is a notable rise in stormwater runoff. Due to the vegetation loss in a wildfire, the soil becomes hydrophobic, which means it repels or doesn't mix with water. In regular circumstances, plant life decelerates the rate of precipitation upon reaching the ground and allows it to slowly penetrate the soil. Following a fire, soil without foliage does not easily soak up water. The burn scars enhance runoff and create a way for debris and sediment to be carried to rivers, lakes, and reservoirs. Nutrient concentrations in waterbodies rise as debris accumulates. Nitrogen and phosphorus levels from scorched leaves flow into the water, causing algal blooms that, in turn, cause changes in oxygen levels or oxygen exhaustion in water bodies downstream. Finally, flash floods after a fire bring large amounts of heavy metals from ash, soil, and geological sources into rivers (Paul, M. J, 2022).

GLOBAL EFFORTS IN TACKLING WILDFIRES

The main duty for stopping spontaneous wildfires and controlling scheduled fires rests with authorities. Nevertheless, international organizations and nongovernmental organizations can assist countries in their endeavours. Multiple UN agencies and programs collaborate on issues concerning the management of wildfires:

- FAO: sustainable forest management, fire management and community involvement, promotion of international cooperation in wildland fire management; jointly with GFMC conducting Global Forest Fire Assessments (FRA 2000 and 2005), publication of Fire Management Guidelines for Temperate and Boreal Forests, Wildland Fire Management Terminology
- Office for the Coordination of Humanitarian Affairs (OCHA) with United Nations Environment Programme (UNEP): coordination of international response to wildland fire emergencies
- World Health Organization (WHO): protection of human health against adverse effects of vegetation fire, smoke pollution
- World Meteorological Organization (WMO): early warning of precursors leading to critical fire situations

Furthermore, a number of international agreements, such as the Ramsar Convention on Wetlands, the Convention on Biological Diversity (CBD), the Convention to Combat Desertification (UNCCD), and the Framework Convention on Climate

Change (UNFCCC), are mandated to counsel and support nations in safeguarding the world's vegetation cover and ecosystem functioning. Due to their limited funding, UN organisations and convention secretariats typically aren't able to accommodate every request from nations seeking support and technology transfer to advance sustainable fire management. Governments and civil society must work together, both bilaterally and multilaterally, through cooperative initiatives. A global policy conversation was facilitated by the creation of an international platform, given the multiplicity of duties both inside and outside the UN system. Under the United Nations International Strategy for Disaster Reduction (ISDR), a Working Group on Wildland Fire was formed in 2001 as part of the Inter-Agency Task Force for Disaster Reduction (Kust, G. 2022).

An international wildland fire summit was held in Sydney, Australia, in October 2003 as a follow-up to the World Summit for Sustainable Development (WSSD) in Johannesburg, South Africa, in 2002, which laid the foundation for the creation of an action plan to lessen the detrimental effects of wildland fires on the environment and people. The summit's subject was "Sustainable Development and Fire Management: Enhancing International Collaboration to Mitigate the Adverse Effects of Fire on Humanity and the Global Environment." The Summit advocated for agreements on shared global strategy, protocols, and guiding principles to strengthen international cooperation in wildland fire management. International collaboration can take place through a number of channels, including conventions, resolutions passed by the UN General Assembly, and voluntary agreements.

The UN-ISDR Working Group on Wildland Fire produced recommendations, and these suggestions, together with other outputs, led to the creation of the Global Wildland Fire Network, which facilitated regional and international discourse. Facilitating this process are the UN-ISDR Wildland Fire Advisory Group (the replacement for the Working Group of Wildland Fire) and the Global Fire Monitoring Centre (which functions as convener and secretariat).

A strategic "Framework for the Development of the International Wildland Fire Accord" was agreed upon by FAO, ISDR, and GFMC in May 2004. Several regional conferences and consultations were held in 2004 to support this process, including the Pan-American Conference on Wildland Fire, the ministerial-level consultations of the Association of South East Asian Nations (ASEAN), and events in Northeast Asia, Eastern Mediterranean/Balkans/Near East/Central Asia, Baltic, Sub-Saharan Africa, and South America. The final regional consultation before the FAO Ministerial Meeting on Forests, which will take place in Rome on March 14, 2005, with a focus on wildland fire, is being held on the premises of the Central Asian Forest Congress. Ministers in charge of forestry will take into account a request for support made by the regions to the international community (Goldammer, J., 2021).

There is increasing recognition that fire management should involve an integrated approach with five key elements (also known as the five Rs):

1. **Review (monitoring and analysis):** In an integrated approach to fire management, an understanding of the causes of fire will help focus prevention and preparedness activities. Data on fires should be recorded over time to determine the main causes and identify trends. The main questions to be answered in a fire investigation are: How do fires start? Who is involved?

The European Union's harmonized fire causes classification scheme has six categories:

a) Unknown: wildfires with no cause found
b) Natural: wildfires of natural origin, with no human involvement
c) Accident: wildfires that were unintentionally and indirectly caused by humans without the use of fire
d) Negligence: wildfires unintentionally caused by humans using fire or glowing objects
e) Deliberate: wildfires caused intentionally by humans with the use of fire
f) Rekindle: wildfires caused by the re-ignition of a previous fire, due to latent heat or embers

2. **Risk reduction (prevention):** However, eradicating or at least reducing the occurrences or the effects of wildfires requires that various methods be employed so that they fit the modern conservationist methodologies as well as traditional ones. Another aspect is strictly regulating fire for what may be highly advisable for its conservation and Natural purposes such as target shooting. This includes conducting burns when it is appropriate and the weather conditions are good, as well as making sure there are adequate measures in case of an eventuality. Besides prescribed burns, there are other methods, such as conservation and climate-smart agriculture, which can decrease the application of fire in land preparation. Effective grazing can also assist in controlling the built-up of fuel loads, and subsequently minimize the use of fire needed for recycling vegetation growth on the pastures. Other important prevention measures include fuel breaks around structures such as houses or villages and non-combustible building materials. This approach, together with measures that include the establishment of fire separation lines and the use of fire-resistant species in plantations, assist in controlling fires. It is also important for the various target groups including the youths in the community to be educated on the same. How by raising awareness

of the threats of fire and methods of fighting fire within the new generation of land managers, we ensure a better prepared and knowledgeable future.

However, preventing wildfires sometimes aggravates the problem of deforestation, and encouraging other economic activities can go a long way in stopping this act. Sometimes, people set fire to the bushes in order to use them for cultivation, and this is answered by ensuring that employment opportunities are created from forests. Last but not least, regulatory policies and legislation that are harmonized and logical are required. The above activities should be supported by these frameworks in the following ways: By ensuring that the rules governing the use of fire are clear and harmonized. The frameworks should facilitate inter-sectorial collaboration in addressing issues related to fire risks.

3. **Readiness (preparedness):** Communities and government agencies must be ready to respond when wildfires begin. It is important to establish clear responsibility for coordinating action in firefighting, as multiple groups may be involved. In numerous nations, local entities are given the authority to coordinate if the fire is under a specific size. Coordination mechanisms at the subnational or national level might be necessary for combating big fires; occasionally, international coordination could also be essential. The Incident Command System, widely utilized, aids in the coordination and organization of fire suppression efforts. A key aspect of being prepared for fires is making sure that enough individuals, both in local areas and within specific regional or national organizations, are educated in fire prevention and management methods and possess effective teamwork abilities. Proper preparedness for fires means ensuring all individuals involved in fire suppression, including foresters, community members, civil defence personnel, and professional firefighters, have appropriate tools and receive training on how to use them effectively. All staff must receive training in safety protocols and possess a minimum level of proficiency in first aid.

Fire danger rating systems utilize simple daily weather information to assess the possibility of wildfires; they are frequently employed to gauge the extent of fire risk and issue advance alerts about the potential for severe wildfires. Forecasts enable issuing fire danger warnings several days before severe fire weather arrives. Providing timely alerts with detailed spatial and temporal information, along with uncertainty indicators and the probability of extreme events, allows forest and land management agencies, landowners, and communities to proactively address fire risks through prevention, detection, and preparedness plans.

4. **Response (suppression):** Different methods, including satellite imagery, fire observation towers, aerial surveillance, lightning detection systems, and reports from community members, can be used to identify forest fires. Local people who are aware of the dangers of wildfires and take part in community fire management initiatives can be very valuable in detecting fires. Good communication is crucial in informing firefighters and managers about the location, size, and burning conditions of fires once they are detected. In a perfect system, dispatch centres with backup energy sources receive information about fire starts and locations to alert and send fire suppression personnel to individual fires. Dispatch centres also furnish firefighters with routine updates on alterations in weather predictions, fire activity, tactics, and the incident management framework. They oversee the fire status and take direction from the incident controller or commander regarding extra resources and support. Efficient, timely communication with the public is crucial to make sure people are well-informed about fire updates and to help them steer clear of dangerous situations. Local media, including radio, television, and newspapers, along with traditional and new information technologies, should be included in communication strategies.

The first stage of fire suppression, known as the initial attack, plays a crucial role in determining the outcome of a fire management program. If the first assault is effective, then the majority of other components of the program will also succeed. Without proper planning, policies, prevention, fuel management, community involvement, and detection, the initial attack phase will fail.

5. **Recovery:** Recovery includes restoring lost assets and livelihoods, as well as repairing the environment. Reconstructing homes, generating additional jobs, and assisting areas in revitalizing their agricultural and wooded areas could form a vital part of the post-wildfire recovery efforts. Replanting and reseeding in ecologically sensitive areas can prevent invasive alien species from taking root in the vast areas of bare soil left by wildfires. It may be feasible to recover timber from trees killed by fire in planted and natural forests where commercial activities are scheduled, and a comprehensive reforestation strategy could be necessary.

REVIEW OF THE FRAMEWORK PERTAINING TO THE WILDFIRES

A. The United States of America

One of the duties of the federal government of the United States is to manage wildland fires. It is responsible for managing public lands where controlled burns and wildfires take place, helping to put out wildfires, and conducting research to better understand the effects of fires. This measure is a proactive approach to addressing major fires, with strategies for both preventing and managing them (De Abreu, S. J. 2021).

The Cohesive Strategy has become an effective policy for interagency wildfire management. However, some scholars have raised concerns about the effectiveness of the governance system in the U.S. Changes in fire behaviour pose a greater threat to houses in the Wildland-Urban Interface, leading to an increased emphasis on suppression response at the expense of prevention efforts to mitigate risk. In the U.S., one of the key drivers in the social-ecological system is funding priorities that favours suppression activities over prevention and preparedness activities. There is a need for policies and practices that are better aligned with the social and ecological realities in the areas where wildfires are expected to be a problem. Therefore, gaining a deeper understanding of how current policies and institutions correspond with the socio-ecological system can result in more sustainable, long-term alternatives to the wildfire issue (Fillmore, S. D, 2024).

B. Canada

In Canada, wildfire management agencies have developed over time and continue to operate independently as individual organizations focused solely on controlling wildfires. At a national scale, the federal government is in charge of emergency management and has set up the Canadian Interagency Forest Fire Center (CIFFC) to oversee fire management and responses nationwide. Every province and territory has its own agency and policies for managing wildfires. Additionally, Indigenous populations in Canada are exposed to higher dangers due to wildfires. These communities hold important fire management knowledge and practices. Numerous obstacles, including legal, political, and attitudinal factors, impede the revival of Indigenous fire management, impacting the wildfire management system in the nation. Due to the rising number and severity of wildfires in Canada, the wildfire season has transformed as well. Recent wildfires that are unprecedented have exceeded the ability to control them, underlining the critical requirement for improved methods and resources to reduce the risk of wildfires. The shift to a risk-based strategy for responding appropriately is a result of severe wildfire seasons, reviews, policy recommendations, and acknowledgement of wildfires' ecological importance and cost control necessity. The Wildfires Act is the main federal law that controls how forest fires are managed on federal lands. It sets rules for regulating fires, consequences for not following them, and permission to use fire for managing forests.

The Canadian government has implemented an all-hazard approach to handle both natural and human-caused emergencies in order to effectively respond to all types of disasters (Public Safety Canada). Public Safety Canada helps coordinate federal agencies to implement a comprehensive emergency management approach on federal lands and properties.

C. Australia

In the past, Australia changed from traditional aboriginal-controlled fires to a firefighting strategy for managing wildfires. Nevertheless, in recent times, serious fires have led to a shift towards combining wildfire prevention and management of wildfire risks in order to enhance Australia's ability to withstand fires (Santos, J. L., 2022).

The current focus is mainly on safeguarding life and property while also preserving the ecosystem services of Australian ecosystems (Forest et al. Group, 2014). Nevertheless, widespread land fragmentation in Australian towns has led to multiple wildland-urban interface zones, consequently putting many lives and properties at risk of wildfires. In order to reach the intended goal, existing wildfire strategies aim to combine wildfire control and spatial planning strategies by including performance-based strategies, fire risk education, and self-protection measures, with a specific emphasis on wildland-urban interface regions. All states and territories have signed The National Bushfire Management Policy Statement (Forest Fire Management Group, 2014) as a commitment to carry out efficient fire management practices. Yet, the progress has been sluggish and incomplete, resulting in recent revisions and the incorporation of essential performance metrics to aid in its implementation (Forestry Australia and Forest Fire Management Group, 2022). The intricacy of implementing policies can be seen in the example of prescribed burns. Moreover, the National Construction Code (NCC), an essential regulatory tool, is continuously reviewed and adjusted to suit the requirements of various states and territories. Finally, the strategy of “Stay and Defend or Leave Early” demonstrates both collective and personal accountability in risk management protocols. Yet, following the destructive fires of 2009, the approach was modified to “Get ready, take action, stay alive”.

D. India

Forests are included in the concurrent list of the Indian Constitution. Both the Central Government and State Governments have the authority to pass laws on the matter. The Government of India is primarily responsible for policy planning and finance matters. The management of the forests falls under the jurisdiction of the different state governments. The state Government is directly responsible for managing the country's forest resources. Consequently, the state forest departments implement fire prevention and control measures. Every State and Union Territory has their individual forest department. The Inspector General of Forests and Special Secretary to the Government of India leads the professional forest service at the national level in the Government of India. The Inspector General of Forests and special Secretary are supported by an additional Inspector General of Forest and Deputy Inspector Generals. Forest Fire prevention is overseen by the Forest Protection Division in the Ministry, led by a Deputy Inspector General of Forests. The Ministry is carrying out a program named “Modern Forest Fire Control Methods” in India, through which financial aid is given to state governments to prevent and control fires (Attri, V., 2020).

India has been utilizing scientific forest management techniques for more than 130 years. Forestry techniques have been created for various types of forests and species in India. Forest management involves carefully crafted forest working plans, with fire prevention and control being a crucial aspect of these plans. Primarily, the guidelines focus on using customary methods such as establishing and upkeeping firebreaks, fire lanes, controlled burns, and hiring fire guards during fire seasons. Residents living near forested regions are also required by law to help forest department personnel in putting out fires. These techniques were effective in managing forest fires in India initially but became less successful over time due to increasing population pressure on forests, resulting in conflicts and resource depletion. Increasing biotic pressure caused more frequent fires, leading to decreased regeneration in forested areas. Considering this, it was deemed essential to establish a contemporary system for managing forest fires in the nation. Between 1985 and 1990, a UNDP initiative was carried out in the country to tackle the issue of resource devastation caused by unchecked forest fires. The main emphasis of the project was on implementing a structured method to address forest fire destruction by utilizing the expertise of certain advanced nations in fire prevention, detection, and control and transferring this knowledge to India. A pilot project was initiated in two states, namely Uttar Pradesh and Maharashtra, as part of this project due to the significant impact of fires on approximately half of the forest area. The two states had distinct ecological and physical features, providing a great chance to test forest fire prevention and control technologies (Dandabathula, G., 2021)

The project produced outstanding outcomes in terms of technical feasibility and economic effectiveness. It was shown clearly that fire incidents can be significantly reduced, with some areas experiencing a 90% decrease in incidents. Since 1992-93, the Government of India, Ministry of Environment & Forests, has introduced a program called “Modern Forest Fire Control Methods” after seeing the success of this project. The program was initiated in 11 states during the 8th Plan period, targeting areas with frequent forest fires. The project included 60% of the country's forest areas. Financial aid was given by the Government to state forest departments for purchasing hand tools, fire-resistant clothing, firefighting equipment, wireless sets, building fire watch towers and fire finders, making fire lines, and for fire fighting research, training, and promotion. An air operation wing was upheld by the Central Government. The project has been ongoing throughout the ninth plan period (1997-2002), and an additional four states are being included in the program. The main part of the plan involves shutting down the Air Operation Wing to focus on traditional and cost-effective methods while also implementing a system to closely monitor forest fires and gather data through the Forest Survey of India, research institutes, and other agencies to improve planning and management. By engaging village forest protection committees established under the Joint Forest Management (JFM) program, an additional aspect is being incorporated into the project. The JFM scheme is being carried out in 22 states with the help of 35,000 village forest protection committees, covering 7 million hectares of land (Attri, V.,2020).

CONCLUSION

An analysis of the patterns of occurrence and scale of fire activity globally clearly points to the need to step up international cooperation in the area of management of wildfires. Indeed, with the changing climate contributing to the development of fires, one cannot avoid realising that no nation can tackle these rising complications on its own. National measures concerning wildfires and their prevention are very important and enshrined in the respective policies and frameworks of each country, but due to the cross-boundary effects noted in circumstances like in the case of smoke or environmental destruction, there is a need for regional strategies and policies. Some of these challenges can be mitigated by organizations such as the United Nations and other regional bodies that have started implementing some measures towards these issues; however, more comprehensive and enhanced international cooperation is needed in order to support such measures and achieve the best results in preventing and combating wildfires as well as in helping people recover their losses.

International organizations, including the FAO, WHO, and the UNEP, have played an essential part in influencing various global strategies for combating wildfires. Such organizations can also provide key assistance in early warning systems, public health and safer forests. Nevertheless, due to inadequate funding and resources within these bodies, they sometimes lack sufficient capability to adequately address the needs of nations, especially the vulnerable ones with high susceptibilities to wild-fire. This has clearly illustrated the need for member states and the private sector to increase their financial contributions and call for better ways of utilising these funds and transferring technologies. Building their capacity will prove critical to ensuring that they will be in a position to deliver the assistance needed to countries with high risks of wildfire events.

The last of the factors that can be positively attributed to the fight against fires is regional collaboration. It shows that regional meetings can be instrumental in setting a platform of discussion and cooperation, such as the ASEAN meeting or a Continent-to-continent initiative like the Pan-American Conference on Wildland Fire. These provide a forum through which the dissemination of the best practices, training and coordination of response interventions is done, which are critical in areas that experience wildfire. However, such processes are not without problems concerning political differences, developmental disparities, and resources among the member nations. These are factors that will have to be addressed in the future to promote regional solidarity and the creation of institutions that will ensure fairness in the sharing of benefits among key players.

Furthermore, local knowledge and people-centred fire management systems should be incorporated into national and international frameworks in order to increase the effectiveness of wildfire prevention and suppression. Aboriginal people have always used fire in the management of the occupations that they controlled in ways that can still be learned today about the prevention of fires and the strengthening of ecosystems. It would be very helpful to understand these traditional methods in practice to be able to incorporate them into the existing methods of extinguishing fires to have a better, more comprehensive way of handling fires. Aboriginal people speak for the self-governing systems used in managing wildfires, and their rights should be respected as regional, national and international capacities are considered appropriate for managing wildfire disasters.

All in all, one should state that the effects of the wildfire problem are increasing, and the measures for its regulation should be global and united. Six priorities that support the effective management of wildfires are the reinforcement of international legal tools, improving cooperation in fire-prone regions, raising financial and human resources, and incorporating knowledge systems. Wildfires remain a major menace to human life, the environment, and the global economy. Thus, they call for increased international cooperation, backed by intergovernmental organisations and civil society, to enhance policies and measures for preventing or mitigating these disasters. The effects of these fires can only be managed through concerted efforts, and if all stakeholders come together, we can ensure that the earth is preserved for use in future.

REFERENCES

Adams, M. A. (2013). Mega-fires, tipping points and ecosystem services: Managing forests and woodlands in an uncertain future. *Forest Ecology and Management*, 294, 250–261. DOI: 10.1016/j.foreco.2012.11.039

Akhtar-Schuster, M., Stringer, L. C., Metternicht, G., Barger, N. N., Chotte, J. L., & Kust, G. (2022). Assessing the impact of science in the implementation of the United Nations Convention to Combat Desertification. *Land (Basel)*, 11(4), 568. DOI: 10.3390/land11040568

Arshad, A., Azhar Ali, A., & Anjali, K. S. (2022). Impact of forest fire on forest ecosystem. *Journal of Agriculture and Technology*, 9(1&2), 18–29.

Attri, V., Dhiman, R., & Sarvade, S. (2020). A review on status, implications and recent trends of forest fire management. *Archives of Agriculture and Environmental Science*, 5(4), 592–602. DOI: 10.26832/24566632.2020.0504024

Bargali, H., Calderon, L. P. P., Sundriyal, R. C., & Bhatt, D. (2022). Impact of forest fire frequency on floristic diversity in the forests of Uttarakhand, western Himalaya. *Trees, Forests and People*, 9, 100300. DOI: 10.1016/j.tfp.2022.100300

Bera, B., Shit, P. K., Sengupta, N., Saha, S., & Bhattacharjee, S. (2022). Forest fire susceptibility prediction using machine learning models with resampling algorithms, Northern part of Eastern Ghat Mountain range (India). *Geocarto International*, 37(26), 11756–11781. DOI: 10.1080/10106049.2022.2060323

Bhattarai, N., Dahal, S., Thapa, S., Pradhananga, S., Karky, B. S., Rawat, R. S., Windhorst, K., Watanabe, T., Thapa, R. B., & Avtar, R. (2022). Forest fire in the hindu kush Himalayas: A major challenge for climate action. *Journal of Forest and Livelihood*, 21(1), 14–31. DOI: 10.3126/jfl.v21i1.56576

BINJOLA, R., DOBHAL, S., & LAL, P. (2022). A REVIEW ON THE CURRENT STATUS OF FOREST FIRE IN UTTARAKHAND STATE OF INDIA. *Journal of Global Ecology and Environment*, 16(4), 26–33.

Bowman, D. M., Balch, J. K., Artaxo, P., Bond, W. J., Carlson, J. M., Cochrane, M. A., ... & Pyne, S. J. (2009). Fire in the Earth system. *science, 324*(5926), 481-484.

Box, R. C. (2023). Policy Options for Minimizing Wildfire Damage in the Western USA. In *International Handbook of Disaster Research* (pp. 1–16). Springer Nature Singapore.

Brockerhoff, E. G., Barbaro, L., Castagneyrol, B., Forrester, D. I., Gardiner, B., González-Olabarria, J. R., Lyver, P. O. B., Meurisse, N., Oxbrough, A., Taki, H., Thompson, I. D., van der Plas, F., & Jactel, H. (2017). Forest biodiversity, ecosystem functioning and the provision of ecosystem services. *Biodiversity and Conservation*, 26(13), 3005–3035. DOI: 10.1007/s10531-017-1453-2

Chandra, K. K., & Bhardwaj, A. K. (2015). Incidence of forest fire in India and its effect on terrestrial ecosystem dynamics, nutrient and microbial status of soil. *International Journal of Agriculture and Forestry*, 5(2), 69–78.

Dandabathula, G., Chintala, S. R., Ghosh, S., Balakrishnan, P., & Jha, C. S. (2021). Exploring the nexus between Indian forestry and the Sustainable Development Goals. *Regional Sustainability*, 2(4), 308–323. DOI: 10.1016/j.regsus.2022.01.002

De Abreu, S. J. (2021). Toward a holistic approach: Considerations for improved collaboration in wildfire management. *Open Journal of Forestry*, 12(1), 107–121. DOI: 10.4236/ojf.2022.121006

De Barros, A. E., Morato, R. G., Fleming, C. H., Pardini, R., Oliveira-Santos, L. G. R., Tomas, W. M., Kantek, D. L. Z., Tortato, F. R., Fragoso, C. E., Azevedo, F. C. C., Thompson, J. J., & Prado, P. I. (2022). Wildfires disproportionately affected jaguars in the Pantanal. *Communications Biology*, 5(1), 1028. DOI: 10.1038/s42003-022-03937-1 PMID: 36229543

Fillmore, S. D., McCaffrey, S., Bean, R., Evans, A. M., Iniguez, J., Thode, A., Smith, A. M. S., & Thompson, M. P. (2024). Factors influencing wildfire management decisions after the 2009 US federal policy update. *International Journal of Wildland Fire*, 33(1). Advance online publication. DOI: 10.1071/WF23129

Gajendiran, K., Kandasamy, S., & Narayanan, M. (2023). Influences of wildfire on the forest ecosystem and climate change: A comprehensive study. *Environmental Research*, •••, 117537. PMID: 37914016

Geraskina, A. P., Tebenkova, D. N., Ershov, D. V., Ruchinskaya, E. V., Sibirtseva, N. V., & Lukina, N. V. (2021). Wildfires as a factor of the loss of biodiversity and functions of forest ecosystems. *For. Sci*, 4(2), 4. DOI: 10.31509/2658-607x-202142-11

Geraskina, A. P., Tebenkova, D. N., Ershov, D. V., Ruchinskaya, E. V., Sibirtseva, N. V., & Lukina, N. V. (2022). WILDFIRES AS A FACTOR OF LOSS OF BIODIVERSITY AND FOREST ECOSYSTEM FUNCTIONS. *Вопросы лесной науки, 5*(1), 1-70.

Goldammer, J. (2021). Thirty years international wildland fire conferences: Review and achievements of a circumglobal journey from Boston to Campo Grande. *Biodiversidade Brasileira*, 11(2), 6–52. DOI: 10.37002/biobrasil.v11i2.1743

Goldammer, J. G. (2013). *Vegetation fires and global change: Challenges for concerted international action. A white paper directed to the united nations and international organizations*. Kessel.

Hemery, L. G., Garavelli, L., Copping, A. E., Farr, H., Jones, K., Baker-Horne, N., Kregting, L., McGarry, L. P., Sparling, C., & Verling, E. (2024). Animal displacement from marine energy development: Mechanisms and consequences. *The Science of the Total Environment*, 917, 170390. DOI: 10.1016/j.scitotenv.2024.170390 PMID: 38286287

Jones, G. M., & Tingley, M. W. (2022). Pyrodiversity and biodiversity: A history, synthesis, and outlook. *Diversity & Distributions*, 28(3), 386–403. DOI: 10.1111/ddi.13280

Kamali Lassem, N., Gaafar, O. M. H. A., & Ali, S. A. (2023, October). Capitalizing the Predictive Potential of Machine Learning to Detect Various Fire Types Using NASA's MODIS Satellite Data for the Mediterranean Basin. In *Proceedings of the 2023 7th International Conference on Advances in Artificial Intelligence* (pp. 24-28). DOI: 10.1145/3633598.3633603

Kumar, P., Kumar, K., Mathur, S., & Singh, M. (2021). Aligning Developmental Programs to Implement SDGs in India: Challenges and the Way Forward. *Journal of Contemporary Issues in Business and Government*, 27(6), 55–64.

Mo, M., Roache, M., Reid, T., Oliver, D. L., Broome, L., Fawcett, A., Howard, K., Thomas, P., Tracey, S., Andersen, G., & Lowry, R. (2021). Corporate support for threatened species recovery efforts: Three case studies from the 2019–20 Australian bushfire season. *Australian Zoologist*, 41(2), 186–193. DOI: 10.7882/AZ.2020.031

Ostad-Ali-Askari, K. (2022). Review of the effects of the anthropogenic on the wetland environment. *Applied Water Science*, 12(12), 260. DOI: 10.1007/s13201-022-01767-4

Pacheco, R. M., & Claro, J. (2023). Characterising wildfire impacts on ecosystem services: A triangulation of scientific findings, governmental reports, and expert perceptions in Portugal. *Environmental Science & Policy*, 142, 194–205. DOI: 10.1016/j.envsci.2023.02.008

Paul, M. J., LeDuc, S. D., Lassiter, M. G., Moorhead, L. C., Noyes, P. D., & Leibowitz, S. G. (2022). Wildfire induces changes in receiving waters: A review with considerations for water quality management. *Water Resources Research, 58*(9), e2021WR030699.

Pausas, J. G., & Keeley, J. E. (2019). Wildfires as an ecosystem service. *Frontiers in Ecology and the Environment*, 17(5), 289–295. DOI: 10.1002/fee.2044

Pengshuai, S. H. A. O., Hongyan, H. A. N., Jingkuan, S. U. N., & Hongtu, X. I. E. (2023). Effects of global change and human disturbance on soil carbon cycling in boreal forest: A review. *Pedosphere*, 33(1), 194–211. DOI: 10.1016/j.pedsph.2022.06.035

Romps, D. M., Seeley, J. T., Vollaro, D., & Molinari, J. (2014). Projected increase in lightning strikes in the United States due to global warming. *Science*, 346(6211), 851–854. DOI: 10.1126/science.1259100 PMID: 25395536

Saha, S., Bera, B., Shit, P. K., Bhattacharjee, S., & Sengupta, N. (2023). Prediction of forest fire susceptibility applying machine and deep learning algorithms for conservation priorities of forest resources. *Remote Sensing Applications: Society and Environment*, 29, 100917. DOI: 10.1016/j.rsase.2022.100917

Santos, J. L., Hradsky, B. A., Keith, D. A., Rowe, K. C., Senior, K. L., Sitters, H., & Kelly, L. T. (2022). Beyond inappropriate fire regimes: A synthesis of fire-driven declines of threatened mammals in Australia. *Conservation Letters*, 15(5), e12905. DOI: 10.1111/conl.12905

Sathishkumar, V. E., Cho, J., Subramanian, M., & Naren, O. S. (2023). Forest fire and smoke detection using deep learning-based learning without forgetting. *Fire Ecology*, 19(1), 9. DOI: 10.1186/s42408-022-00165-0

Sazawa, K., Yoshida, H., Okusu, K., Hata, N., & Kuramitz, H. (2018). Effects of forest fire on the properties of soil and humic substances extracted from forest soil in Gunma, Japan. *Environmental Science and Pollution Research International*, 25(30), 30325–30338. DOI: 10.1007/s11356-018-3011-1 PMID: 30159838

Schmerbeck, J., & Fiener, P. (2015). Wildfires, ecosystem services, and biodiversity in tropical dry forest in India. *Environmental Management*, 56(2), 355–372. DOI: 10.1007/s00267-015-0502-4 PMID: 25900601

Sharma, N., & Parameswaran, K. (2023). An appraisal of the measures taken by the Indian Government to attain sustainable development goals and to meet the commitments of the Paris Agreement. *Passagens: International Review of Political History & Legal Culture, 15*(2).

Sharma, U., Kaushal, R., & Sharma, Y. P. (2022). Effect of forest fires on soil carbon dynamics in different land uses under NW Himalayas. *Indian Journal of Ecology*, 49(6), 2322–2329.

Shi, K., & Touge, Y. (2022). Characterization of global wildfire burned area spatiotemporal patterns and underlying climatic causes. *Scientific Reports*, 12(1), 644. DOI: 10.1038/s41598-021-04726-2 PMID: 35022560

Singha, A. P., Kumarb, J., Chauhanc, I., Saxenaa, P. K., Gaurd, P. K., Sharmae, S. K., ... & Singha, P. K. (2022). Role of climate change in disasters occurrences: Forecasting and management options. *Visualization Techniques for Climate Change with Machine Learning and Artificial Intelligence*, 149.

Souza-Alonso, P., Saiz, G., García, R. A., Pauchard, A., Ferreira, A., & Merino, A. (2022). Post-fire ecological restoration in Latin American forest ecosystems: Insights and lessons from the last two decades. *Forest Ecology and Management*, 509, 120083. DOI: 10.1016/j.foreco.2022.120083

Tedim, F., & Leone, V. (2020). The dilemma of wildfire definition: What it reveals and what it implies. *Frontiers in Forests and Global Change*, 3, 553116. DOI: 10.3389/ffgc.2020.553116

Vasudeva, S. P. (2018). Disastrous Forest Fires: Management and Control. *The Indian Journal of Public Administration*, 64(2), 237–253. DOI: 10.1177/0019556117750900

Wang, S. W., Lim, C. H., & Lee, W. K. (2021). A review of forest fire and policy response for resilient adaptation under changing climate in the Eastern Himalayan region. *Forest Science and Technology*, 17(4), 180–188. DOI: 10.1080/21580103.2021.1979108

Weiskopf, S. R., Rubenstein, M. A., Crozier, L. G., Gaichas, S., Griffis, R., Halofsky, J. E., Hyde, K. J. W., Morelli, T. L., Morisette, J. T., Muñoz, R. C., Pershing, A. J., Peterson, D. L., Poudel, R., Staudinger, M. D., Sutton-Grier, A. E., Thompson, L., Vose, J., Weltzin, J. F., & Whyte, K. P. (2020). Climate change effects on biodiversity, ecosystems, ecosystem services, and natural resource management in the United States. *The Science of the Total Environment*, 733, 137782. DOI: 10.1016/j.scitotenv.2020.137782 PMID: 32209235

Chapter 2
Advancing Artificial Intelligence for Wildfire Prediction and Control: New Edges in Environmental Safety and Fire Ecology

Tarun Kumar Kaushik
https://orcid.org/0000-0002-3279-8307
Sharda University, India

Ravish
Meerut College, India

Anurag Singh
https://orcid.org/0009-0005-1322-7380
Meerut College, India

Anjali Raghav
https://orcid.org/0009-0004-0248-7956
Sharda University, India

Bhupinder Singh
https://orcid.org/0009-0006-4779-2553
Sharda University, India

Kittisak Jermsittiparsert
https://orcid.org/0000-0003-3245-8705
Shinawatra University, Thailand

ABSTRACT

The advancements in data science, along with the advancements in digital and satellite technologies, have increased the capabilities for using artificial intelligence (AI) in the fields of forestry and wildlife. However, the rapid expansion of construction projects, agricultural activities, and urban regions presents a substantial risk to biodiversity worldwide. Therefore, using cutting-edge technology like artificial intelligence (AI) in the domains of forests and biodiversity might enhance the effective monitoring, management, and conservation of biodiversity and forest resources. This chapter

DOI: 10.4018/979-8-3693-7565-5.ch002

aims to provide a thorough overview of the global use of AI and machine learning (ML) algorithms in the forestry sector and biodiversity protection. Moreover, this study investigates the challenges faced during the use of AI technology in the domains of forestry and biodiversity. The results of this study would motivate forest officials, scientists, researchers, and conservationists to explore the possibilities of using AI technology for forest management and biodiversity protection.

INTRODUCTION

Currently, Artificial Intelligence (AI) or Machine Learning (ML) is emerging as a crucial element that influences several aspects of our life. The United Nations shares the view that Artificial Intelligence (A.I.) can play a crucial role in achieving the aims outlined in the Sustainable Development Goals (SDGs). Effective forest management is crucial for the preservation and protection of our environment, making it a vital role in ensuring the survival of mankind on Earth (Oliveira Fiorini et al., 2023). Given the wide expanse of the forest, conducting patrols throughout the entire region would be an insurmountable undertaking. Artificial intelligence (A.I.) can provide significant assistance in this context. A.I. is currently being utilized in several forest management initiatives, mostly to monitor and combat unlawful activities and efficiently handle data related to endangered species (Lofqvist et al., 2023). Increasing the accessibility of comprehensive data on forests and biodiversity, together with the use of cloud computing and digital and satellite technologies, can promote the broader adoption and use of AI technology. This research explores the application of artificial intelligence (A.I.) in the field of forest management. Artificial intelligence (AI) is a wide-ranging discipline in computer science that aims to develop intelligent machines capable of augmenting, automating, and accelerating crucial everyday tasks traditionally dependent on human intellect. The method involves extracting patterns, predicting future conditions, and spotting anomalies. The progress in artificial intelligence has resulted in its extensive adoption in diverse industries, such as e-commerce, social networks, agriculture, environmental sustainability, education, healthcare, data manipulation prevention, social care, urban design, public safety, transportation, and environmental conservation (Wicki et al., 2023). This also include its utilization in forest and wildlife sectors. The progress has been driven by several reasons. First and foremost, the prompt dissemination of information has significantly risen, offering a chance to develop impactful insights and expedite the process of making decisions (Potner et al., 2023).

Furthermore, the expenses associated with data storage and computation have diminished as a result of the accessibility of cloud technologies. This allows us to create complex solutions on a big scale, therefore speeding up testing and research

(Watson et al., 2023). Furthermore, the advancements are further enhanced by the presence of extensive data sources, including high-resolution satellite imagery, sensors and telemetry technologies, drone and camera technologies, and data sources that focus on individuals (such as social networks, citizen science, and other open data sources). However, the impact of AI technology has been uneven, predominantly benefiting companies that provide substantial economic gains, while being less utilized in the fields of forestry and biodiversity protection. Although the use of AI in forest and natural resources management started three decades ago, the progress and adoption of AI technology in the forestry industry have been somewhat sluggish in comparison to other industries such as healthcare, transportation, and agriculture (Kennedy et al., 2023). Forests comprise almost 30% of the Earth's surface area, making them the most widespread ecosystems on land and hosting 90% of the planet's species. Forests are vital for sustaining the proper operation of our planet by offering necessary activities that support life, including as protective functions and environmental services (Singh & Kaunert, 2024). They aid in the purification of the environment by stabilizing greenhouse gases, therefore maintaining the availability of clean and breathing air. In addition, they possess the ability to assimilate a substantial proportion of the annual worldwide atmospheric carbon dioxide (CO2) emissions, comprising around 30%, corresponding to 2 billion tons annually (Dhali et al., 2023). Their contribution to global food security is substantial since they provide support to pollinators, natural predators of agricultural pests, and the hydrological cycle. They function as a substantial repository of medicinal plants and account for around 40% of the global renewable energy supply in the form of biofuels. Wetlands are essential for maintaining the hydrological equilibrium of many ecosystems and improving their resilience to extreme events such as floods and droughts (Singh, 2023).

Moreover, over 1.6 billion people depend on forests as their main source of nutrition. Unfortunately, woods are faced with a multitude of challenges. Globally, forests are rapidly deteriorating due to the logging industry, expansion of agriculture, and urban development. Climate change is intensifying the global deterioration of forests by causing events like wildfires (Hansen et al., 2023). In the last 25 years, about 129 million hectares of forest land have been lost globally, resulting in a reduction of the global carbon storage by 17.4 billion tons. Continued forest degradation and carbon depletion are expected to continue globally in the near future. While the forest industry still relies on conventional methods to handle the benefits and risks of forests, it is crucial to recognize the potential rewards and difficulties they bring (Batisha, 2023).

Indeed, it is one of the few industries that are slow to adopt modern technologies. For example, in many countries, forest authorities still use conventional pen and paper techniques to conduct forest inventories (Singh, 2024). Conventional methods have several drawbacks, including the incorporation of subjective viewpoints, the slowing down of data collection and analysis, and the limited scalability of the strategy. In contrast to the forestry industry, other industries, such as agriculture, have rapidly embraced technological alternatives, while being a significant source to deforestation (Qwaider et al., 2023). Precision agriculture utilizes robotic devices to carry out activities such as weed eradication, fertilization, and pesticide administration, leading to enhanced agricultural production per unit of land (Singh, 2024). Forests, in contrast to agricultural systems, are known for their ever-changing nature, requiring proper management, especially since many nations are currently looking for ways to achieve significant changes in order to meet their nationally determined contributions (NDCs) (Singh, 2023).

Technology may successfully overcome the constraints of conventional approaches to data collection and analysis, therefore playing a crucial role in improving forest management and conservation. Access to extensive and superior datasets, a network infrastructure for Internet-of-Things (IoT), cutting-edge technology like high-resolution cameras, satellite technology, sensors, drones, and unmanned aerial vehicles (UAVs), along with computational space and storage, are indispensable prerequisites for the advancement and application of AI in any domain (Sobha, 2023). Research on AI applications in biodiversity preservation and forestry has been motivated by the requirement to access a combination of these characteristics. The domains encompassed in this context are forest inventory, identification of illegal wildlife, illegal wood trade, and tree cutting. Furthermore, these study efforts have significantly altered the prompt use of AI technology in the domains of biodiversity protection and forestry. This is supported by the increasing quantity of start-ups that employ AI technology in various industries (Muhmad Kamarulzaman et al., 2023).

Given the aforementioned rationale, the preservation and safeguarding of forests and animals have become crucial, since they directly impact the survival of life on Earth. In order to ensure the safeguarding, preservation, and sustainable use of forests, it is imperative that we employ highly effective forest management practices. In terms of forest management, we possess a diverse array of legal frameworks, encompassing both local and international laws, as we do not consider forests to be owned by any individual or entity (Bogers, 2023). While certain states may own exclusive rights to utilize something, they do not have the authority to destroy it. Under these conditions, Forest Management has become an imperative and vital activity that must be executed with great efficiency (Singh, 2023). Forest management encompasses the control and oversight of many activities related to forests, including surveillance, administrative, legal, economic, and social issues (Nahar, 2024).

The utilization of artificial intelligence technologies is quite recent. Forest management utilizing A.I.-based technology is a recent addition to the use of A.I. methodologies. The manual forest management is not yielding the expected outcomes (Desai et al., 2023). The forest cover is diminishing, while the incidence of forest fires is rising, despite the presence of stringent rules against unlawful hunting. Furthermore, there has been a noticeable surge in the illicit trade of wild animals. In order to discover effective methods of forest management, it is necessary to investigate the use of artificial intelligence in this field (Sharma & Singh, 2022). At now, we possess several methods that utilize artificial intelligence, such as Artificial Neural Network, for the purpose of managing forests. These methods include automated forest fire prediction, bird's eye view analysis, and AI-based carbon cycle mapping, all of which contribute to more effective forest management. However, it is imperative that we provide answers to the issues that are associated with this. It need to thoroughly analyze and document all the legal implications associated with the use of these artificial intelligence (A.I.) based methods for forest management and legislation (Ashukem & Sama, 2023).

Necessity for A.I. Based Forest Management

While many individuals prioritize the improvement of the environment, only a few occasionally contemplate the importance of forest management (Cornford et al., 2023). However, it is widely acknowledged that forest management is a crucial and fundamental undertaking for safeguarding our ecosystem and ensuring our survival on this planet. The lumber harvested from the forest is used to manufacture our homes, the paper cup that holds your morning coffee, and the cardboard box that contains your latest internet delivery. All of these products are derived from trees growing in the forests. Assessing the capacity of forests to provide those resources has been expensive, time-consuming, and lacking advanced technology. The top ranger service organizations in the United States annually provide millions of dollars to compensate workers for the task of accurately counting and measuring trees (Singh, 2022).

Regarding forest management, there are several dimensions associated with it. However, traditional methods of forest management are inadequate for monitoring and controlling forest-related activities, natural or man-made hazards, and wildlife smuggling. The outcomes of these manual approaches have been unsatisfactory (Koh et al., 2023). In contrast, forest management techniques based on artificial intelligence have demonstrated some promising results. Due to the vast expanse of forest cover, the current methods employed to address the aforementioned issue are both costly and ineffective. The impact of human activities on the environment, caused by climate change, is leading to an increase in both natural and man-made risks (Sheng & Lian, 2023). Consequently, our forests are being depleted and wildlife is being adversely affected. We enforce stringent regulations to save forests and animals. However, due to rampant corruption and a lack of an adequate number of forest guards, we are still unable to effectively curb the criminal operations of smuggling and poaching of wild animals (Singh, 2022).

Forest Correlated Actions

We have several companies that rely on trees for their primary resources. The enormous financial worth of the forest has led to enterprises seeking to over exploit it. However, in order to ensure the sustainable use of forest resources, the government has implemented a range of rules and regulations aimed at limiting excessive exploitation of these resources and designated authorities for the effective enforcement of prescribed legislation for the preservation of forests (Ashukem, 2024). Despite completing all these procedures, the commercial worth of the forest remains quite high, resulting in a substantial amount of money being invested. Consequently, corruption has become a prevalent issue in this context. These enterprises engage in corrupt practices by offering bribes to local officials and over exploiting forest resources. Due to the vast expanse of woods, conducting manual monitoring is exceedingly challenging and costly (Singh & Kaunert, 2024). In light of this, it is imperative that we implement an automated kind of monitoring that is devoid of prejudices and corruption. Therefore, artificial intelligence (A.I.) approaches can offer more effective methods for regulating forest-related activities and improving forest management.

Monitoring, Managing and Forest Hazards

Experts believe that forest dangers are natural phenomena that contribute to the improvement of the life cycle of plants and creatures in the forest. However, it must be acknowledged that the aforementioned perspective is only partially accurate (Prasanna et al., 2023). This is because the occurrence of forest dangers has experienced a significant and rapid rise as a direct consequence of human activities, leading to extensive deforestation. For instance, let's consider the case of a forest fire. Traditionally, this event falls within the category of natural processes. However, due to human activity, we have observed a rise in both the average temperature and the peak temperature from one season to another (Terry et al. 2024). Consequently, there has been a rapid increase in the frequency of forest fire incidents, accompanied by an expanding coverage area. Consequently, this leads to a reduction in the extent of forested areas. The similar scenario applies to other dangers such as forest floods and landslides. The frequency of these dangers has grown as a result of human activities, leading to a detrimental impact on the forest cover. If we are able to accurately forecast these dangers in a timely manner, we can effectively reduce their consequences. Therefore, the forecasting of these forest dangers may be carried out effectively by utilizing A.I. powered technologies. In order to make this forecast, we must analyze an extensive volume of data using a designated method. Currently, we possess two algorithms for forecasting forest fires. However, the processing of data at such a vast scale is a significant challenge due to the nascent stage of these artificial intelligence-based technologies. However, artificial intelligence (A.I.) technologies have the capability to accurately forecast forest risks to prevent these dangers from causing deforestation (Ashukem & Sama, 2023).

Trafficking of Wild Animals

Wild animals play a crucial role in the ecosystem of forests, as they are home to billions and trillions of creatures. Forest management encompasses the control of the wildlife inhabiting the forest region as well (Fopa et al., 2023). Over the past 20-30 years, significant efforts have been made to prohibit the illicit hunting and trading of wild animals. These efforts have included the establishment of legislative frameworks and the creation of specialized authorities at various levels to preserve these species in their natural habitats (Ozkavaf Senalp, 2023). Although the aforementioned factors have brought about a significant and observable transformation in this matter, there still persists a substantial illicit market involved in the trade of wild animals and items derived from them. The market for medicines derived from wild animals nevertheless persists, particularly in relation to fertility-related medications. Despite our extensive endeavors, we continue to fall short in safeguarding the biodiversity of our wildlife. Therefore, artificial intelligence (A.I.) based technologies can also assist in this matter. Artificial intelligence has the capability to evaluate and handle vast quantities of data obtained from live cameras, sensors, and satellite pictures that are received on a continuous basis. Therefore, it is capable of providing alerts for any abnormal or unrecognized motions. This can also aid in the illicit transportation of timber from the forest and has significant potential for curbing the smuggling of both wildlife and timber from the forest (Cumming et al., 2023).

A.I. CONNECTED WAYS FOR FOREST SUPERVISION

As previously said, artificial intelligence-based technology is necessary for effective forest management and regulation. This technology removes bias and has a significantly greater data processing capability compared to human processes (Cork et al. 2023). Now we shall explore many significant artificial intelligence (A.I.) methods that contribute to various elements of forest management (Gadsden et al., 2023).

Using Artificial Neural Network for Forest Management

Artificial Neural Network (ANN) is a significant focus of study in the field of Artificial Intelligence. Artificial Neural Networks (ANN) are considered to be one of the most crucial techniques in the field of machine learning. Before analyzing and processing the data, it is displayed in a manner similar to how a human brain functions, although at a far faster pace. When discussing advanced techniques for forest management and regulation, a big challenge arises in processing the vast amount of data and information linked with it. ANN perceives the data and evaluates or processes it using the supplied algorithms (Lamalle & Stoett, 2023). The forest management and regulatory data are of a non-linear nature, and the most effective method for processing such data is through the use of artificial neural networks.

Automated Forest Fire Calculation

Artificial Intelligence technologies and techniques have the potential to effectively manage and mitigate forest fire dangers. The product can be prepared by inputting data on areas that are particularly susceptible to combustion. It has the ability to detect vulnerable areas inside a forest and promptly alert the forest management team, enabling them to swiftly respond to these areas and extinguish any fires. As a result, they will need less time and may conserve more natural resources by easily extinguishing the fire. It has the potential to save a significant amount of money (Dunlap, 2023).

One possible use of A.I.-based technology is the prompt detection and management of forest fires, which may be highly beneficial for effectively controlling forest dangers and maintaining forest cover. If firefighters can promptly respond or appropriate measures can be implemented to conserve forests, it would be beneficial when considering the utilization of energy by the utilized calculations (Newman et al. 2024).

Regarding Forest hazard management, a significant danger that has a huge impact is 'Forest Fire'. Quantifying the extent of damage caused by recent forest fires, such as the Amazon Forest fire, Arizona Forest fire, and Australia forest fire, is exceedingly challenging. Therefore, it is preferable to accurately forecast such forest fires in order to prevent them and save our forest resources. By analyzing the climatic conditions, ecological data, satellite images, and sensor signals of a specific forest using prescribed algorithms, we can obtain a more precise prediction of the severity of a particular event. Natural processes in the forest environment might provide threats. Therefore, we must make a prediction of the potential negative impact in relation to the benefits that will be gained for the forest, and then manage the situation accordingly (Ghosh et al., 2023).

Future Perspective

This is a highly popular method for forest management that is based on artificial intelligence. Using A.I.-based technologies, data, photos, or any other information obtained from satellites, cameras, or sensors will be processed. High-resolution satellite photos are valuable for monitoring the precise location of every individual tree within the forest. Drones have been employed to survey and document the activities occurring within the forest (Shinde et al., 2023). However, if we were to accurately document the state and actions occurring in the forest on a continuous basis, the volume of data obtained would be substantial. Humans lack the capacity to digest the data. In this scenario, artificial neural networks (ANN) or other artificial intelligence (AI) methodologies may be employed to handle the extensive data for the purpose of efficiently mapping the state and activities of the forest (Cardwell, 2023).

A.I. Based Carbon Cycle Mapping

Researchers now utilize artificial intelligence (A.I.) based tools to map the carbon cycle, aiming to enhance forest management practices. The conventional method was associated with the empirical manipulation and parameterization or identification of input or output data. However, utilizing artificial intelligence (A.I.) in carbon cycle mapping, the artificial neural network (ANN) may be employed with splines, neural networks, and kernel approaches to accurately forecast carbon dioxide (CO2) and energy flow. The carbon cycle mapping has several applications, including hazard prediction and lifetime mapping in forests. Overall, this also contributes to improved forest management by addressing the vast undiscovered areas under its purview (Aryal et al., 2023).

Automated Accuracy

The precision of agricultural practices may significantly contribute to forest management by reducing deforestation and promoting afforestation. A significant proportion of deforestation is attributed to the food supply chain (Ngadze, 2023). Utilizing AI-based technology in all aspects of agriculture has the potential to provide us with numerous benefits and unlimited possibilities. By adopting sustainable practices, we can conserve water resources and reduce the number of pesticides utilized. Precision in agriculture may lead to reduced pesticide usage and more efficient water utilization through the integration of technology, machinery, and intelligent application of algorithms. Artificial intelligence gadgets designed for strategy producers and agronomists may also help facilitate climate-friendly initiatives (Sofi et al., 2023).

Automation in Forestry and Afforestation

The earth has a maximum capacity of 1.5 trillion trees. This has the potential to counterbalance a period of carbon emissions. Mechanization can be beneficial in this context. The two newly established companies, Bio Carbon Engineering and Drone Seed, are mentioned in this way. Artificial intelligence may be employed to identify suitable planting locations, monitor plant health, evaluate weed presence, and analyze trends. Although many instances of deforestation can be attributed to the expansion of agriculture or urban development, the bulk of deforestation is caused by the logging industry (Shukla et al. 2023). Clear-cutting is a very destructive and widespread activity worldwide. Strategy designers should seek advice on the consequences of deforestation. Rainforest Connection has implemented outdated, sophisticated cell phones powered by solar panels in the jungle. At that juncture, a machine learning computation is capable of identifying the sounds of cutting tools within a radius of one kilometre and promptly transmitting this information to a nearby wireless antenna. Artificial intelligence (A.I.) technology may also be utilized for coordination and transportation, but this might have adverse effects and should be accompanied by appropriate solutions (Malekpour et al., 2023).

Legal Impression of AI-Based Ways for Forest Organization

Across the globe, legal professionals are collaborating with Artificial Intelligence specialists to establish regulations for Artificial Intelligence and A.I.-based technology. In addition, we have yet to discover a precise and comprehensive definition for artificial intelligence. Therefore, we can see the challenges associated with regulating artificial intelligence (Quamar et al., 2023).

Regulation is essential due to the inherent duality of everything, encompassing both good and bad qualities. Currently, Artificial Intelligence is in its nascent stage, although it has the potential to cause significant harm to human existence. Despite the implementation of AI-based technologies in forest management, it is important to acknowledge that these technologies can have both beneficial and detrimental effects. However, we must acknowledge the practicality of employing these artificial intelligence (A.I.) methods for the purpose of forest management and control. In my view, it is imperative to regulate these matters due to the significant potential advantages they offer (Basu, 2024).

The preceding chapter has examined the benefits and effects of utilizing artificial intelligence for forest management. Artificial intelligence (A.I.) methods may effectively aid in limiting and controlling illicit activities, as well as ensuring the enforcement of rules in forest management. One component of forest management pertains to the legal dimension of forest management. The potential of A.I.-based technologies is very great, since they have the capability to eradicate many forms of illicit activity, such as illegal hunting and the unlawful clearance of timber (Banerjee et al., 2023). Furthermore, it has the potential to deter illicit activity associated with industries such as mining, which may have detrimental effects on forests, leading to deforestation and the loss of species that rely on these ecosystems (Escap, 2023).

The management of the forest cover is not only a mechanical process, and innovation alone cannot account for all factors. Currently, we have an understanding that forests are experiencing loss, and it is imperative to address this issue. Therefore, it seems that designing forest holdings is just one component of the whole arrangement. Political determination and the understanding of the grave consequences by a larger population are essential (Jarvelainen, 2023). If change does not come from a single source or individual, but rather emerges from a collective effort, then it can be expected that every action taken to improve circumstances for our forests will have significance. Technologists and innovation enthusiasts can undoubtedly have a significant influence. Therefore, why should we not embrace technological advancements for the sake of humanity? One may contemplate the optimal utilization of artificial intelligence to provide improved air quality and increased natural carbon sequestration by effectively managing our forest cover and fauna (Sliuzas et al., 2023).

The disadvantages of using A.I.-based technology for forest management are as follows: The majority of countries with significant forest resources are emerging nations, but industrialized nations own exclusive control of modern artificial intelligence technology. Moreover, these advanced nations are unwilling to share their expertise on AI-based technology, and the developing nations are equally unwilling to grant access to monitoring and mapping of their forest resources. While it may have negative implications for the privacy of any country, the primary worry lies in relation to the national security of a State (Platjouw & Pozdnakova, 2023).

The autonomy of a developing nation will be subject to doubt. Another significant problem is the issue of responsibility in the event of misconduct or unfortunate incidents. None of the parties are prepared to assume the obligation. In order to implement AI-based forest management techniques in all forests, it is necessary to establish appropriate regulations and determine responsibilities in the event of any accidents or misconduct. Additionally, developing nations need assurance that their sovereignty and national security will not be compromised (Santarsiero, 2023).

Foremost Inventiveness

The expanded use of A.I. technologies has greatly enhanced the interaction and performance of these technologies, leading to remarkable improvements. On a conceptual level, it consistently offered opportunities for remarkable potential

(Wellmann, 2023). We have used the phrase "On the conceptual level" since we are aware that the integration of A.I. based technologies with Forest management is not a novel notion, even at a theoretical level. This occurred as a result of advancements in the methodology of developing tools and technologies that are based on artificial intelligence, as well as the approach to general development in our society and environment. As a result of this shift in our attitude towards the environment and ecology, we may consider incorporating artificial intelligence (AI) technology and tools into one of the most challenging, complex, and crucial jobs for human existence on this planet (Neimark, 2023).

Company Plan

One of the prominent and often debated projects in this field is Microsoft's "A.I. for Earth Project". This project aims to address four key areas - climate, agriculture, biodiversity, and water - in order to achieve sustainability (Zhang, 2023). It utilizes A.I.-based innovation and technology to pave the way for a better and more sustainable future. The key reasons for selecting these options are due to the detrimental impact of the changing environment on human well-being, infrastructure, and natural systems (De Vries, 2023. Artificial intelligence (A.I.) driven innovation can provide individuals with more accurate weather forecasts to assist mitigate any impacts. In the next two or three decades, we will need to enhance food production on limited arable land while minimizing the ecological impact in order to sustain the growing global population. Artificial intelligence (A.I.) technology can let users monitor the health of farms in real-time. Species are facing an alarming rate of extinction. Artificial intelligence methods can aid in accelerating the discovery, monitoring, and protection of biodiversity across our world (Anekwe et al., 2024).

Over the next two decades, demand for fresh water is expected to greatly exceed supply. Artificial intelligence (A.I.) can aid persons in modelling the distribution of Earth's water resources to help us conserve and save fresh water (Sarker et al., 2023).

This research utilizes geophysics and artificial intelligence to aid individuals in comprehending hyper-local climate-related hazards. This strategy employs artificial intelligence-based advancements to predict weather-related hazards. The project aimed to show the fire threat for any location by collecting and analyzing quickly spreading fire information, doing mathematical recreations, and utilizing satellite imagery. Fire-related statistical information may be accessed by anybody using APIs and graphical tools. Collecting agricultural data on the cloud and providing ranchers with access to enhance precision farming. This initiative use sensors to collect soil, cultivation, and crop data for specific areas of agriculture. The data is stored in Microsoft Azure and made available to farmers using user-friendly APIs to help them reduce costs, enhance yields, and minimize the environmental impact of agriculture (Bustamante, 2023).

Employing satellites and artificial intelligence (A.I.) technology to protect biodiversity, secure livelihoods, and prevent slavery in the fishing industry. The oceanic component of the initiative collaborates with government experts to prevent illegal, unreported, and uncontrolled fishing by continually monitoring vessel movements (Gill et al., 2023). Artificial intelligence algorithms identify suspicious behaviour and communicate it to authorities to enhance the coordination of surveillance vessels. This research is revolutionizing the methods we use to quantify and assess our forest canopy. It is essential for conservationists, governments, and landowners to replenish forests for the purpose of promoting environmental, social, and economic well-being. Through the utilization of artificial intelligence (AI) and cloud programming, these meetings may collaborate to analyze the consequences of climate change and enhance areas. Using high-definition satellite images and field data from the US Forest Service, we are developing artificial intelligence models to accurately map the forest cover (Khan, 2024).

Artificial intelligence (A.I.) based innovations and technologies collaborate with local researchers to combat elimination. Earth's artificial intelligence is employing AI technology and advanced cloud software to identify animal species that are on the brink of extinction. It employs artificial intelligence algorithms and deep learning algorithms to manipulate. They are utilizing innovation to examine and differentiate among animals and species (Yousuf, 2023).

Researchers are employing continuous data mapping to enhance precision in privacy. As part of our ongoing effort to protect intricate ecosystems, we collaborated with Chesapeake Conservancy to develop a specialized system for generating one-meter resolution land cover data over the whole United States. Our meticulous preparation utilizes artificial intelligence and comprehensive databases to disrupt precision confidentiality (Papas, 2023). The utilization of artificial intelligence (A.I.) has aided in combating poaching through the implementation of A.I.-based methods, including A.I. analysis and organisation, as well as the execution of demonstrations. They collect data from previous poaching operations and then employ AI-based methods and social modelling to provide predictions on poaching hotspots and optimal patrol routes. The outcome is the production of more effective timepieces and enhanced use of resources in the fight against poaching (Luger, 2023).

Institute of Data Science Project

The project undertaken by A.I. professionals and other specialists at this institute provides valuable insights into the importance of precise, definitive, and state-of-the-art data (van Dam et al., 2024). They have been considering the impact of unusual weather conditions on the growth and restoration of forest cover, specifically focusing on how this affects the ability of forests to store carbon. It has been shown that younger and less dense forests are less effective in carbon sequestration compared to older and denser areas. During their study, they discovered significant data on the damage to the forest caused by Hurricane Maria. In the future, it will utilize the collected data, together with remote sensing images and measurements, to generate a precise estimate of the damage caused by the storm. In the absence of up-to-date and accurate data, as well as a proactive outlook on the future potential of forest stock, planners may underestimate the value or countries may overestimate the capacity for carbon sequestration (Irvine et al., 2023).

Planning and conducting research strategies like this are essential for developing comprehensive and globally significant knowledge. We acknowledge that the world requires more organizations that are able to achieve greater results in this manner, which is why we are excited about the rapid growth of the AI for Earth grantees network. This project has provided us with a comprehensive understanding of the innovative combination of human ingenuity with AI, and the significant impact it is anticipated to have on the rapid, impactful, and global progress that our planet need (Singh et al., 2024). Fortunately, the current technology already exists to collect and analyze the data coming in from sensors, satellites, drones, local researchers, camera traps, and many other sources (Singh & Kaunert, 2024). Utilizing cloud computing and artificial intelligence, we can efficiently and cost-effectively perform tasks that were previously considered difficult and expensive to the point of being impossible (Singh & Kaunert, 2024).

The progress was witnessed on land, and the spread planning serves as an exemplary example. In the United States, the most comprehensive and easily accessible information repositories on land cover, spanning a distance of thirty metres, were last updated nearly seven years ago. Utilizing artificial intelligence (A.I.) for innovation can provide a clear benefit, as it simplifies and enhances the effectiveness of decision-making processes. Additionally, it reduces the vulnerability to political manipulation, particularly when it comes to understanding the current state of affairs on Earth, including the timing and location of events. The pace of growth is one of the only factors keeping up with climate change. Utilizing artificial intelligence to monitor the impact of our current land use practices and to simulate scenarios allows us to get accurate information, which may help us manage our properties, watersheds, and ecosystems more effectively and sustainably (Singh & Kaunert, 2024).

CONCLUSION

Since the whole ecosystem, climate, and our oxygen cycle depend on forests, forest management is a top priority if people are to continue existing on our planet. Despite the existence of regulations pertaining to forest management, we rely on artificial intelligence (AI) based methods such as automated forest fire prediction, bird's eye view, and AI-based carbon cycle mapping to ensure the sustainable, scientific, and efficient management of the forest. However, there is a downside to every benefit. Since developing nations possess a disproportionate amount of forest resources, the use of artificial intelligence (AI) based technology to manage these resources raises concerns about privacy and national security. Therefore, these concerns need to be resolved in order to have better methods of forest management and regulation; otherwise, emerging nations will continue to have inefficient management and control of their forest resources. In terms of Microsoft and DSI's research projects, those dealing with forest management using AI-based technologies are in their early stages but show promise. As a result, we may conclude that forest management approaches based on artificial intelligence have the potential to be the most important and promising for achieving sustainable development and effective forest management.

REFERENCES

Anekwe, I. M. S., Zhou, H., Mkhize, M. M., & Akpasi, S. O. (2024). Addressing Climate Change Challenges in South Africa: A Study in KwaZulu Natal Province. *Climate Crisis: Adaptive Approaches and Sustainability*, 475-496.

Aryal, K., Maraseni, T., & Apan, A. (2023). Examining policy– institution– program (PIP) responses against the drivers of ecosystem dynamics. A chronological review (1960–2020) from Nepal. *Land Use Policy*, 132, 106789. DOI: 10.1016/j.landusepol.2023.106789

Ashukem, J. C. N. (2024). Land grabbing in the Anthropocene and the "questionable" pursuit of sustainability in Africa–highlighting ecological disproportionality. In *Human Rights and the Environment in Africa* (pp. 276–294). Routledge.

Ashukem, J. C. N., & Sama, S. M. (Eds.). (2023). *Domestic and Regional Environmental Laws and Policies in Africa: A Research Companion*. Taylor & Francis. DOI: 10.4324/9781003382256

Banerjee, O., Cicowiez, M., Malek, Ž., Verburg, P. H., Vargas, R., Goodwin, S., Bagstad, K. J., & Murillo, J. Á. (2023). Banking on strong rural livelihoods and the sustainable use of natural capital in post-conflict Colombia. *Environment, Development and Sustainability*, 26(10), 1–22. DOI: 10.1007/s10668-023-03740-w

Basu, M. (2024). *The Routledge Handbook of International Environmental Policy*. Taylor & Francis. DOI: 10.4324/9781003440574

Batisha, A. (2023). A lighthouse to enhance the quality of life in the Nile River basin. *Environmental Economics and Policy Studies*, 1–29. DOI: 10.1007/s10018-023-00380-2

Bogers, M. (2023). *Integrating global sustainability governance: How the Sustainable Development Goals impact institutional and policy integration at the global level* (Doctoral dissertation, Utrecht University).

Bustamante, M., Roy, J., Ospina, D., Achakulwisut, P., Aggarwal, A., Bastos, A., ... & Zscheischler, J. (2023). Ten new insights in climate science 2023/2024. *Global Sustainability*, 1-58.

Cardwell, M. (2023). Results-based agri-environmental scheme design: Legal implications. *Environmental Law Review*, 25(4), 260–288. DOI: 10.1177/14614529231185678

Cork, S., Alexandra, C., Alvarez-Romero, J. G., Bennett, E. M., Berbés-Blázquez, M., Bohensky, E., & Wyborn, C. (2023). Exploring Alternative Futures in the Anthropocene. *Annual Review of Environment and Resources*, 48(1), 25–54. DOI: 10.1146/annurev-environ-112321-095011

Cornford, R., Spooner, F., McRae, L., Purvis, A., & Freeman, R. (2023). Ongoing over-exploitation and delayed responses to environmental change highlight the urgency for action to promote vertebrate recoveries by 2030. *Proceedings of the Royal Society B, 290*(1997), 20230464. DOI: 10.1098/rspb.2023.0464

Cumming, G. S., Adamska, M., Barnes, M. L., Barnett, J., Bellwood, D. R., Cinner, J. E., Cohen, P. J., Donelson, J. M., Fabricius, K., Grafton, R. Q., Grech, A., Gurney, G. G., Hoegh-Guldberg, O., Hoey, A. S., Hoogenboom, M. O., Lau, J., Lovelock, C. E., Lowe, R., Miller, D. J., & Wilson, S. K. (2023). Research priorities for the sustainability of coral-rich western Pacific seascapes. *Regional Environmental Change*, 23(2), 1–15. DOI: 10.1007/s10113-023-02051-0 PMID: 37125023

De Vries, B. J. (2023). *Sustainability science*. Cambridge University Press. DOI: 10.1017/9781009300230

Desai, B. H. (2023). Land-Soil Sustainability: India's International Obligations. In *Soil Law and Governance in India* (pp. 3–24). Springer International Publishing. DOI: 10.1007/978-3-031-32360-7_1

Dhali, M., Hassan, S., & Subramaniam, U. (2023). Comparative analysis of oil and gas legal frameworks in Bangladesh and Nigeria: A pathway towards achieving sustainable energy through policy. *Sustainability (Basel)*, 15(21), 15228. DOI: 10.3390/su152115228

Dunlap, A. (2023). The Structures of Conquest: Debating Extractivism (s), Infrastructures and Environmental Justice for Advancing Post-Development Pathways. *International Development Policy| Revue internationale de politique de développement.*

ESCAP. (2023). The changing landscape of regional cooperation in Asia and the Pacific.

Fopa, V. K., Bayir, N., & Özdal, D. (2023). Assessing the status and spatial-temporal dynamics of the Bamenda Mountains (BM), North West region of Cameroon. *Environmental Monitoring and Assessment*, 195(9), 1053. DOI: 10.1007/s10661-023-11630-x PMID: 37589789

Gadsden, G. I., Golden, N., & Harris, N. C. (2023). Place-Based Bias in Environmental Scholarship Derived from Social–Ecological Landscapes of Fear. *Bioscience*, 73(1), 23–35. DOI: 10.1093/biosci/biac095 PMID: 36643594

Ghosh, S., Dinda, S., Chatterjee, N. D., & Bera, D. (2023). Linking ecological vulnerability and ecosystem service value in a fast-growing metropolitan area of eastern India: A scenario-based sustainability approach. *Environment, Development and Sustainability*, •••, 1–31. DOI: 10.1007/s10668-023-03966-8

Gill, D. A., Blythe, J., Bennett, N., Evans, L., Brown, K., Turner, R. A., Baggio, J. A., Baker, D., Ban, N. C., Brun, V., Claudet, J., Darling, E., Di Franco, A., Epstein, G., Estradivari, , Gray, N. J., Gurney, G. G., Horan, R. P., Jupiter, S. D., & Muthiga, N. A. (2023). Triple exposure: Reducing negative impacts of climate change, blue growth, and conservation on coastal communities. *One Earth*, 6(2), 118–130. DOI: 10.1016/j.oneear.2023.01.010

Hansen, H. H., Bergman, E., Cowx, I. G., Lind, L., Pauna, V. H., & Willis, K. A. (2023). Resilient rivers and connected marine systems: a review of mutual sustainability opportunities. *Global Sustainability*, 1-71.

Irvine, K., Katsanou, K., Spears, B. M., Carvalho, L., Elelman, R., Free, G., & Warner, S. (2023). *Embedding lakes into the global sustainability agenda*. United Nations Environment Programme (UNEP).

Jarvelainen, M. (2023). Evaluating the Effectiveness of Forest Conservation as a CO2 Offset Measure for Urban Development: A Case Study. *Forest Ecology and Management*.

Kennedy, C. M., Fariss, B., Oakleaf, J. R., Garnett, S. T., Fernández-Llamazares, Á., Fa, J. E., Baruch-Mordo, S., & Kiesecker, J. (2023). Indigenous Peoples' lands are threatened by industrial development; conversion risk assessment reveals need to support Indigenous stewardship. *One Earth*, 6(8), 1032–1049. DOI: 10.1016/j.oneear.2023.07.006

Khan, I. A. (Ed.). (2024). *Security and Global Governance*. INTERDISCIPLINARY INSTITUTE OF HUMAN SECURITY & GOVERNANCE.

Koh, T., Lye, L. H., & Lum, S. (Eds.). (2023). *Peace With Nature: 50 Inspiring Essays On Nature And The Environment*. World Scientific. DOI: 10.1142/13561

Lamalle, S., & Stoett, P. (Eds.). (2023). *Representations and Rights of the Environment*. Cambridge University Press. DOI: 10.1017/9781108769327

Löfqvist, S., Kleinschroth, F., Bey, A., de Bremond, A., DeFries, R., Dong, J., Fleischman, F., Lele, S., Martin, D. A., Messerli, P., Meyfroidt, P., Pfeifer, M., Rakotonarivo, S. O., Ramankutty, N., Ramprasad, V., Rana, P., Rhemtulla, J. M., Ryan, C. M., Vieira, I. C. G., & Garrett, R. D. (2023). How social considerations improve the equity and effectiveness of ecosystem restoration. *Bioscience*, 73(2), 134–148. DOI: 10.1093/biosci/biac099 PMID: 36896142

Luger, J. (2023). *Governing the High Seas: effective institutional arrangements for the conservation and sustainable use of marine biodiversity beyond national jurisdiction* (Doctoral dissertation, The University of St Andrews).

Malekpour, S., Allen, C., Sagar, A., Scholz, I., Persson, Å., Miranda, J. J., Bennich, T., Dube, O. P., Kanie, N., Madise, N., Shackell, N., Montoya, J. C., Pan, J., Hathie, I., Bobylev, S. N., Agard, J., & Al-Ghanim, K. (2023). What scientists need to do to accelerate progress on the SDGs. *Nature*, 621(7978), 250–254. DOI: 10.1038/d41586-023-02808-x PMID: 37704766

Muhmad Kamarulzaman, A. M., Wan Mohd Jaafar, W. S., Mohd Said, M. N., Saad, S. N. M., & Mohan, M. (2023). UAV Implementations in Urban Planning and Related Sectors of Rapidly Developing Nations: A Review and Future Perspectives for Malaysia. *Remote Sensing (Basel)*, 15(11), 2845. DOI: 10.3390/rs15112845

Nahar, S. (2024). Modeling the effects of artificial intelligence (AI)-based innovation on sustainable development goals (SDGs): Applying a system dynamics perspective in a cross-country setting. *Technological Forecasting and Social Change*, 201, 123203. DOI: 10.1016/j.techfore.2023.123203

Neimark, B. (2023). *Hottest of the Hotspots: The Rise of Eco-precarious Conservation Labor in Madagascar*. University of Arizona Press. DOI: 10.2307/jj.7941382

Newman, R. J., Enns, C., Capitani, C., Thorn, J. P., Courtney-Mustaphi, C. J., Buckton, S. J., Om, E. S., Fazey, I., Haji, T. A., Nchimbi, A. Y., Kariuki, R. W., & Marchant, R. A. (2024). 'Kesho'Scenario Development for Supporting Water-Energy Food Security under Future Conditions in Zanzibar. *Land (Basel)*, 13(2), 195. DOI: 10.3390/land13020195

. Ngadze, F. (2023). Applying the Safe and Just Operating Space (SJOS) framework to Sustainable Development in Zimbabwe.

Oliveira Fiorini, A. C., Rua Rodriguez Rochedo, P., Angelkorte, G., Diuana, F. A., Império, M., Silva Carvalho, L., Nogueira Morais, T., Vasquez-Arroyo, E., & Schaeffer, R. (2023). Investigating biodiversity trends in different mitigation scenarios with a national integrated assessment model. *Journal of Integrative Environmental Sciences*, 20(1), 2239323. DOI: 10.1080/1943815X.2023.2239323

Özkavaf Şenalp, S. I. L. A. (2023). Exploring resilience of socio-ecological productive landscape through understanding change, impact and response among farmers: the case of northwestern Ankara.

Papas, M. (2023). *Capacity-Building and the Water-Energy-Food Nexus: Rethinking Integration in the Asia-Pacific*. Taylor & Francis. DOI: 10.4324/9780429342905

Platjouw, F. M., & Pozdnakova, A. (Eds.). (2023). *The Environmental Rule of Law for Oceans: Designing Legal Solutions*. Cambridge University Press. DOI: 10.1017/9781009253741

Pörtner, H. O., Scholes, R. J., Arneth, A., Barnes, D. K. A., Burrows, M. T., Diamond, S. E., Duarte, C. M., Kiessling, W., Leadley, P., Managi, S., McElwee, P., Midgley, G., Ngo, H. T., Obura, D., Pascual, U., Sankaran, M., Shin, Y. J., & Val, A. L. (2023). Overcoming the coupled climate and biodiversity crises and their societal impacts. *Science*, 380(6642), eabl4881. DOI: 10.1126/science.abl4881 PMID: 37079687

Prasanna, S., Ameen, A., Saurav, K., Singh, D. B., Tripathy, A., Sharma, P., ... Sanjaya, K. (2023). *Empowerment and equality navigating human rights law in a complex world*. Institute of Legal Education.

Quamar, M. M., Al-Ramadan, B., Khan, K., Shafiullah, M., & El Ferik, S. (2023). Advancements and Applications of Drone-Integrated Geographic Information System Technology—A Review. *Remote Sensing (Basel)*, 15(20), 5039. DOI: 10.3390/rs15205039

Qwaider, S., Al-Ramadan, B., Shafiullah, M., Islam, A., & Worku, M. Y. (2023). GIS-Based Progress Monitoring of SDGs towards Achieving Saudi Vision 2030. *Remote Sensing (Basel)*, 15(24), 5770. DOI: 10.3390/rs15245770

Santarsiero, V. (2023). Spatial analyzes and remote sensing for land cover change dynamics: Assessing in a spatial planning.

Sarker, M. N. I., Hossain, B., Shi, G., & Firdaus, R. R. (2023). Promoting net-zero economy through climate-smart agriculture: Transition towards sustainability. *Sustainability Science*, 18(5), 2107–2119. DOI: 10.1007/s11625-023-01379-0

Sharma, A., & Singh, B. (2022). Measuring Impact of E-commerce on Small Scale Business: A Systematic Review. *Journal of Corporate Governance and International Business Law*, 5(1).

Sheng, S., & Lian, H. (2023). The spatial pattern evolution of rural settlements and multi-scenario simulations since the initiation of the reform and opening up policy in China. *Land (Basel)*, 12(9), 1763. DOI: 10.3390/land12091763

Shinde, V. R., Mishra, R. R., Bhonde, U., & Vaidya, H. (Eds.). (2023). *Managing Urban Rivers: From Planning to Practice*. Elsevier.

Shukla, G., Bhat, J. A., Chakravarty, S., Almutairi, A., Li, M., & Iyer-Raniga, U. (2023). *Floristic Diversity: Biology and Conservation*. BoD–Books on Demand. DOI: 10.5772/intechopen.100674

Singh, B. (2022). Relevance of Agriculture-Nutrition Linkage for Human Healthcare: A Conceptual Legal Framework of Implication and Pathways. *Justice and Law Bulletin*, 1(1), 44–49.

Singh, B. (2022). COVID-19 Pandemic and Public Healthcare: Endless Downward Spiral or Solution via Rapid Legal and Health Services Implementation with Patient Monitoring Program. *Justice and Law Bulletin*, 1(1), 1–7.

Singh, B. (2023). Blockchain Technology in Renovating Healthcare: Legal and Future Perspectives. In *Revolutionizing Healthcare Through Artificial Intelligence and Internet of Things Applications* (pp. 177-186). IGI Global.

Singh, B. (2023). Federated Learning for Envision Future Trajectory Smart Transport System for Climate Preservation and Smart Green Planet: Insights into Global Governance and SDG-9 (Industry, Innovation and Infrastructure). *National Journal of Environmental Law*, 6(2), 6–17.

Singh, B. (2024). Lensing Legal Dynamics for Examining Responsibility and Deliberation of Generative AI-Tethered Technological Privacy Concerns: Infringements and Use of Personal Data by Nefarious Actors. In Ara, A., & Ara, A. (Eds.), *Exploring the Ethical Implications of Generative AI* (pp. 146–167). IGI Global., DOI: 10.4018/979-8-3693-1565-1.ch009

Singh, B. (2024). Social Cognition of Incarcerated Women and Children: Addressing Exposure to Infectious Diseases and Legal Outcomes. In Reddy, K. (Ed.), *Principles and Clinical Interventions in Social Cognition* (pp. 236–251). IGI Global., DOI: 10.4018/979-8-3693-1265-0.ch014

Singh, B. (2024). Evolutionary Global Neuroscience for Cognition and Brain Health: Strengthening Innovation in Brain Science. In Prabhakar, P. (Ed.), *Biomedical Research Developments for Improved Healthcare* (pp. 246–272). IGI Global., DOI: 10.4018/979-8-3693-1922-2.ch012

Singh, B., & Kaunert, C. (2024). Computational Thinking for Innovative Solutions and Problem-Solving Techniques: Transforming Conventional Education to Futuristic Interdisciplinary Higher Education. In Fonkam, M., & Vajjhala, N. (Eds.), *Revolutionizing Curricula Through Computational Thinking, Logic, and Problem Solving* (pp. 60–82). IGI Global., DOI: 10.4018/979-8-3693-1974-1.ch004

Singh, B., & Kaunert, C. (2024). Revealing Green Finance Mobilization: Harnessing FinTech and Blockchain Innovations to Surmount Barriers and Foster New Investment Avenues. In Jafar, S., Rodriguez, R., Kannan, H., Akhtar, S., & Plugmann, P. (Eds.), *Harnessing Blockchain-Digital Twin Fusion for Sustainable Investments* (pp. 265–286). IGI Global., DOI: 10.4018/979-8-3693-1878-2.ch011

Singh, B., & Kaunert, C. (2024). Computational Thinking for Innovative Solutions and Problem-Solving Techniques: Transforming Conventional Education to Futuristic Interdisciplinary Higher Education. In Fonkam, M., & Vajjhala, N. (Eds.), *Revolutionizing Curricula Through Computational Thinking, Logic, and Problem Solving* (pp. 60–82). IGI Global., DOI: 10.4018/979-8-3693-1974-1.ch004

Singh, B., & Kaunert, C. (2024). Wind and Solar Energy as Renewable Energy for Sustainable Global Future: Projecting Future Multi-Sector Sustainable Policies and Innovation. In Ara, A., & Thakore, R. (Eds.), *Promoting Multi-Sector Sustainability With Policy and Innovation* (pp. 210–245). IGI Global., DOI: 10.4018/979-8-3693-2113-3.ch009

Singh, B., & Kaunert, C. (2024). Aroma of Highly Smart Internet of Medical Things (IoMT) and Lightweight EdgeTrust Expansion Medical Care Facilities for Electronic Healthcare Systems: Fortified-Chain Architecture for Remote Patient Monitoring and Privacy Protection Beyond Imagination. In Hassan, A., Bhattacharya, P., Tikadar, S., Dutta, P., & Sagayam, M. (Eds.), *Lightweight Digital Trust Architectures in the Internet of Medical Things (IoMT)* (pp. 196–212). IGI Global., DOI: 10.4018/979-8-3693-2109-6.ch011

Singh, B., Vig, K., Dutta, P. K., & Kaunert, C. (2024). Unraveling Agile Transformation for Customer Satisfaction in Changing Market Conditions: Roadmap for Industry Embracing Change in Project Management. In Misra, S., Jadeja, R., & Mittal, M. (Eds.), *Practical Approaches to Agile Project Management* (pp. 305–321). IGI Global., DOI: 10.4018/979-8-3693-3318-1.ch017

Sliuzas, R., Mwesigye, P., Kisembo, T., Wamboga, J., & Ajambo, E. (2023). Land Governance for Climate Resilience.

Sobha, P. (2023). *Future Energy Landscapes in Northern Sweden: Sustainable Transition Scenarios for Municipalities* (Doctoral dissertation, Luleå University of Technology).

Sofi, I. I., Shah, M. A., & Ganie, A. H. (2023). Integrating human footprint with ensemble modelling identifies priority habitats for conservation: A case study in the distributional range of Arnebia euchroma, a vulnerable species. *Environmental Monitoring and Assessment*, 195(8), 914. DOI: 10.1007/s10661-023-11528-8 PMID: 37395941

Terry, N., Castro, A., Chibwe, B., Karuri-Sebina, G., Savu, C., & Pereira, L. (2024). Inviting a decolonial praxis for future imaginaries of nature: Introducing the Entangled Time Tree. *Environmental Science & Policy*, 151, 103615. DOI: 10.1016/j.envsci.2023.103615

van Dam, A. A., Robertson, H. A., Prieler, R., Dubey, A., & Finlayson, C. M. (2024). Recognizing diversity in wetlands and farming systems to support sustainable agriculture and conserve wetlands.

Varughese, C., Henry, L., Morris, A., Bickerton, S., Rattenbury, N., Mankelow, C., Gorman, A., Katavich-Barton, S., & Dhopade, P. (2023). The intersection of space and sustainability: The need for a transdisciplinary and bi-cultural approach. *Acta Astronautica*, 211, 684–701. DOI: 10.1016/j.actaastro.2023.07.009

Watson, J. E., Ellis, E. C., Pillay, R., Williams, B. A., & Venter, O. (2023). Mapping Industrial Influences on Earth's Ecology. *Annual Review of Environment and Resources*, 48(1), 289–317. DOI: 10.1146/annurev-environ-112420-013640

Wellmann, T. (2023). Remote Sensing for sustainable and resilient. *Cities (London, England)*.

Wicki, S., Black, B., Kurmann, M., & Grêt-Regamey, A. (2023). Archetypes of social-ecological-technological systems for managing ecological infrastructure. *Environmental Research Letters*, 19(1), 014038. DOI: 10.1088/1748-9326/ad1080

Yousuf, A. H. (2023). Integrating ecosystem–based management and marine spatial planning for sustainable ocean governance in the Bay of Bengal.

Zhang, Z. (2023). *Mitigating Investment Risks in Nature-Based Solutions: A Strategic Approach Towards Sustainable Project Implementation* (Doctoral dissertation, Massachusetts Institute of Technology).

Chapter 3
AI and ML Applications in Wildlife Conservation and Forest Management: A Comprehensive Review

Hind Hammouch
https://orcid.org/0000-0002-5897-1649
Faculty of Legal, Economic, and Social Sciences of Fez, Morocco

Anjali Raghav
https://orcid.org/0009-0004-0248-7956
Sharda University, India

ABSTRACT

AI and ML have opened new avenues into wildlife conservation, linking efforts worldwide to improve forest management. However, a comprehensive review highlights the manifold applications of these new technologies for biodiversity conservation and natural resource management. These technologies are automated species identification and population monitoring to predict habitat degradation and anti-poaching strategies, all contributing immensely to the preservation of Wildlife. This chapter outlines how AI-powered drones and sensor networks can quickly monitor environments, find illegal actions as they happen, and collect reams of data on wildlife behavior. It also explores the opportunities for ML algorithms to study intricate ecological inter-relationships, predict climate change effects & optimize forest restoration and management activities. AI and ML provide a roadmap of the great promise that emergent technologies hold in providing new configurations in Wildlife Conservation and Forest Management to promote a Sustainable future.

DOI: 10.4018/979-8-3693-7565-5.ch003

1. INTRODUCTION

Protecting the environment in the world includes wildlife conservation and forest management as an aspect of environmental stewardship to maintain biodiversity on Earth so that natural resources will not be used up. Wildlife conservation seeks to protect wild species and their habitats, paying attention to harmony between wildlife, ecosystems, and human activities (Wicki et al., 2023). This is crucial to the conservation successes that ensure viable future populations, as human pressure on natural habitats only increases with time (Potner et al., 2023). Forest management is a complex of administrative, legal, economic, ecological, and technical activities that ensure the maintenance or conservation of all functions available in forests (Watson et al., 2023). Though invaluable, traditional conservation practices have proven to be fraught with difficulty. Deforestation and urbanization are significant issues regarding forests, which lead to habitat loss, creating biodiversity and ecological imbalance (Kennedy et al., 2023). Climate change is an additional threat to wildlife and forest ecosystems as it works in changing habitats for the better or worse but at a cost of survival rates. Old-fashioned methods often bend and break under these pressures, with minimal resources running on archaic mechanics. Compounded by inter-specific competition and exacerbated, conflict is prompted when animals come into human settlements for food or shelter, leading to dangerous interactions that could threaten wildlife populations and humans. The ever-highest resource lure is in jeopardy (Singh & Kaunert, 2024). However, technology has been used more proactively over the last few years to solve these issues. Advancements in Artificial Intelligence (AI) and Machine Learning (ML) help revolutionize wildlife conservation and forest management with an efficient toolset that aids researchers in developing tools for monitoring ecosystems, predicting trends, and optimizing resource management. AI algorithms could collect and analyze countless satellite image sensor network data to monitor wildlife preserves where animals move, poaching, and illegal logging (Dhali et al., 2023).

By utilizing past data, machine learning models can predict the potential effects of climate change on a particular species or habitat and push conservationists to take preventive action. These tech solutions increase conservation efficiency by allowing real-time data management and evidence-based decision-making. It assesses how AI and ML could be added to existing frameworks to obtain better wildlife conservation and forest management outcomes. The characterizing state-of-the-art technological applications in these domains, recognizing information or deployment gaps, and ii recommending that best practices be followed by means for implementation to improve conservatoire effectiveness (Singh, 2023). Also, investigating case studies of technology-based conservation to ameliorate circumstances builds a foundation for potential solutions that can be applied globally (Hansen et al., 2023).

2. TECHNOLOGICAL ADVANCEMENTS IN WILDLIFE CONSERVATION

The application of Artificial Intelligence (AI) and Machine Learning (ML) in wildlife conservation has brought new perspectives, improving several folds for monitoring, anti-poaching measures, and habitat analysis (Singh et al., 2022). The potential of this technology is game-changing for conservation, enabling access to real-time data and the adoption of predictive analytics and advanced monitoring capabilities. Some of those advancements are as follows:

Monitoring and Surveillance: Historically, monitoring wildlife necessitated manual field observations that were inefficient in time and cost-consuming; additionally, this generated datasets that were restrained in scale. The transformative technologies that have emerged for real-time wildlife monitoring are camera traps and AI-powered drones (Batisha, 2023). Motion-sensitive camera traps can be placed intelligently in the field to capture images or video shots of movement, typically done prior over a range of habitats. These images are analyzed by AI algorithms, which identify species of interest and keep an eye on animal behavior while estimating population sizes. Drones or unmanned aerial vehicles (UAV) have increased monitoring. They can fly for long distances and snap high-resolution images of wildlife in their natural environments. Drones are also invaluable for surveilling remote areas like dense forests or hilly terrains. Their infrared cameras help spot wildlife wandering at night and inform about potential poachers lurking after dark. Computer vision algorithms can automatically identify species, track animal movements, and assess habitat health with AI-driven drones. Conservationists, too, use drones to track orangutan populations in Borneo or illegal loggers in the Amazon Rainforest (Singh, 2024).

Efficient species identification and population estimation are significant obstacles in wildlife conservation. Machine learning (ML) algorithms are introduced to perform these tasks automatically. Deep learning models such as Convolutional Neural Networks (CNNs) can process images from camera traps and drones to identify species based on their key characteristics. They are trained on massive labeled image data and know to distinguish between species even when light or angle changes (Qwaider et al., 2023). This includes Wildbook, an AI-equipped ML platform that can identify individual animals through characteristic markings and patterns. This identification allows us to track individual animals over time, understand where they go when not in the rivers, and attempt population size estimates. This detailed data enables conservationists to make better-informed decisions about how best to protect endangered species (Singh, 2024).

Illegal hunting continues to kill rare animals. AI and ML are effectively delivering predictive analytics and real-time monitoring solutions to combat this illegitimate activity. Based on historical poaching incident data, such as date, time, location, and other environmental factors, predictive models can forecast where further activities may occur. This data helps park authorities dispatch rangers intelligently, allocating resources where needs are the greatest (Singh, 2023).

For example, an AI system called the Protection Assistant for Wildlife Security (PAWS) has been used to help counter poaching. The game theory algorithms use historical data, animal migratory patterns, and environmental factors to predict hot spots for poaching. PAWS Based on poachers' movements discovered by the system, it suggests patrolling some routes to maximize the chances of catching them cost-effectively (Sobha, 2023). This method has proven successful in Uganda, Cambodia, and Botswana, decreasing poaching events at an impressive rate.

Many AI implementations have been used to combat poaching in the real world. Conservationists in South Africa's Kruger National Park are using AI-powered acoustic sensors that pick up gunshots, a hallmark of would-be poachers. The sensors can distinguish gunshots from noises similar to the human ear, like thunder or vehicles passing by, so you know with certainty what triggered the detection. The sensors alert the rangers instantly when a gunshot is detected so they can act as soon as possible. This includes using facial recognition technology with rhinos. The AI algorithms use images of the rhinos' faces, which are as unique to each animal as human fingerprints. The rangers have a database of each implanted rhino, letting them know if one goes missing or is killed by poachers (Muhmad Kamarulzaman et al., 2023).

Modern satellite imagery combined with ML has created enormous potential for wildlife habitat mapping and analysis. Previously, habitat assessment was conducted via manual surveys, achieving limited scope on a commercial scale. However, satellites can take pictures of large expanses, giving an overview of entire ecosystems and their evolution over time. Conservationists can use ML algorithms to analyze these images and identify patterns a human eye might fail to recognize. For example, small variations in vegetation or water levels suggest environmental trouble is ahead (Singh, 2023). An example is the tracking of deforestation in real-time, as seen by this platform, using satellite imagery and ML on a massive scale. By processing satellite imagery and then comparing it with historical results, the system can also identify cases of illegal logging. That has allowed authorities to act fast, stopping more of the wilderness from going up in flames. Likewise, it has enabled mapping coral reefs and management actions on water temperature and, more broadly, assessing marine ecosystem health using satellite data water (Nahar, 2024).

Migration ecology is critical to conserving large herbivores as it highlights essential habitats and migration routes that must be conserved. Technology such as AI and ML has further enabled monitoring migration patterns by using data from GPS collars, satellite tags, etc., to trap animals. Data mining algorithms can automatically analyze vast data to discern patterns and derive future trajectories (machine learning), empowering conservationists to design interventions proactively. One example is the Elephant Listening Project, which employs AI-decoded sounds recorded by acoustic sensors in rainforests. The sensors detect the elephants' low-frequency rumbles, which are too deep for human ears to hear but can travel many kilometers. Animals' 3D models, by capturing these noises, researchers can piece together critical aspects of elephant behavior without waiting for them to appear through the brush where they cannot be visually monitored. This information is essential for forming safe migration corridors) for those who migrate through a secure passage (Desai et al., 2023).

Satellite imagery has also been examined using ML models to gauge seasonal changes in vegetation. These models predict food availability, which herbivores rely on, and hence migration patterns such as deer or antelope. This knowledge will be helpful to conservationists, as they may now know how to predict where wildlife conflict is likely due to its overlap with human activity (Sharma & Singh, 2022).

3. AL AND ML FOREST MANAGEMENT

Artificial Intelligence (AI) and Machine Learning (ML) in forestry have revolutionized how we monitor, preserve, and maintain forest ecosystems. AI and ML can generate real-time data analysis and predictions, contribute to the sustainable management of resources (especially wood) for landowners, and remain useful tools in climate change mitigation practices. Here is a breakdown of what AI and ML are up to in the world of forest health, sustainable forestry practices, and climate change mitigation (Ashukem & Sama, 2023).

3.1 Forest Health Monitoring

Diseases and pest invasions are serious threats to forests worldwide, sometimes causing substantial ecological and economic losses. Early detection of such problems may prevent them from spreading and causing further damage. Machine learning (ML) algorithms have been highly efficient in preventing these scenarios from happening by detecting premonitory signs that usually occur before a disease or pest overcomes an entire field (Cornford et al., 2023).

ML models can process extensive and frequent volumes of data coming from different sources such as satellite images, drone footage, and even ground sensors, to find anomalies in the health of vegetation. For instance, they can detect perceivable differences in hue, reflectance, and canopy structure indicative of disease or pest infestation. By working with data collected from infrared and multispectral sensors, machine learning models can identify pre-symptomatic stress in trees (including a lack of water or imbalances among nutrients) that often appear months before outward manifestations are visible to human observers. In Canada, AI and ML technologies are being used to prevent the spread of a mountain pine beetle pest that has taken its toll on millions of hectares (hector) of Pine forests (Singh, 2022). Using satellite data and meteorological trends, ML models can foresee areas prone to beetle infestations so that foresters do not have to wait for dead trees before taking action. For example, cyclone monitoring in the Indian Ocean and tropical disease detection affecting rubber and palm oil plantations with early warnings to farmers on farm-related gutters.

Deforestation still accounts for an essential part of the environmental problems, including those related to biodiversity loss, climate change, and soil erosion. AI technologies are crucial in tracing deforestation patterns based on monitoring forests by satellite imagery and analyzing historical data to detect naturally endangered logging activities and forest regeneration over time. Unlike the manual survey and delayed reporting of traditional methods, AI can send immediate alerts, which can help law enforcement agencies take action against illegal activities at an early stage (Koh et al., 2023).

Deforestation in real-time by analyzing satellite pictures and using these machine learning algorithms. It can tell the difference between natural changes, such as seasonal leaf shedding, and human-caused deforestation due to logging or agricultural expansion. Using AI, the same models can predict where deforestation is likely to occur by incorporating variables like distance from roads, population density, and historical land use. Policymakers can take preventative action by predicting where deforestation is expected, such as establishing protected areas or encouraging sustainable land use (Sheng & Lian, 2023).

Indeed, as David Erlendson of Rainforest Connection (RFCx) has already shown, in the Amazon rainforest where AI is used to thwart illegal logging operations, AI can also hear chainsaws or the rev of a truck engine and then inform authorities not only that there is potential unlawful activity happening in some remote area. Illegal logging is complex and challenging to manage, but new technologies used by satellites can reduce deforestation and provide crucial information on conservation policies' effectiveness (Singh, 2022). Using technology, precision forestry results in managing forest resources more efficiently, allowing balanced, sustainable use while achieving higher productivity. In precision forestry, AI and ML make data-based predictions that help foresters decide about planting new trees, thinning, or harvesting existing ones. Topher also explained that those companies and technologies provide some insight into soil type information. Still, when it comes to best forestry practices, there is no replacement for experience/practice on the ground measuring trees (Ashukem, 2024).

For example, when planting trees, you should plant them according to weather conditions like soil moisture and temperature, along with sufficient nutrients. During the growing stages, ML algorithms can take constant readings of tree health and determine when thinning is necessary. In the harvest phase, AI can evaluate tree size and market demand to determine when trees are at their peak for cutting schedules so that fully matured ones get cut down while less developed ones remain (Singh & Kaunert, 2024).

Countries like Finland are leading the way in forest management with AI integrated into precision forestry practices to improve and enable sustainable wood production. When armed with AI and sensors, drones flown at low altitudes can allow foresters to measure the health of trees or estimate their growth rates without disturbing natural ecosystems. Such technologies have provided for the more efficient management of resources, less waste, and higher local community economic returns.

3.2 Examples Of Improved Forest Management Outcomes Through Case Studies

Multiple testimonies show how AI and ML applications have successfully enhanced forest management. The Forest Service in the United States has used AI for tree density monitoring and wildfire risk prediction. Using satellite images and historical fire data, AI models can pinpoint where vegetation is dense, dry, and ripe for a wildfire. This allows authorities to take preventative steps, including reducing hazard burns that reduce fuel loads and decrease the risk of dangerous fires (Prasanna et al., 2023).

AI can map rainforests like Brazil with the Amazon Conservation Team. The rainforests of South America and Southeast Asia are two ecosystems that could benefit from an increased use of drones. On the other hand, AI models process data from satellites to monitor land use changes, such as forests being cut down and converted into crop fields and revealing some of the drivers behind deforestation. These findings inform policy and Conservation action, such as reforestation projects or the defense of indigenous territories (Terry et al. 2024). Forests are essential for combating climate change as they function as carbon sinks, which help to remove carbon dioxide from the atmosphere and store it in trees and soil. Quantifying how much carbon forests take up is essential to fight climate change. AI models now predict the rate at which individual trees sequester carbon by training them on things like tree species, age, soil composition, and climatic factors (Ashukem & Sama, 2023). These models give policymakers a way to predict the amount of carbon each forest can store and determine if any offset project will do enough good for it to be worth supporting.

Some of its applications could include the likelihood of one tree species or another growing under different conditions and environments, allowing foresters in reforestation projects to choose the most adaptive type. It was also seen planting billions of trees nationwide successfully; hence, it was the approach taken to create China's "Green Great Wall" against desertification. Artificial intelligence models were then implemented to determine what species would grow best in arid areas and contribute more successful results to the reforestation project (Fopa et al., 2023).

The climate of forests can substantially affect the growth and biodiversity of forest ecosystems, and those changes depend on climatic factors such as temperature and precipitation that directly relate to the ecosystem's health status. When AI is combined with climate models, researchers can articulate the long-term health of forests and predict how they might react to a changing climate. By systematically integrating these data, this partnership provides deep insights into the resilience AI has been employed to develop "digital twins" of forests virtual replicas that model real-world ecosystems (Ozkavaf Senalp, 2023). Different scenarios are tested on a digital model to explore the impacts of deforestation or climate change due to conservation interventions. The bottom line is that the knowledge from these simulations' guides forest managers in their decisions about how they can most effectively hold on to forests as climate changes (Cumming et al., 2023).

4. BENEFITS AND CHALLENGES OF AI & ML IN CONSERVATION EFFORTS

Artificial Intelligence (AI) and Machine Learning (ML) benefit wildlife conservation and forest management, but it also presents some critical challenges that must be tackled for sustainable implementation with ethics. The following subsection delves into these technologies' significant benefits and challenges, illustrating how closely intertwined AI and ML can be when applied to conservation on a field study level (Cork et al. 2023).

The great thing about AI and ML in conservation is the improved accuracy and increased efficiency through better monitoring and data analysis. Conventional methods of monitoring wildlife, such as statistical surveys or trail cameras, are more labor-intensive and expensive. On the other hand, AI tools are used to ingest vast amounts of data very quickly and identify species, track their populations, deduce trends in ecosystems, etc., without much human intervention. In particular, AI algorithms enable the automatic analysis of images from camera traps to recognize and even count different species down to the individual level with much higher precision than human experts could ever hope for while managing not only similar-looking animals but subtle differences between them as well. This automates an aspect of the analysis and means conservationists can spend more time thinking about what to do rather than having their work dominated by data manipulation (Gadsden et al., 2023).

Furthermore, AI and ML technology have made tracking areas far from our sight, like dense forests or hilly terrain, easier. AI-powered drone cameras quickly survey large areas in high-quality resolution, and machine learning algorithms analyze images to detect trends or anomalies. This allows us to track animal migration, monitor forest health, and detect poaching or logging. In addition, because AI can analyze data in real-time and trigger instant alerts, it also enables faster response by the authorities to any threats to wildlife or natural resources (Lamalle & Stoett, 2023).

AI-based conservation means that it is cost-effective and easily scalable. The setup cost of AI systems can be expensive at first, but the long-term benefits are worth it. The systems can work 24×7 without fatigue through intelligent built-in cameras, reducing the amount of field worker personnel needed. Once you train an AI model, it is easy to scale- up. They can be distributed over many other regions, which makes them an exciting application for conservation needs that demand monitoring tools all over habitats. In other illustrative use cases, AI algorithms process satellite imagery to track deforestation on multiple continents and offer a holistic view of forest health that helps inform worldwide conservation projects (Dunlap, 2023).

AI and ML in conservation present challenges. One of the biggest concerns is privacy. AI systems process data, particularly from far-away sensors like drones and satellites; this can lead to accidentally including potentially sensitive local in the analysis stream. The most vital aspect of this data is that even if it falls into the wrong hands, others will not have access to exploit these resources, causing just as much damage for years or decades, which should be necessary when communities spend so much time, money and energy in doing everything they can to protect their respective environments (Newman et al. 2024).

A fourth challenge is that AI algorithms can carry inherent bias. AI systems learn from data, and when used with biased or incomplete information (used to train them), these algorithms might produce inaccurate output. One such example is using an AI model trained on data broadly representative of one region, and hence behavior well in that area but not so suitable to other regions with different species or environmental conditions. This creates the potential for biased monitoring and conservation strategies, which could be counterproductive (Ghosh et al., 2023). AI needs valid and complete data sets representing all walks of life and constant scrutiny and adaptation algorithms whose accuracy holds in various contexts.

This also places technical constraints on how AI can be best harnessed for conservation efforts. AI algorithms might be the most advanced, but they are not error-free. Many factors affect the effectiveness of AI-driven tools, such as poor-quality images and extreme weather conditions or lighting. If, for instance, a drone's camera fails to get high-quality photos for any reason, such as fog or rain, this will cause mistakes in species identification or habitat assessment. Moreover, most conservation areas are remote and may lack internet connectivity to avail of cloud-based AI solutions that require real-time data transfer. There needs to be a more flexible AI system that can service capabilities across the broad spectrum of arduous environments beyond these specific limitations (Shinde et al., 2023).

Another critical thing to consider when using AI and ML for conservation is ethical concerns. Although such technologies provide valuable tools for wildlife protection, their application has also posed ethical considerations over surveillance and data-gathering privacy issues. For example, using drones to detect wildlife is intrusive and may disturb animals in their natural environment (Cardwell, 2023). However, conservationists are also required by ethical and scientific standards and have tremendous respect for the ecosystems they study to minimize harm while ensuring data is appropriately collected. While AI can be an excellent tool in conservation, any application of this technology must occur within the realms of transparency and accountability so that various agencies use this powerful innovation ethically and responsibly (Aryal et al., 2023). This involves making transparent how data is gathered, filtered, and used and engaging local communities trustfully at every stage of decision-making processes regarding conservation.

Conservation problems are too complicated to tackle with AI alone (Ngadze, 2023). Therefore, partnerships between different sectors will achieve the most influential tech solutions. Conservation problems are not truly conservation problems because they do not occur in a vacuum (Sofi et al., 2023). These issues exist at the nexus of political, economic, and social influences that all feed into one another ad infinitum until nothing is left to compete over (Shukla et., 2023). Solving these challenges is something that AI models should work on in collaboration with technologists, conservationists, policymakers, and local communities. Implementing such a system would force the AI to consider not only environmental factors but also socio-economic indicators of deforestation, as many cases occur in connection with illegal logging or agricultural expansion. Changes addressed through this level of collaboration can then guarantee AI solutions that are both technically and socially/politically viable, as well as more long-lasting conservation results (Malekpour et al., 2023).

5. FUTURE DIRECTIONS AND RECOMMENDATIONS

The future of wildlife conservation and forest management lies in the continued integration of Artificial Intelligence (AI) with other emerging technologies, such as the Internet of Things (IoT) and big data analytics. The combination of AI, IoT, and big data has the potential to create a more interconnected and intelligent system for monitoring ecosystems. For instance, IoT devices like sensors, drones, and camera traps can continuously collect vast amounts of environmental data. At the same time, AI algorithms analyze this data to provide real-time insights into wildlife activity, forest health, and ecosystem changes. By connecting these devices through IoT networks, conservationists can create a robust and comprehensive system that offers enhanced capabilities for tracking animal movements, detecting illegal activities, and identifying environmental threats. Big data analytics can further help by processing and interpreting these large datasets, revealing patterns and trends that might not be apparent through traditional methods. This integration could lead to a new era of precision conservation, where resources are allocated more efficiently, and interventions are targeted more accurately.

To make the most of these technological advancements, policymakers, researchers, and technologists should consider several recommendations. Policymakers should focus on creating frameworks that encourage the use of AI and related technologies in conservation while addressing data privacy and ethical considerations. Clear regulations that outline the responsible use of AI, data sharing, and collaboration across borders can enhance the effectiveness of conservation efforts. Researchers should prioritize the development of AI models that are adaptable to different ecosystems and capable of functioning under various environmental conditions. This includes building more robust machine learning algorithms that can be learned from diverse data sets, minimizing the risk of bias, and ensuring accuracy across different regions. Moreover, interdisciplinary collaboration between conservation scientists, data scientists, and AI experts is essential to develop technologically sound and ecologically sensitive solutions.

Technologists can contribute by focusing on the creation of more user-friendly and cost-effective AI solutions for conservation. Many organizations in this field operate with limited budgets, so scalable and affordable technologies are crucial for widespread adoption. Furthermore, developers should work towards improving the energy efficiency of AI systems to reduce the carbon footprint of conservation technologies, ensuring that the tools used to protect the environment do not inadvertently harm it. Involving local communities in deploying these technologies is also critical, as their knowledge and experience can provide valuable insights into the local ecosystems and improve the accuracy and effectiveness of AI-driven conservation projects.

6. CONCLUSION

Integrating AI and ML in wildlife conservation and forest management has shown promising results, offering more efficient, accurate, and scalable solutions for monitoring and protecting ecosystems. From real-time species identification to predictive analytics for anti-poaching initiatives, these technologies have transformed how conservationists approach biodiversity conservation challenges. However, there are still challenges, including data privacy, bias, technical limitations, and ethical considerations. Moving forward, the continued development of AI, combined with IoT and big data and a focus on collaboration between stakeholders, will be key to harnessing the full potential of these technologies. By addressing the existing challenges and promoting responsible, ethical, and innovative use of AI, we can pave the way for a future where technology and nature work hand in hand to sustain our planet's ecosystems.

REFERENCES

Anekwe, I. M. S., Zhou, H., Mkhize, M. M., & Akpasi, S. O. (2024). Addressing Climate Change Challenges in South Africa: A Study in KwaZulu Natal Province. *Climate Crisis: Adaptive Approaches and Sustainability*, 475-496.

Aryal, K., Maraseni, T., & Apan, A. (2023). Examining policy– institution– program (PIP) responses against the drivers of ecosystem dynamics. A chronological review (1960–2020) from Nepal. *Land Use Policy*, 132, 106789. DOI: 10.1016/j.landusepol.2023.106789

Ashukem, J. C. N. (2024). Land grabbing in the Anthropocene and the "questionable" pursuit of sustainability in Africa–highlighting ecological disproportionality. In *Human Rights and the Environment in Africa* (pp. 276–294). Routledge.

Ashukem, J. C. N., & Sama, S. M. (Eds.). (2023). *Domestic and Regional Environmental Laws and Policies in Africa: A Research Companion*. Taylor & Francis. DOI: 10.4324/9781003382256

Banerjee, O., Cicowiez, M., Malek, Ž., Verburg, P. H., Vargas, R., Goodwin, S., Bagstad, K. J., & Murillo, J. Á. (2023). Banking on strong rural livelihoods and the sustainable use of natural capital in post-conflict Colombia. *Environment, Development and Sustainability*, 26(10), 1–22. DOI: 10.1007/s10668-023-03740-w

Basu, M. (2024). *The Routledge Handbook of International Environmental Policy*. Taylor & Francis. DOI: 10.4324/9781003440574

Batisha, A. (2023). A lighthouse to enhance the quality of life in the Nile River basin. *Environmental Economics and Policy Studies*, 1–29. DOI: 10.1007/s10018-023-00380-2

Bogers, M. (2023). *Integrating global sustainability governance: How the Sustainable Development Goals impact institutional and policy integration at the global level* (Doctoral dissertation, Utrecht University).

Bustamante, M., Roy, J., Ospina, D., Achakulwisut, P., Aggarwal, A., Bastos, A., ... & Zscheischler, J. (2023). Ten new insights in climate science 2023/2024. *Global Sustainability*, 1-58.

Cardwell, M. (2023). Results-based agri-environmental scheme design: Legal implications. *Environmental Law Review*, 25(4), 260–288. DOI: 10.1177/14614529231185678

Cork, S., Alexandra, C., Alvarez-Romero, J. G., Bennett, E. M., Berbés-Blázquez, M., Bohensky, E., & Wyborn, C. (2023). Exploring Alternative Futures in the Anthropocene. *Annual Review of Environment and Resources*, 48(1), 25–54. DOI: 10.1146/annurev-environ-112321-095011

Cornford, R., Spooner, F., McRae, L., Purvis, A., & Freeman, R. (2023). Ongoing over-exploitation and delayed responses to environmental change highlight the urgency for action to promote vertebrate recoveries by 2030. *Proceedings of the Royal Society B, 290*(1997), 20230464. DOI: 10.1098/rspb.2023.0464

Cumming, G. S., Adamska, M., Barnes, M. L., Barnett, J., Bellwood, D. R., Cinner, J. E., Cohen, P. J., Donelson, J. M., Fabricius, K., Grafton, R. Q., Grech, A., Gurney, G. G., Hoegh-Guldberg, O., Hoey, A. S., Hoogenboom, M. O., Lau, J., Lovelock, C. E., Lowe, R., Miller, D. J., & Wilson, S. K. (2023). Research priorities for the sustainability of coral-rich western Pacific seascapes. *Regional Environmental Change*, 23(2), 1–15. DOI: 10.1007/s10113-023-02051-0 PMID: 37125023

De Vries, B. J. (2023). *Sustainability science*. Cambridge University Press. DOI: 10.1017/9781009300230

Desai, B. H. (2023). Land-Soil Sustainability: India's International Obligations. In *Soil Law and Governance in India* (pp. 3–24). Springer International Publishing. DOI: 10.1007/978-3-031-32360-7_1

Dhali, M., Hassan, S., & Subramaniam, U. (2023). Comparative analysis of oil and gas legal frameworks in Bangladesh and Nigeria: A pathway towards achieving sustainable energy through policy. *Sustainability (Basel)*, 15(21), 15228. DOI: 10.3390/su152115228

Dunlap, A. (2023). The Structures of Conquest: Debating Extractivism (s), Infrastructures and Environmental Justice for Advancing Post-Development Pathways. *International Development Policy| Revue internationale de politique de développement*.

ESCAP. (2023). The changing landscape of regional cooperation in Asia and the Pacific.

Fopa, V. K., Bayir, N., & Özdal, D. (2023). Assessing the status and spatial-temporal dynamics of the Bamenda Mountains (BM), North West region of Cameroon. *Environmental Monitoring and Assessment*, 195(9), 1053. DOI: 10.1007/s10661-023-11630-x PMID: 37589789

Gadsden, G. I., Golden, N., & Harris, N. C. (2023). Place-Based Bias in Environmental Scholarship Derived from Social–Ecological Landscapes of Fear. *Bioscience*, 73(1), 23–35. DOI: 10.1093/biosci/biac095 PMID: 36643594

Ghosh, S., Dinda, S., Chatterjee, N. D., & Bera, D. (2023). Linking ecological vulnerability and ecosystem service value in a fast-growing metropolitan area of eastern India: A scenario-based sustainability approach. *Environment, Development and Sustainability*, •••, 1–31. DOI: 10.1007/s10668-023-03966-8

Gill, D. A., Blythe, J., Bennett, N., Evans, L., Brown, K., Turner, R. A., Baggio, J. A., Baker, D., Ban, N. C., Brun, V., Claudet, J., Darling, E., Di Franco, A., Epstein, G., Estradivari, , Gray, N. J., Gurney, G. G., Horan, R. P., Jupiter, S. D., & Muthiga, N. A. (2023). Triple exposure: Reducing negative impacts of climate change, blue growth, and conservation on coastal communities. *One Earth*, 6(2), 118–130. DOI: 10.1016/j.oneear.2023.01.010

Hansen, H. H., Bergman, E., Cowx, I. G., Lind, L., Pauna, V. H., & Willis, K. A. (2023). Resilient rivers and connected marine systems: a review of mutual sustainability opportunities. *Global Sustainability*, 1-71.

Jarvelainen, M. (2023). Evaluating the Effectiveness of Forest Conservation as a CO2 Offset Measure for Urban Development: A Case Study. *Forest Ecology and Management*.

Kennedy, C. M., Fariss, B., Oakleaf, J. R., Garnett, S. T., Fernández-Llamazares, Á., Fa, J. E., Baruch-Mordo, S., & Kiesecker, J. (2023). Indigenous Peoples' lands are threatened by industrial development; conversion risk assessment reveals need to support Indigenous stewardship. *One Earth*, 6(8), 1032–1049. DOI: 10.1016/j.oneear.2023.07.006

Khan, I. A. (Ed.). (2024). *Security and Global Governance*. INTERDISCIPLINARY INSTITUTE OF HUMAN SECURITY & GOVERNANCE.

Koh, T., Lye, L. H., & Lum, S. (Eds.). (2023). *Peace With Nature: 50 Inspiring Essays On Nature And The Environment*. World Scientific. DOI: 10.1142/13561

Lamalle, S., & Stoett, P. (Eds.). (2023). *Representations and Rights of the Environment*. Cambridge University Press. DOI: 10.1017/9781108769327

Malekpour, S., Allen, C., Sagar, A., Scholz, I., Persson, Å., Miranda, J. J., Bennich, T., Dube, O. P., Kanie, N., Madise, N., Shackell, N., Montoya, J. C., Pan, J., Hathie, I., Bobylev, S. N., Agard, J., & Al-Ghanim, K. (2023). What scientists need to do to accelerate progress on the SDGs. *Nature*, 621(7978), 250–254. DOI: 10.1038/d41586-023-02808-x PMID: 37704766

Muhmad Kamarulzaman, A. M., Wan Mohd Jaafar, W. S., Mohd Said, M. N., Saad, S. N. M., & Mohan, M. (2023). UAV Implementations in Urban Planning and Related Sectors of Rapidly Developing Nations: A Review and Future Perspectives for Malaysia. *Remote Sensing (Basel)*, 15(11), 2845. DOI: 10.3390/rs15112845

Nahar, S. (2024). Modeling the effects of artificial intelligence (AI)-based innovation on sustainable development goals (SDGs): Applying a system dynamics perspective in a cross-country setting. *Technological Forecasting and Social Change*, 201, 123203. DOI: 10.1016/j.techfore.2023.123203

Neimark, B. (2023). *Hottest of the Hotspots: The Rise of Eco-precarious Conservation Labor in Madagascar*. University of Arizona Press. DOI: 10.2307/jj.7941382

Newman, R. J., Enns, C., Capitani, C., Thorn, J. P., Courtney-Mustaphi, C. J., Buckton, S. J., Om, E. S., Fazey, I., Haji, T. A., Nchimbi, A. Y., Kariuki, R. W., & Marchant, R. A. (2024). 'Kesho' Scenario Development for Supporting Water-Energy Food Security under Future Conditions in Zanzibar. *Land (Basel)*, 13(2), 195. DOI: 10.3390/land13020195

Ngadze, F. (2023). Applying the Safe and Just Operating Space (SJOS) framework to Sustainable Development in Zimbabwe.

Özkavaf Şenalp, S. I. L. A. (2023). Exploring resilience of socio-ecological productive landscape through understanding change, impact and response among farmers: the case of northwestern Ankara.

Papas, M. (2023). *Capacity-Building and the Water-Energy-Food Nexus: Rethinking Integration in the Asia-Pacific*. Taylor & Francis. DOI: 10.4324/9780429342905

Platjouw, F. M., & Pozdnakova, A. (Eds.). (2023). *The Environmental Rule of Law for Oceans: Designing Legal Solutions*. Cambridge University Press. DOI: 10.1017/9781009253741

Pörtner, H. O., Scholes, R. J., Arneth, A., Barnes, D. K. A., Burrows, M. T., Diamond, S. E., Duarte, C. M., Kiessling, W., Leadley, P., Managi, S., McElwee, P., Midgley, G., Ngo, H. T., Obura, D., Pascual, U., Sankaran, M., Shin, Y. J., & Val, A. L. (2023). Overcoming the coupled climate and biodiversity crises and their societal impacts. *Science*, 380(6642), eabl4881. DOI: 10.1126/science.abl4881 PMID: 37079687

Prasanna, S., Ameen, A., Saurav, K., Singh, D. B., Tripathy, A., Sharma, P., ... Sanjaya, K. (2023). *Empowerment and equality navigating human rights law in a complex world.* Institute of Legal Education.

Quamar, M. M., Al-Ramadan, B., Khan, K., Shafiullah, M., & El Ferik, S. (2023). Advancements and Applications of Drone-Integrated Geographic Information System Technology—A Review. *Remote Sensing (Basel)*, 15(20), 5039. DOI: 10.3390/rs15205039

Qwaider, S., Al-Ramadan, B., Shafiullah, M., Islam, A., & Worku, M. Y. (2023). GIS-Based Progress Monitoring of SDGs towards Achieving Saudi Vision 2030. *Remote Sensing (Basel)*, 15(24), 5770. DOI: 10.3390/rs15245770

Santarsiero, V. (2023). Spatial analyses and remote sensing for land cover change dynamics: assessing in a spatial planning.

Sarker, M. N. I., Hossain, B., Shi, G., & Firdaus, R. R. (2023). Promoting net-zero economy through climate-smart agriculture: Transition towards sustainability. *Sustainability Science*, 18(5), 2107–2119. DOI: 10.1007/s11625-023-01379-0

Sharma, A., & Singh, B. (2022). Measuring Impact of E-commerce on Small Scale Business: A Systematic Review. *Journal of Corporate Governance and International Business Law*, 5(1).

Sheng, S., & Lian, H. (2023). The spatial pattern evolution of rural settlements and multi-scenario simulations since the initiation of the reform and opening up policy in China. *Land (Basel)*, 12(9), 1763. DOI: 10.3390/land12091763

Shinde, V. R., Mishra, R. R., Bhonde, U., & Vaidya, H. (Eds.). (2023). *Managing Urban Rivers: From Planning to Practice*. Elsevier.

Shukla, G., Bhat, J. A., Chakravarty, S., Almutairi, A., Li, M., & Iyer-Raniga, U. (2023). *Floristic Diversity: Biology and Conservation*. BoD–Books on Demand. DOI: 10.5772/intechopen.100674

Singh, B. (2022). Relevance of Agriculture-Nutrition Linkage for Human Healthcare: A Conceptual Legal Framework of Implication and Pathways. *Justice and Law Bulletin*, 1(1), 44–49.

Singh, B. (2022). COVID-19 Pandemic and Public Healthcare: Endless Downward Spiral or Solution via Rapid Legal and Health Services Implementation with Patient Monitoring Program. *Justice and Law Bulletin*, 1(1), 1–7.

Singh, B. (2023). Blockchain Technology in Renovating Healthcare: Legal and Future Perspectives. In *Revolutionizing Healthcare Through Artificial Intelligence and Internet of Things Applications* (pp. 177-186). IGI Global.

Singh, B. (2023). Federated Learning for Envision Future Trajectory Smart Transport System for Climate Preservation and Smart Green Planet: Insights into Global Governance and SDG-9 (Industry, Innovation and Infrastructure). *National Journal of Environmental Law*, 6(2), 6–17.

Singh, B. (2024). Lensing Legal Dynamics for Examining Responsibility and Deliberation of Generative AI-Tethered Technological Privacy Concerns: Infringements and Use of Personal Data by Nefarious Actors. In Ara, A., & Ara, A. (Eds.), *Exploring the Ethical Implications of Generative AI* (pp. 146–167). IGI Global., DOI: 10.4018/979-8-3693-1565-1.ch009

Singh, B. (2024). Social Cognition of Incarcerated Women and Children: Addressing Exposure to Infectious Diseases and Legal Outcomes. In Reddy, K. (Ed.), *Principles and Clinical Interventions in Social Cognition* (pp. 236–251). IGI Global., DOI: 10.4018/979-8-3693-1265-0.ch014

Singh, B. (2024). Evolutionary Global Neuroscience for Cognition and Brain Health: Strengthening Innovation in Brain Science. In Prabhakar, P. (Ed.), *Biomedical Research Developments for Improved Healthcare* (pp. 246–272). IGI Global., DOI: 10.4018/979-8-3693-1922-2.ch012

Singh, B., & Kaunert, C. (2024). Computational Thinking for Innovative Solutions and Problem-Solving Techniques: Transforming Conventional Education to Futuristic Interdisciplinary Higher Education. In Fonkam, M., & Vajjhala, N. (Eds.), *Revolutionizing Curricula Through Computational Thinking, Logic, and Problem Solving* (pp. 60–82). IGI Global., DOI: 10.4018/979-8-3693-1974-1.ch004

Singh, B., & Kaunert, C. (2024). Revealing Green Finance Mobilization: Harnessing FinTech and Blockchain Innovations to Surmount Barriers and Foster New Investment Avenues. In Jafar, S., Rodriguez, R., Kannan, H., Akhtar, S., & Plugmann, P. (Eds.), *Harnessing Blockchain-Digital Twin Fusion for Sustainable Investments* (pp. 265–286). IGI Global., DOI: 10.4018/979-8-3693-1878-2.ch011

Singh, B., & Kaunert, C. (2025). Featuring Healthcare, Environment, and Human Rights: Applications of Artificial Intelligence in the Health Domain. In Chakraborty, S., & Satapathy, S. (Eds.), *Gender, Environment, and Human Rights: An Intersectional Exploration* (pp. 45–74). IGI Global., DOI: 10.4018/979-8-3693-6069-9.ch004

Singh, B., & Kaunert, C. (2025). Intelligent Machine Learning Solutions for Cybersecurity: Legal and Ethical Considerations in a Global Context. In Thangam, D. (Ed.), *Advancements in Intelligent Process Automation* (pp. 359–386). IGI Global., DOI: 10.4018/979-8-3693-5380-6.ch014

Singh, B., & Kaunert, C. (2025). Innovative Business Models in Digital and Virtual Reality for the Gig Economy: Space-Age Sustainable Development Projecting Smart Business Practices. In Taherdoost, H., Drazenovic, G., Madanchian, M., Khan, I., & Arshi, O. (Eds.), *Business Transformation in the Era of Digital Disruption* (pp. 127–156). IGI Global., DOI: 10.4018/979-8-3693-7056-8.ch005

Singh, B., Kaunert, C., & Ravesangar, K. (2025). Artificial Intelligence and Employee Well-Being: Projecting Psychological and Physical Healthcare. In Yadav, P. (Ed.), *Human Resource Strategies in the Era of Artificial Intelligence* (pp. 109–134). IGI Global., DOI: 10.4018/979-8-3693-6412-3.ch005

Singh, B., Singh, A., Kaunert, C., Arora, M. K., Lal, S., & Ravesangar, K. (2025). Unleashing the Ethical and Legal Implications of E-Business Revolution: Consumer Privacy and Security Concerns in Phase of Digital Disruption. In Taherdoost, H., Drazenovic, G., Madanchian, M., Khan, I., & Arshi, O. (Eds.), *Business Transformation in the Era of Digital Disruption* (pp. 157–180). IGI Global., DOI: 10.4018/979-8-3693-7056-8.ch006

Sliuzas, R., Mwesigye, P., Kisembo, T., Wamboga, J., & Ajambo, E. (2023). Land Governance for Climate Resilience.

Sobha, P. (2023). *Future Energy Landscapes in Northern Sweden: Sustainable Transition Scenarios for Municipalities* (Doctoral dissertation, Luleå University of Technology).

Sofi, I. I., Shah, M. A., & Ganie, A. H. (2023). Integrating human footprint with ensemble modelling identifies priority habitats for conservation: A case study in the distributional range of Arnebia euchroma, a vulnerable species. *Environmental Monitoring and Assessment*, 195(8), 914. DOI: 10.1007/s10661-023-11528-8 PMID: 37395941

Terry, N., Castro, A., Chibwe, B., Karuri-Sebina, G., Savu, C., & Pereira, L. (2024). Inviting a decolonial praxis for future imaginaries of nature: Introducing the Entangled Time Tree. *Environmental Science & Policy*, 151, 103615. DOI: 10.1016/j.envsci.2023.103615

Varughese, C., Henry, L., Morris, A., Bickerton, S., Rattenbury, N., Mankelow, C., Gorman, A., Katavich-Barton, S., & Dhopade, P. (2023). The intersection of space and sustainability: The need for a transdisciplinary and bi-cultural approach. *Acta Astronautica*, 211, 684–701. DOI: 10.1016/j.actaastro.2023.07.009

Watson, J. E., Ellis, E. C., Pillay, R., Williams, B. A., & Venter, O. (2023). Mapping Industrial Influences on Earth's Ecology. *Annual Review of Environment and Resources*, 48(1), 289–317. DOI: 10.1146/annurev-environ-112420-013640

Wellmann, T. (2023). Remote Sensing for sustainable and resilient. *Cities (London, England).*

Yousuf, A. H. (2023). Integrating ecosystem–based management and marine spatial planning for sustainable ocean governance in the Bay of Bengal.

Zhang, Z. (2023). *Mitigating Investment Risks in Nature-Based Solutions: A Strategic Approach Towards Sustainable Project Implementation* (Doctoral dissertation, Massachusetts Institute of Technology).

Chapter 4
Artificial Intelligence and Machine Learning Technological Tools in Enhancing Efficiency in Wildlife and Forest Conservation

Tarun Kumar Kaushik
https://orcid.org/0000-0002-3279-8307
Sharda University, India

Ravish
Meerut College, India

Anurag Singh
https://orcid.org/0009-0005-1322-7380
Meerut College, India

Anjali Raghav
https://orcid.org/0009-0004-0248-7956
Sharda University, India

Bhupinder Singh
https://orcid.org/0009-0006-4779-2553
Sharda University, India

DOI: 10.4018/979-8-3693-7565-5.ch004

ABSTRACT

General enthusiasm surrounding the possibilities of AI, there are persistent concerns regarding its negative impacts, including substantial energy consumption and issues of ethics. The evaluation has looked at several uses in fields including building, transportation, healthcare, manufacturing, agriculture, and water. Among the numerous techniques used in sustainability regression, DSS-based (Decision Support System) AI models and RL (Reinforcement Learning) are the most often used ones. The assessment also provides direction on the industrial sectors using strategies to include ideas of sustainable development into their operations.

1. INTRODUCTION

Given the widespread impact of climate change on multiple countries, the significance of environmental sustainability in global economic operations is growing. Sustainability is a concept that emphasizes the pursuit of fairness and equality, and is typically characterized by three dimensions: environmental, economic, and social. The coexistence of flourishing economies and environmental deterioration is incompatible, as attaining environmental sustainability involves risks that can affect corporate operations and prospects in a highly competitive market. This concept can be applied to make immediate judgements at several levels, including the global, national, and individual consumer levels. The World Meteorological Organisation estimates that there is an equal chance, around 50/50, that world average temperatures may reach pre-industrial levels (Potner et al., 2023).

The temperature is projected to increase by more than 1.5 degree Celsius within the next five years, according to Alex Knapp in 2022. To address the situation, the UNFCCC establishes fundamental criteria for organizing a conference called COP, which includes the objective of achieving carbon neutrality in the meeting's agenda (Watson et al., 2023). The UK Government has pledged to adopt the International Norm for Event Sustainability Management Systems (ISO20121) and become the first country to reach PAS 2060, the internationally accepted benchmark for carbon neutrality, which is the global norm for achieving zero carbon emissions. The COP26 Sustainability Report provides details on the actions we implemented to ensure an environmentally conscious summit, mitigate and reduce greenhouse gas emissions, and extract valuable insights for future COPs and significant gatherings. This demonstrates the leadership and ambition of countries in the field of sustainable event management (Kennedy et al., 2023).

The growing demand for AI in the pursuit of sustainability leads to a growth in data volume, which in turn results in higher energy requirements for infrastructure in the future. Therefore, it necessitates methodologies for the storage, analysis, and visualization of data to integrate artificial intelligence into a study on sustainable development. On the other hand, some argue that artificial intelligence either significantly influences environmental sustainability issues or is only a component of them (Singh & Kaunert, 2024). The study underlined in terms of resources uses the financial effects of artificial intelligence. Still unsolved, nevertheless, are issues about the sustainability of artificial intelligence and how it affects environmental conditions. Focusing on one facet of sustainability and debating the several challenges faced in that subject, several scholars have carried out a literature study (Dhali et al., 2023). This study intends to clarify the several approaches, challenges, and strategies required in the application of artificial intelligence (AI) for sustainable development. Our goal in providing pertinent insights is to address the following research agendas:

- To explore the many techniques employed in the fields of artificial intelligence and sustainability
- To have a deeper comprehension of the difficulties, possibilities, and obstacles that are considered significant for sustainable development when utilizing artificial intelligence
- To determine the influence of artificial intelligence (AI) on sustainable development (SD) and the specific characteristics of this influence (Singh, 2023)

The literature review section has been partitioned into two stages. First, we underline the exhaustive review of 287 research publications analyzing the challenges faced and the strategies applied to get above such constraints. We also underline the numerous fields of the environment in which artificial intelligence has been applied, including healthcare and supply chains (Hansen et al., 2023). Moreover, we present findings about many AI-integrated solutions applied in many fields. Furthermore, our work has made important additions on two different levels to the corpus of knowledge already in publication. Using PRISMA-based literature, our study of the literature will find and examine the several elements influencing artificial intelligence in the framework of sustainability as well as possible challenges. Moreover, it points forth other possible directions for next studies for academic researchers. This literature focuses on the several strategies used to improve sustainable practices throughout several levels, from local to big, through the usage of artificial intelligence (AI). The research will then concentrate mostly on answering the questions (Batisha, 2023).

Furthermore, the given document will have the following structure:

Particularly about their impact on the environment, Section 2 offers a thorough summary of the body of current research on the junction between artificial intelligence (AI) and sustainability. Section 3 offers comprehensive details on the study technique used here. Section 4 offers the results of the research together with the descriptive data analysis and the field-based approach applied here. We will now go over the possible ramifications of our results in great detail. The constraints of our study and recommendations for additional investigation will come next. At last, Section 6 offers the conclusion (Singh, 2024).

2. LITERATURE REVIEW

Developing a Simple Linear Regression (SLR) model requires an awareness of the scope of the research and the several elements connected to the intended use. Finding the criteria for inclusion and exclusion depends on a clear characterization of the extent of the evaluation. Table 1 offers the exact definitions of some key terms connected to the chosen area.

Environmental Sustainability

Sustainable development practices are challenged by population rates, geographical considerations, infrastructure, health conditions, and infrastructure in many nations (Kuenzel et al., 2016). Moreover, these countries encounter financial barriers while adopting sustainable management methods (Qwaider et al., 2023).

Therefore, we need a thorough evaluation that can help researchers in doing studies in this field and pointing up further actions to reduce the difficulties and barriers related to artificial intelligence in sustainable development. The word "sustainability" has been used in several contexts and is relevant to several industrial sectors, including those engaged in the palm oil industry. From being seen as a constraint, it is now acknowledged as a major tool for greatly raising output (Singh, 2024).

Artificial Intelligence (AI) in Sustainable Development

Through analysis of elements including energy emissions and reductions, carbon dioxide absorption, development of greener transportation networks, monitoring deforestation, weather forecasting, accident prediction, and supply chain management efficiency, artificial intelligence (AI) can accelerate worldwide efforts to protect the environment and conserve resources. The 2020 Vinuesa et al. study looked at how artificial intelligence might either help or hinder the fulfilment of the 169 targets and 17 goals set forth in the 2030 Agenda for Sustainable Development.

In the framework of sustainable development artificial intelligence (AI) is influencing several facets of different sectors. Global energy sectors, environmental results, productivity, and other factors are just a few of the spheres of influence AI is having on several facets of civilization. With an emphasis on including artificial intelligence into many spheres of sustainability and efficiently addressing pragmatic challenges, research in the field of artificial intelligence has lately increased in volume. Observed is the development of artificial intelligence in sectors like mechanics, parallelizing of CPUs, sensors, algorithms, and software. This advancement is enabling the spread of several artificial intelligence applications, so enabling activities such environmental scanning, knowledge exchange, monitoring, and goal development about adaptation. In industrial sectors, artificial intelligence offers a complete assessment of equipment performance and condition along with great monitoring and management capacity (Singh, 2023).

Estupin' and Alvarez (2016) clarified the difficulties related to sustainable development (SD) and have provided large datasets on sustainability that might be carefully investigated to handle the problems of SD (Sachs, 2015).

3. METHODOLOGY FOR CONDUCTING RESEARCH

3.1. A Systematic Literature Review is a Comprehensive and Methodical Examination of Existing Literature

In the sciences, both physical and medicinal, systematic review is a widely used technique. In computer science, it is presently being used more and more to examine the enormous amount of internet-available data. Using a quantitative, repeatable, methodical, and precisely defined technique, one hopes to have a complete and thorough awareness of a certain issue. Using the Preferred Reporting Items for Systematic Reviews and Meta-Analyzes (PRISMA) methodology, which is well-known for its thorough reviewing approach we systematically reviewed the literature. The PRISMA approach provides a complete list of elements that let other researchers do a reproducible evaluation and produce exact data for next projects. The PRISMA approach covers several procedures relevant for field research in the domains of physical science and medicine. Still at the exploratory stage, nonetheless, it is being developed for use in computer science research. By means of the PRISMA-based approach, systematic literature reviews ensure their quality, help readers to evaluate their strengths and shortcomings, enable the replication of review approaches, structure, and format using PRISMA, and so set a benchmark for others in the same field (Sobha, 2023).

We searched the Scopus database thoroughly for papers in our methodical review. Review is important since it provides a larger range of scholarly knowledge, thereby enabling us to better grasp the research we wish to undertake. Research papers included in the Scopus database have been carefully selected, suggesting that we might trust them for academic research to have a certain set of keywords and concepts, we searched the database with an eye towards article titles to perform our study. Using the terms “AI,” “artificial intelligence,” and “machine learning” in combination with Boolean logic specifically, the operators AND and NOT and the terms “sustainability,” “environment,” a search was done to collect pertinent content about “AI” and “sustainability.” By employing these criteria, certain restrictions were imposed: it is possible that general articles pertaining to 'neural network,' 'cognitive,' 'sustainable development,' 'sustainability,' 'sustainably,' or 'sustainable' may have been overlooked. However, these restrictions resulted in a total of 7753 articles in the category of 'Article and Review', which appeared to be like other reviews in this subject (Muhmad Kamarulzaman et al., 2023).

Subsequently, all these publications underwent a manual review as part of the screening phase to ensure their relevance to the subject matter, and any articles that did not meet this criterion were excluded. We have excluded conference papers, books, and conference reviews from this literature assessment. We employed a content analysis methodology to identify themes and assure their trustworthiness. Prior to reporting in section 3.1, we conducted an intercoder reliability test on specific themes. We conducted a search for the terms on Scopus separately and quantitatively assessed the level of agreement between the resulting data. The titles and keywords of the relevant database were reviewed twice to verify the accuracy of the search results. Subsequently, the full-text articles were evaluated for their pertinence, and any duplicates were eliminated. Figure 1 denotes the selection of a flowchart for the purpose of searching publications (Bogers, 2023).

Figure 1. Methodology for choosing research publications

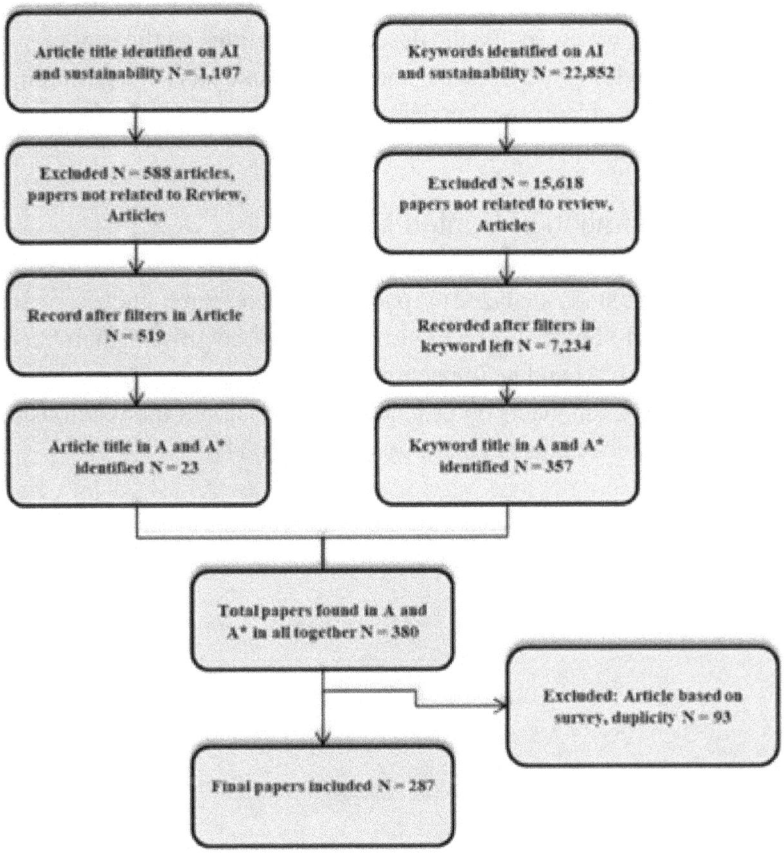

We obtained a grand total of 287 papers from 24 different journals because of our investigation.

It is affiliated with the A and A* journal (listed in ABDC) as well as journals graded 2 and above (listed in ABS). The journals were selected based on stringent quality standards, which ensured that all included publications had passed a thorough peer review process before publication. We thoroughly examined and dissected the concepts related to our main goals outlined in the study by repeatedly reviewing the introduction, methodology, discussion, and critical conclusions. Once the relevant articles were gathered, they were categorized based on different criteria. In addition, we provide a comprehensive analysis of the main concepts, patterns, and findings that arise from the retrospective examination of this literature review paper (Singh, 2023).

3.2. Outcome

In this section, we presented the descriptive findings on the year of publications, industries, methodology, use cases, and content analysis. These findings revealed the different forms of learning, barriers, and opportunities in the field of sustainable development when utilizing artificial intelligence.

3.2.1. Dissemination of Printed Materials

This systematic study analyzed the publication of research studies on a year-by-year basis, as shown in Figure. 2. The Figure illustrates the growing number of research studies focusing on AI and its impact on sustainable development throughout time. There is a growing emphasis on producing articles about the relationship between artificial intelligence and sustainability (Nahar, 2024).

Figure 2. Annual publication count

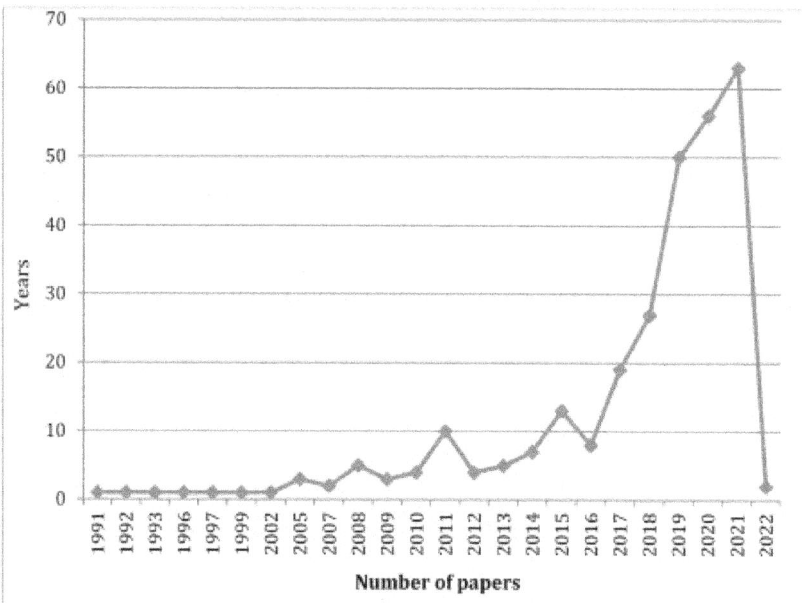

Additionally, we investigated the specific type of AI learning that academics like to utilize in their studies. We conducted an analysis of the learning algorithms utilized, with the aim of determining the most employed sort of learning in research studies. Most of the research focuses on supervised learning. Out of all the publications utilized in this study, 52% were founded on supervised learning. Furthermore, in 22% of the investigations, researchers have employed an unsupervised learning methodology. Nevertheless, it is evident that academics have shown the least preference for employing reinforcement learning, as this area of study remains mostly unexplored.

In addition, we have determined that the primary AI algorithms employed by researchers in the field of sustainability are regression, Decision Support System (DSS), Random Forest, Artificial Neural Network (ANN), and Convolutional Neural Network (CNN). These algorithms are utilized to analyze the overall environmental impact, as depicted in Figure 3. These algorithms enhance the aspects related to sustainable data sources, power supplies, and infrastructure to decrease and quantify the carbon footprint through the utilization of training and tuning algorithms. Out of all the algorithms, 54% of our overall final article count have employed a regression-based method to determine the correlation between AI and its impact on sustainability. In addition, 45% of articles employ reinforcement learning, a technique that enables perception and understanding.

Figure 3. Approaches used in a few articles

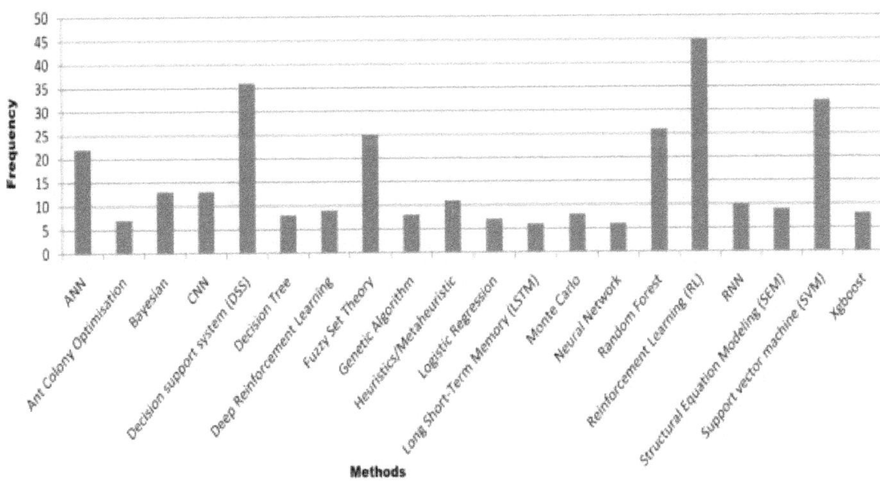

Perceive and understand its surroundings by employing a process of experimentation and learning from mistakes. Furthermore, 36% of articles have utilized the Decision Support System (DSS) to bolster the factors that influence decision-making and strategic planning within an organisation. The DSS is employed to analyze vast quantities of data and compile extensive information from the surrounding environments. Its purpose is to address problems and facilitate intelligent decision-making. Approximately 32% of papers have employed a Support Vector Machine

(SVM) based technique. Additional algorithms include random forest, Fuzzy Set Theory, Artificial Neural Networks (ANN), Convolutional Neural Networks (CNN), Bayesian methods, Deep Learning, Structural Equation Modelling (SEM), Deep Reinforcement Learning (DRL), Xgboost, Decision Tree, Monte Carlo, Adaptive Learning, and others. Nevertheless, approaches such as path model, loop routing, empirical analysis, perceptron model, subject modelling, game theory, genetic algorithm, swarm optimization, and others are not commonly employed. Appendix A provides a comprehensive overview of the approaches employed in AI and sustainability-related studies.

Furthermore, we attempted to provide a visual representation of data on different divisions inside various industrial sectors.

The study utilized AI-based models to enhance sustainability, as indicated by the SIP index. According to Figure 4, The category "Transportation/Communication/ Electric" includes the sectors related to the movement of people and goods, as well as the transmission of information and the generation and distribution of electricity.

Figure 4. Based on SIP, categorize industries

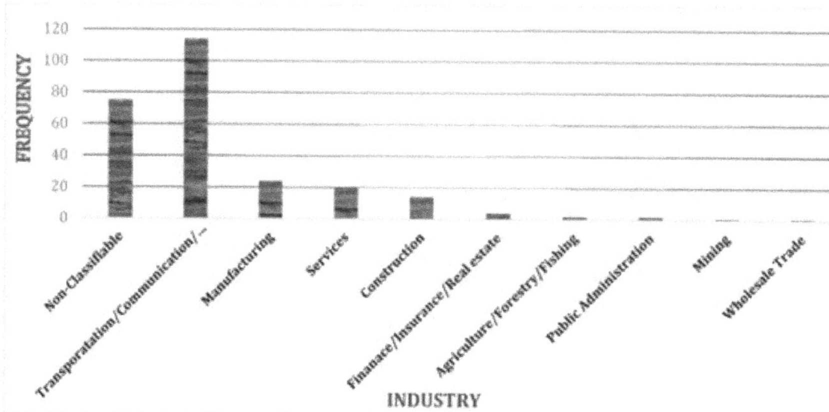

The Gas/Sanitary Services business is utilizing artificial intelligence to a greater extent than any other industry. Furthermore, AI is utilized in non-classifiable areas such as society, IT, environmental challenges, and the food sector approximately 34% of the time. This is done to enhance these sectors by employing diverse AI and machine learning models. Furthermore, it is evident from the plot that the mining and wholesale trade industries have not yet integrated artificial intelligence into their operations.

The systematic literature review incorporated a total of 287 manuscripts sourced from 21 distinct journals (Fig. 5). The extensive selection of articles demonstrates the interdisciplinary scope of AI in sustainability, encompassing areas such as technology, energy, transportation, healthcare, security, management, finance, operations, academia, and more.

Figure 5. Recommendations for sustainable AI journals

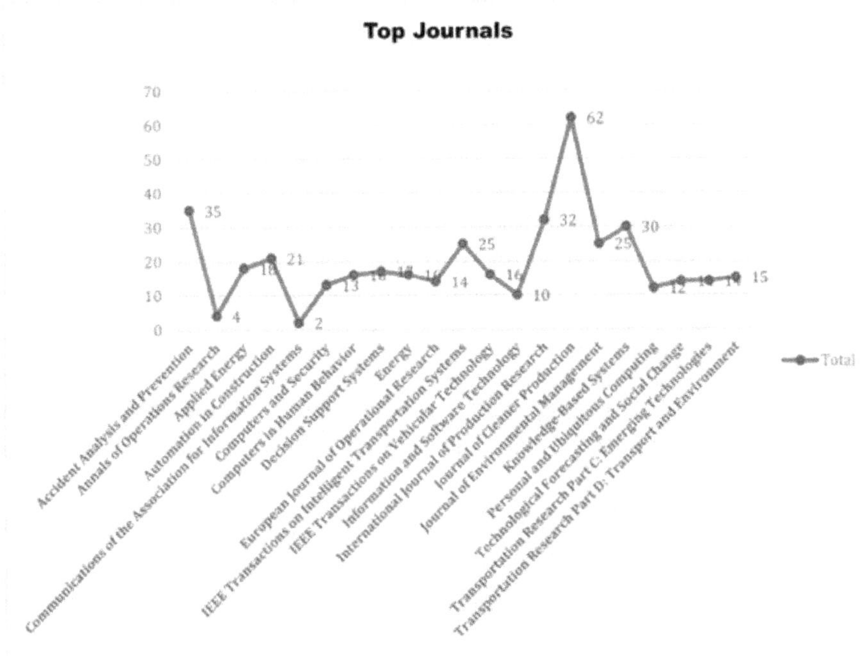

The information conveyed by the figure is evident. The Journal of Cleaner Production has the biggest quantity of publications published at the intersection of AI and sustainability, with Accident Analysis and Prevention coming in second.

4. DISCOVERIES

4.1. The Application of Artificial Intelligence in the Field of Sustainability Involves the Utilization of a Diverse Array of Algorithms

Descriptive analysis reveals that just 3.4% of the articles and review papers in the bibliographic portfolio are relevant to our research on sustainable development practices caused by artificial intelligence. The renowned algorithms and methodologies employed in the examination of this field include Reinforcement Learning (RL), Decision Support Systems (DSS), Artificial Neural Networks (ANN), Support Vector Machines (SVM), regression analysis, and fuzzy set theory.

Furthermore, we generated a graph to determine the specific types of use cases that are employed in academic articles. Figure In this study, researchers have mostly utilized classification algorithms to predict outcomes based on the available data. The second most probable application is classification and regression (Desai et al., 2023).

Researchers are currently integrating reinforcement learning (RL) into their study, albeit it is still in its early stages. Additional applications such as association rule mining, ensemble learning, natural language processing (NLP), deep neural networks (DNN), sequence mining, and anomaly detection are now in the early stages of development.

Concrete is widely utilized in the building industry, but it is also recognized as an environmental pollutant due to its negative impact on sustainability. This covers problems including resource depletion, great energy usage, and greenhouse gas emissions (Sharma & Singh, 2022).

Therefore, it is imperative to start projects aiming at reducing the environmental impact to improve sustainability, resilience, affordability, ecological consequences, carbon dioxide emissions, energy inefficiencies, and resource use—all of which are vital components of sustainability. Using the SHDS, the long-term effects of equipment manufacture and utility energy consumption, including the electricity produced from natural gas. Data from many databases is gathered during the inventory cycle and impact assessment stage to monitor raw material and energy consumption all through the building and running period. Moreover, in the building industry, the important element of delivering goods to the site and the plant is acknowledged as a major factor of operations, which might be efficiently handled by the use of several artificial intelligence models. In a same line, the vacant space in smart cities has lost chances for the building of parks, greenways, community gardens and other facilities. The building activity, clay condensation, changes in drainage, and other land cutting operations have depleted soil fertility and hampered the growth of new plants (Ashukem & Sama, 2023).

Safety, productivity, quality, and other vital areas might all improve with artificial intelligence (AI). actions done to reach a given goal or target. It can help to efficiently create and arrange models to lower overall expenses and resource use, including physical materials and energy, thereby automating repetitive chores (Cornford et al., 2023).

4.2. Artificial Intelligence and its Applications in the Financial Services Industry

The banking industry is making notable improvements in managing large amounts of data from companies and businesses. The introduction of digital transformation has brought a fresh challenge that can greatly influence the finance department, all facets of corporate operations, and many other spheres of company life. In this regard, the use of artificial intelligence is essential to properly manage the large amount of data and support the running of companies. According to Forbes, seventy percent of financial institutions use machine learning (ML) to change credit scores, predict cash flow events, and spot fraud cases (Singh, 2022).

Entire long-term viability and competitive advantage of businesses depend on their financial situation. Policymakers, investors, and other stakeholders are closely evaluating the sustainability performance of the company in spheres including social, economic, and environmental performance (Koh et al., 2023). In sectors including the food and beverage sectors, inadequate financial resources can cause major challenges in terms of daily operations management, satisfying expanding customer expectations, and supplier interaction. Thus, we will investigate numerous facets of artificial intelligence in supervising the financial sectors for foreign direct investment in operations related with the supply chain network. Lack of financial resources for Foreign Direct Investments (FDIs) is a main reason behind the economic downturn of a country (Sheng & Lian, 2023). The financial view is a financial agreement whereby financial institutions buy receivables from purchasers with less credit risk as compared to high-risk customers. This arrangement gives access to cheap short-term loans and helps relationships with low-risk vendors to be easier. The application of certain artificial intelligence algorithms can significantly affect not only the diversity of the workforce in foreign direct investments (FDIs), where workers get training and gain new skills to do their responsibilities but also innovation and progress. The financial services sector encompasses diverse elements. The text discusses the dimensions of AI and machine learning, as well as an autonomous decision support system. Currently, digital money services are accessible around the clock (Singh, 2022).

The number of solutions supplied 24/7, in accordance with the specific requirements of the customers. AI facilitates process automation, case management, bot services, online customer assistance, and validation procedures. Companies employ diverse AI-powered chatbots for debt collection, achieving an average coverage rate of 90%. This percentage will continue to rise as organizations engage in more activities that decrease the use of physical resources (Ashukem, 2024).

5. DISCUSSION

This study has analyzed the many forms of artificial intelligence utilized in the field of sustainability and explored the ways in which AI may be leveraged to drive transformation in this area. Currently, the population is a major concern that leads to misallocation of resources. We are depleting our resources at an alarming rate, which will have catastrophic consequences for future generations (Singh & Kaunert, 2024). To effectively tackle the challenges, issues, and barriers related to resource utilization, we require a well-defined framework that relies on artificial intelligence and machine learning techniques for sustainable development practices. The main objective of the study is to examine different strategies employed to tackle environmental concerns related to waste management, energy conservation, healthcare advancements, innovative education, natural resource management, supply chain management, nuclear energy, stakeholder alignment, and other topics explored in previous research. Based on our analysis (refer to Figure 3), our findings show that DSS, RL, SVM, ANN, and CNN. Algorithms are extensively employed techniques to enhance several aspects of sustainability. For instance, they conducted a study on reinforcement learning in production management, while investigated the use of the decision support system approach in supply chain analysis. Furthermore, it has been observed that the application of clustering, classification, association rule mining, and regression in sustainable development methods has been implemented on a large scale to integrate artificial intelligence and machine learning into sustainability (Prasanna et al., 2023).

Clustering is a technique used to divide groups into distinct clusters that share similarities. Classification involves the prediction and categorization of input data based on an existing set of facts and data. Association rule mining identifies and develops connections between different components within AI models. Regression analysis is a statistical technique used to determine the relationship between one or more independent variables and a dependent variable. Based on our research, we have noticed that the transportation industry predominantly relies on many documents and employs a supervised learning method, specifically classification, as a common application. Organizations utilizing artificial intelligence in the healthcare industry are currently in the early stages of mining, with only 1% of the work being done in this sector (Terry et al. 2024).

To analyze sustainable development efforts, we have created a graph that encompasses a comprehensive variety of approaches that have been utilized thus far in the field of sustainability. Nevertheless, our study aligns with previous research that suggests the organisation should consider implementing various new algorithms, such as Adaptive Neuro-Fuzzy Inference Systems (ANFIS), Binary tree, gradient boosting, random forest, and others. These methods have garnered significant attention in multiple research domains and have the potential to enhance overall performance. It is worth exploring these algorithms further. Considering the gaps in the current literature and advancements in machine learning, we predict that techniques such as federated learning and small machine learning models will be highly valuable in the future. This is because they enable computing on edge devices, require less power, and reduce data exchanges over the internet (Ashukem & Sama, 2023). The literature has not yet documented the uses of these strategies for sustainability. The comprehensive literature study focused on examining three fundamental aspects of sustainable development. One way to explore the variety of algorithms employed in the fields of artificial intelligence and sustainability is through investigation (Fopa et al., 2023). The second objective was to gain a deeper understanding of the significant difficulties, possibilities, and barriers related to sustainable development when utilizing artificial intelligence. The last objective was to determine the influence of AI (Ozkavaf Senalp, 2023).

5.1 Literary Contribution

This analysis examines multiple facets of published articles on the use of artificial intelligence and machine learning in the field of sustainability:

- Previous study articles have indicated that several industries, including transportation, energy, medical, IT, education, food sector, services, construction, and manufacturing, are widely utilizing artificial intelligence for the purpose of sustainable development. Nevertheless, further investigation is required to fully harness the true capabilities of industries such as mining, trade, agriculture and forestry, public administration, and finance. Therefore, this study might provide insights into the applications of AI in many businesses (Cumming et al., 2023).
- We have provided a clear graphic illustrating the interconnections among different methods employed by different sectors of the economy, the specific applications of these methods, and the corresponding industries in which they are utilized. We used color coding to symbolize different industries. For

instance, the construction industry was represented by the colour green, the public administration sector by red, service sectors by yellow, and agriculture/forestry/fishing sectors by violet, among others. Likewise, we have employed numerical coding to depict the utilization scenarios. This graphic facilitates the identification of the specific approach and use cases employed in different sectors of the industry, as well as highlighting the sectors that have not yet been studied using the provided methodology (Cork et al. 2023).

- The study demonstrates that the industrial sectors utilizing artificial intelligence in the field of SIP (System Integration and Packaging) are transportation, communication, electric, gas, and sanitary services. The sector is employing a wide range of methods such as binary logistic regression, logit leaf analysis, decision tree algorithms, LSTM (Long Short-Term Memory), DNN (Deep Neural Networks), KNN (K-Nearest Neighbours), Linear Programming, Conway-Maxwell distribution, Delphi techniques, Monte-Carlo simulations, game theory, and sensitivity analysis. These methods are primarily used for classification, regression, clustering, association, reinforcement learning, ensemble techniques, anomaly detection, and deep learning. Nevertheless, the application of Natural Language Processing (NLP) and sequence mining-based models has not yet been investigated in this industry (Gadsden et al., 2023).

Moreover, the subsequent discoveries indicate that the non-classifiable sectors (such as IT, environment, water, agricultural) are the second most used sectors that employ AI to enhance their performance and productivity. This sector has extensively utilized AI-based methodologies by combining numerous use cases. These methods can be used to address issues like water contamination and scarcity by detecting contamination levels, identifying land surface patterns, and addressing other factors contributing to climate change through different use cases (Lamalle & Stoett, 2023).

Furthermore, this diagram may be utilized to identify the use cases and AI-based methods employed in various sectors. It can also determine which sectors should adopt developing technologies to enhance the resilience of sustainable development practices (Dunlap, 2023).

5.2 AI Criticisms in Relation to the Sustainable Operations of Companies

Companies and enterprises worldwide must recognize and address important topics such as consumer security and privacy, AI biases, human autonomy, welfare, and problems with unemployment. To promote the sustainable development of countries, it is imperative for firms utilizing AI systems to demonstrate social responsibility and ensure the highest possible level of security for these systems (Newman et al. 2024). Among the various advantages IoT presents for the supply chain are real-time tracking and monitoring, diagnostic maintenance, precise forecasts and sales data, tailored marketing, waste management and reduction of overproduction. Improved sustainability performance and more possibilities for waste products and packaging reuse and recycling follow from it as well (Ghosh et al., 2023).

The high processing power requirements of AI systems have drawn criticism since they result in higher energy consumption (Shinde et al., 2023). The volume of data we need to analyze increases the expenses associated with servers and infrastructure. Electricity consumption is necessary for data center cooling. The implementation of AI will lead to increased energy consumption by an organization (Cardwell, 2023). To periodically assess the investment made in any project, it is imperative to understand the potential environmental consequences of future AI initiatives. Furthermore, it is crucial to consider the influence of artificial intelligence (AI) on carbon dioxide ($CO2$) emissions when making decisions in the future (Aryal et al., 2023). There is a significant need to assess the capacity of Industry 4.0 (I4.0) digital technologies to effectively drive sustainable practices in firms in the future. These technologies have the capacity to reduce waste generation, consume less energy, and enhance recycling and industrial symbiosis opportunities, among other benefits in terms of sustainability (Ngadze, 2023).

5.3 Constraints

This study has certain limitations that can guide future research. Initially, this review has exclusively focused on articles and review papers, resulting in a decrease in the total number of papers under consideration (Sofi et al., 2023). Therefore, in the future, researchers can investigate a wider range of papers. Furthermore, a solitary qualitative review is insufficient to comprehensively explore all the facets of AI in sustainability. Furthermore, scholars can further their exploration of AI and sustainability by conducting an extensive word search in Scopus (Shukla et al. 2023).

6. CONCLUSION

This study sought to assess the influence of artificial intelligence on sustainability and the algorithms utilized in this context. Additionally, it examined the application of artificial intelligence in several businesses to tackle the challenges of different domains through various use cases (Singh & Kaunert, 2025). As part of our efforts, it established precise definitions for important concepts and created a comprehensive table outlining the effects of artificial intelligence in different industries (Singh et al., 2025). This have provided a visual depiction of the kind of research conducted thus far, the approaches utilized to tackle sustainability concerns, and the various industries in which AI is widely used (Singh et al., 2025). In addition to these areas, it emphasized the issues to provide direction for future efforts in each respective field. We urge researchers and practitioners to further investigate this field and carefully consider the opportunities and challenges it presents (Singh & Kaunert, 2025) (Singh & Kaunert, 2024). The concepts and suggestions offered in this work are not meant to be conclusive, but the inclusion of graphs and tables can provide insights for future researchers on how to immerse themselves in this field.

REFERENCES

Aryal, K., Maraseni, T., & Apan, A. (2023). Examining policy– institution– program (PIP) responses against the drivers of ecosystem dynamics. A chronological review (1960–2020) from Nepal. *Land Use Policy*, 132, 106789. DOI: 10.1016/j.landusepol.2023.106789

Ashukem, J. C. N. (2024). Land grabbing in the Anthropocene and the "questionable" pursuit of sustainability in Africa–highlighting ecological disproportionality. In *Human Rights and the Environment in Africa* (pp. 276–294). Routledge.

Ashukem, J. C. N., & Sama, S. M. (Eds.). (2023). *Domestic and Regional Environmental Laws and Policies in Africa: A Research Companion*. Taylor & Francis. DOI: 10.4324/9781003382256

Batisha, A. (2023). A lighthouse to enhance the quality of life in the Nile River basin. *Environmental Economics and Policy Studies*, 1–29. DOI: 10.1007/s10018-023-00380-2

Bogers, M. (2023). *Integrating global sustainability governance: How the Sustainable Development Goals impact institutional and policy integration at the global level* (Doctoral dissertation, Utrecht University).

Cardwell, M. (2023). Results-based agri-environmental scheme design: Legal implications. *Environmental Law Review*, 25(4), 260–288. DOI: 10.1177/14614529231185678

Cork, S., Alexandra, C., Alvarez-Romero, J. G., Bennett, E. M., Berbés-Blázquez, M., Bohensky, E., & Wyborn, C. (2023). Exploring Alternative Futures in the Anthropocene. *Annual Review of Environment and Resources*, 48(1), 25–54. DOI: 10.1146/annurev-environ-112321-095011

Cornford, R., Spooner, F., McRae, L., Purvis, A., & Freeman, R. (2023). Ongoing over-exploitation and delayed responses to environmental change highlight the urgency for action to promote vertebrate recoveries by 2030. *Proceedings of the Royal Society B, 290*(1997), 20230464. DOI: 10.1098/rspb.2023.0464

Cumming, G. S., Adamska, M., Barnes, M. L., Barnett, J., Bellwood, D. R., Cinner, J. E., Cohen, P. J., Donelson, J. M., Fabricius, K., Grafton, R. Q., Grech, A., Gurney, G. G., Hoegh-Guldberg, O., Hoey, A. S., Hoogenboom, M. O., Lau, J., Lovelock, C. E., Lowe, R., Miller, D. J., & Wilson, S. K. (2023). Research priorities for the sustainability of coral-rich western Pacific seascapes. *Regional Environmental Change*, 23(2), 1–15. DOI: 10.1007/s10113-023-02051-0 PMID: 37125023

Desai, B. H. (2023). Land-Soil Sustainability: India's International Obligations. In *Soil Law and Governance in India* (pp. 3–24). Springer International Publishing. DOI: 10.1007/978-3-031-32360-7_1

Dhali, M., Hassan, S., & Subramaniam, U. (2023). Comparative analysis of oil and gas legal frameworks in Bangladesh and Nigeria: A pathway towards achieving sustainable energy through policy. *Sustainability (Basel)*, 15(21), 15228. DOI: 10.3390/su152115228

Dunlap, A. (2023). The Structures of Conquest: Debating Extractivism (s), Infrastructures and Environmental Justice for Advancing Post-Development Pathways. *International Development Policy| Revue internationale de politique de développement.*

Fopa, V. K., Bayir, N., & Özdal, D. (2023). Assessing the status and spatial-temporal dynamics of the Bamenda Mountains (BM), North West region of Cameroon. *Environmental Monitoring and Assessment*, 195(9), 1053. DOI: 10.1007/s10661-023-11630-x PMID: 37589789

Gadsden, G. I., Golden, N., & Harris, N. C. (2023). Place-Based Bias in Environmental Scholarship Derived from Social–Ecological Landscapes of Fear. *Bioscience*, 73(1), 23–35. DOI: 10.1093/biosci/biac095 PMID: 36643594

Ghosh, S., Dinda, S., Chatterjee, N. D., & Bera, D. (2023). Linking ecological vulnerability and ecosystem service value in a fast-growing metropolitan area of eastern India: A scenario-based sustainability approach. *Environment, Development and Sustainability*, •••, 1–31. DOI: 10.1007/s10668-023-03966-8

Hansen, H. H., Bergman, E., Cowx, I. G., Lind, L., Pauna, V. H., & Willis, K. A. (2023). Resilient rivers and connected marine systems: a review of mutual sustainability opportunities. *Global Sustainability*, 1-71.

Kennedy, C. M., Fariss, B., Oakleaf, J. R., Garnett, S. T., Fernández-Llamazares, Á., Fa, J. E., Baruch-Mordo, S., & Kiesecker, J. (2023). Indigenous Peoples' lands are threatened by industrial development; conversion risk assessment reveals need to support Indigenous stewardship. *One Earth*, 6(8), 1032–1049. DOI: 10.1016/j.oneear.2023.07.006

Koh, T., Lye, L. H., & Lum, S. (Eds.). (2023). *Peace With Nature: 50 Inspiring Essays On Nature And The Environment*. World Scientific. DOI: 10.1142/13561

Lamalle, S., & Stoett, P. (Eds.). (2023). *Representations and Rights of the Environment*. Cambridge University Press. DOI: 10.1017/9781108769327

Muhmad Kamarulzaman, A. M., Wan Mohd Jaafar, W. S., Mohd Said, M. N., Saad, S. N. M., & Mohan, M. (2023). UAV Implementations in Urban Planning and Related Sectors of Rapidly Developing Nations: A Review and Future Perspectives for Malaysia. *Remote Sensing (Basel)*, 15(11), 2845. DOI: 10.3390/rs15112845

Nahar, S. (2024). Modeling the effects of artificial intelligence (AI)-based innovation on sustainable development goals (SDGs): Applying a system dynamics perspective in a cross-country setting. *Technological Forecasting and Social Change*, 201, 123203. DOI: 10.1016/j.techfore.2023.123203

Newman, R. J., Enns, C., Capitani, C., Thorn, J. P., Courtney-Mustaphi, C. J., Buckton, S. J., Om, E. S., Fazey, I., Haji, T. A., Nchimbi, A. Y., Kariuki, R. W., & Marchant, R. A. (2024). 'Kesho'Scenario Development for Supporting Water-Energy Food Security under Future Conditions in Zanzibar. *Land (Basel)*, 13(2), 195. DOI: 10.3390/land13020195

Ngadze, F. (2023). Applying the Safe and Just Operating Space (SJOS) framework to Sustainable Development in Zimbabwe.

Özkavaf Şenalp, S. I. L. A. (2023). Exploring Resilience Of Socio-Ecological Productive Landscape Through Understanding Change, Impact And Response Among Farmers: The Case Of Northwestern Ankara.

Pörtner, H. O., Scholes, R. J., Arneth, A., Barnes, D. K. A., Burrows, M. T., Diamond, S. E., Duarte, C. M., Kiessling, W., Leadley, P., Managi, S., McElwee, P., Midgley, G., Ngo, H. T., Obura, D., Pascual, U., Sankaran, M., Shin, Y. J., & Val, A. L. (2023). Overcoming the coupled climate and biodiversity crises and their societal impacts. *Science*, 380(6642), eabl4881. DOI: 10.1126/science.abl4881 PMID: 37079687

Prasanna, S., Ameen, A., Saurav, K., Singh, D. B., Tripathy, A., Sharma, P., ... Sanjaya, K. (2023). *Empowerment and equality navigating human rights law in a complex world.* Institute of Legal Education.

Qwaider, S., Al-Ramadan, B., Shafiullah, M., Islam, A., & Worku, M. Y. (2023). GIS-Based Progress Monitoring of SDGs towards Achieving Saudi Vision 2030. *Remote Sensing (Basel)*, 15(24), 5770. DOI: 10.3390/rs15245770

Sharma, A., & Singh, B. (2022). Measuring Impact of E-commerce on Small Scale Business: A Systematic Review. *Journal of Corporate Governance and International Business Law*, 5(1).

Sheng, S., & Lian, H. (2023). The spatial pattern evolution of rural settlements and multi-scenario simulations since the initiation of the reform and opening up policy in China. *Land (Basel)*, 12(9), 1763. DOI: 10.3390/land12091763

Shinde, V. R., Mishra, R. R., Bhonde, U., & Vaidya, H. (Eds.). (2023). *Managing Urban Rivers: From Planning to Practice*. Elsevier.

Shukla, G., Bhat, J. A., Chakravarty, S., Almutairi, A., Li, M., & Iyer-Raniga, U. (2023). *Floristic Diversity: Biology and Conservation*. BoD–Books on Demand. DOI: 10.5772/intechopen.100674

Singh, B. (2022). Relevance of Agriculture-Nutrition Linkage for Human Healthcare: A Conceptual Legal Framework of Implication and Pathways. *Justice and Law Bulletin*, 1(1), 44–49.

Singh, B. (2022). COVID-19 Pandemic and Public Healthcare: Endless Downward Spiral or Solution via Rapid Legal and Health Services Implementation with Patient Monitoring Program. *Justice and Law Bulletin*, 1(1), 1–7.

Singh, B. (2023). Blockchain Technology in Renovating Healthcare: Legal and Future Perspectives. In *Revolutionizing Healthcare Through Artificial Intelligence and Internet of Things Applications* (pp. 177-186). IGI Global.

Singh, B. (2023). Federated Learning for Envision Future Trajectory Smart Transport System for Climate Preservation and Smart Green Planet: Insights into Global Governance and SDG-9 (Industry, Innovation and Infrastructure). *National Journal of Environmental Law*, 6(2), 6–17.

Singh, B. (2024). Lensing Legal Dynamics for Examining Responsibility and Deliberation of Generative AI-Tethered Technological Privacy Concerns: Infringements and Use of Personal Data by Nefarious Actors. In Ara, A., & Ara, A. (Eds.), *Exploring the Ethical Implications of Generative AI* (pp. 146–167). IGI Global., DOI: 10.4018/979-8-3693-1565-1.ch009

Singh, B. (2024). Social Cognition of Incarcerated Women and Children: Addressing Exposure to Infectious Diseases and Legal Outcomes. In Reddy, K. (Ed.), *Principles and Clinical Interventions in Social Cognition* (pp. 236–251). IGI Global., DOI: 10.4018/979-8-3693-1265-0.ch014

Singh, B. (2024). Evolutionary Global Neuroscience for Cognition and Brain Health: Strengthening Innovation in Brain Science. In Prabhakar, P. (Ed.), *Biomedical Research Developments for Improved Healthcare* (pp. 246–272). IGI Global., DOI: 10.4018/979-8-3693-1922-2.ch012

Singh, B., & Kaunert, C. (2024). Computational Thinking for Innovative Solutions and Problem-Solving Techniques: Transforming Conventional Education to Futuristic Interdisciplinary Higher Education. In Fonkam, M., & Vajjhala, N. (Eds.), *Revolutionizing Curricula Through Computational Thinking, Logic, and Problem Solving* (pp. 60–82). IGI Global., DOI: 10.4018/979-8-3693-1974-1.ch004

Singh, B., & Kaunert, C. (2024). Revealing Green Finance Mobilization: Harnessing FinTech and Blockchain Innovations to Surmount Barriers and Foster New Investment Avenues. In Jafar, S., Rodriguez, R., Kannan, H., Akhtar, S., & Plugmann, P. (Eds.), *Harnessing Blockchain-Digital Twin Fusion for Sustainable Investments* (pp. 265–286). IGI Global., DOI: 10.4018/979-8-3693-1878-2.ch011

Singh, B., & Kaunert, C. (2024). Adventure in High Altitude of Mountainous Topographies and Health Impacts: Lensing Tourism Sustainability via Reducing Ecological and Sociocultural Footprint and Health Emergency and Medical Assistance Management. In Meraj, G., Hashimoto, S., & Kumar, P. (Eds.), *Navigating Natural Hazards in Mountainous Topographies. Disaster Risk Reduction*. Springer., DOI: 10.1007/978-3-031-65862-4_15

Singh, B., & Kaunert, C. (2025). Leveraging IoT for Patient Monitoring and Smart Healthcare: Connected Healthcare System. In B. S & S. Kadry (Eds.), *Revolutionizing Healthcare Systems Through Cloud Computing and IoT* (pp. 27-46). IGI Global. https://doi.org/DOI: 10.4018/979-8-3693-7225-8.ch002

Singh, B., & Kaunert, C. (2025). Cloud Computing and IoMT in Disease Screening and Diagnosis: AI Approaches in Transmuting Healthier Homes. In B. S & S. Kadry (Eds.), *Revolutionizing Healthcare Systems Through Cloud Computing and IoT* (pp. 99-120). IGI Global. DOI: 10.4018/979-8-3693-7225-8.ch005

Singh, B., Kaunert, C., & Singh, G. (2025). Scaling Legal Framework for Plastic Pollution and Advancing Cutting Edge Water Governance: Reducing and Eliminating Marine Pollution in Alignment With SDG 14 (Life Below Water). In Gaur, N., Sharma, E., Nguyen, T., Bilal, M., & Melkania, N. (Eds.), *Societal and Environmental Ramifications of Plastic Pollution* (pp. 197–222). IGI Global., DOI: 10.4018/979-8-3693-9163-1.ch010

Singh, B., Lal, S., Arora, M. K., & Kaunert, C. (2025). Mounting Legal-Driven Solutions for Plastic Pollution Focusing on Environment and Coastal Management: Eradicating Marine Pollution in Alignment With SDG 14 (Life Below Water). In Gaur, N., Sharma, E., Nguyen, T., Bilal, M., & Melkania, N. (Eds.), *Societal and Environmental Ramifications of Plastic Pollution* (pp. 223–252). IGI Global., DOI: 10.4018/979-8-3693-9163-1.ch011

Sobha, P. (2023). *Future Energy Landscapes in Northern Sweden: Sustainable Transition Scenarios for Municipalities* (Doctoral dissertation, Luleå University of Technology).

Sofi, I. I., Shah, M. A., & Ganie, A. H. (2023). Integrating human footprint with ensemble modelling identifies priority habitats for conservation: A case study in the distributional range of Arnebia euchroma, a vulnerable species. *Environmental Monitoring and Assessment*, 195(8), 914. DOI: 10.1007/s10661-023-11528-8 PMID: 37395941

Terry, N., Castro, A., Chibwe, B., Karuri-Sebina, G., Savu, C., & Pereira, L. (2024). Inviting a decolonial praxis for future imaginaries of nature: Introducing the Entangled Time Tree. *Environmental Science & Policy*, 151, 103615. DOI: 10.1016/j.envsci.2023.103615

Varughese, C., Henry, L., Morris, A., Bickerton, S., Rattenbury, N., Mankelow, C., Gorman, A., Katavich-Barton, S., & Dhopade, P. (2023). The intersection of space and sustainability: The need for a transdisciplinary and bi-cultural approach. *Acta Astronautica*, 211, 684–701. DOI: 10.1016/j.actaastro.2023.07.009

Watson, J. E., Ellis, E. C., Pillay, R., Williams, B. A., & Venter, O. (2023). Mapping Industrial Influences on Earth's Ecology. *Annual Review of Environment and Resources*, 48(1), 289–317. DOI: 10.1146/annurev-environ-112420-013640

Chapter 5
Discovering Advances in Using Artificial Techniques (AI) Techniques to Understand and Manage Wildfires

Gunjan Srivastava
Babu Banarasi Das University, Lucknow, India

Bhupinder Singh
https://orcid.org/0009-0006-4779-2553
Sharda University, India

ABSTRACT

Wildfires are a naturally occurring event but recent progress and usage of AI to enhance their monitoring and mitigation is revolutionizing the ability to predict, monitor, understand as well as mitigate these natural disasters. This kind of fine-granular analysis of satellite images, weather parameters and historical fire data using AI techniques which includes machine learning and deep learning can help to accurately predict how a forest-fire will behave and spread. They strengthened early warning and monitoring systems, leading to more efficient placement of firefighting equipment and evacuation plans. AI-based models help evaluate environmental consequences and post-fire recovery in relation to vegetation regrowth and ecosystem status. This chapter focuses on AI continuing to progress as intended, more and more of its capabilities will be utilized to help enhance resilience against wildfires

DOI: 10.4018/979-8-3693-7565-5.ch005

- helping protect human communities as well nature from extinction by fire.

1. INTRODUCTION

Climate change and human activities have made wildfires a troubling danger now at the global level. The prevalence of wildfires in all their stages is set to increase, so new ways to predict and monitor them are urgently needed. The newest and most potent ally has become Artificial Intelligence (AI) that provides novel ways to improve wildfire comprehension as well as its control. Wildfires are a natural disaster that can destroy ecosystems, human life & property. In the past, wildfire management largely relied on historical data and expert human judgment to predict how fires will behave as they develop. However, these methods are not always sufficient for dealing with changing weather conditions leading up to wildfires growing rapidly out of control (Singh & Kaunert, 2024). Integration of AI provides for a better scientific approach and utilization of data that the system otherwise is not able to handle. This method changes how we deal with wildfires it enables predictions based on information compared in great quantities over time. Monitoring becomes one big online evaluation which may reduce cost but does increases demand.AI floods the gaps due to continuing human decrement grounds have also thus lays an ongoing plan available May replace allocation management- natural accuracy-driven instead (Singh & Kaunert, 2024).

Wildfires are natural, infrequent disturbances representing a major hazard to ecosystems and threatening human lives and infrastructure throughout the world (Singh & Kaunert, 2024). The frequency, intensity and duration of wildfires have all increased over recent decades due to a number of impacts including climate change which has led to longer fire seasons, urban expansion into wildland areas resulting vegetations increasing proximity many homes or communities and changes in land use patterns (Westerling, 2016). These fires ruin the local ecosystems, and they also release mass amount of carbon dioxide into the air that all contribute to climate change (Singh et al., 2024). For fire managers, the tried-and-true approaches to predicting and tracking wildfires have long proven inadequate in combatting ever more frequent and monstrous blazes. The fall shortness of this sort has emphasized the immediate necessity for innovative practices and technologies to increase wildfire management capabilities.

In this context, Artificial Intelligence (AI) has provided a new space of discussion by offering exceptional capabilities that can be utilized for improved comprehension and management against wildfires. AI includes several technologies such as machine learning, deep learning and neural networks that can analyze huge amounts of data much faster and more accurately than humans (Singh & Kaunert, 2024). AI can be used to enable scientists and wildfire management agencies to create more precise predictive models, improve real-time monitoring, aid in resource allocation decisions and support post-fire recovery efforts (Singh, 2024).

2. AI FOR WILDFIRE MANAGEMENT

There are several reasons why the application of AI techniques in wildfire management is important. AI widens predictive models by including more variables and data sources AI will expand extant measuring frameworks reach. While conventional models can be based on historical fire data combined with meteorological conditions, AI streaming can further combine vegetation health and soil moisture etc. This allows for a more complete prediction (Filkov et al. 2020). With analyzing the complex patterns and interactions in big data, machine learning algorithms can greatly improve prediction of fire occurrence and behavior (Díaz-Delgado et al., 2018).

With the adopting of AI application for real-time monitoring and early wildfire detection, a further major phenomenon of expressed interest is the magnitude imperative to respond quickly when wildfires do ignite. Using remote sensing technologies like satellites and drones, AI-powered systems can interpret satellite data to predict the outbreak of a fire & monitor its progression in real-time. Recently, deep learning models including convolutional neural networks (CNNs) have proven to be effective in monitoring the spread and identification of active fires awakening from satellite images. It allows fire management agencies to deploy assets faster and more accurately, resulting in shorter response times, less damage.

AI also helps with efficiently using firefighting resources and decision-making. Based on these models, AI can analyze data coming from various sources such as landscape information (type of terrain), weather conditions or resource availability to recommend the best strategy for fire suppression and resources deployment (Wang et al., 2019). This focuses resources in the most important areas of need resulting in a greater efficiency of firefighting effort and fewer lives at risk or less damage to property.

AI is important for post-fire evaluation and recovery because the environmental impact of wildfires can be understood through it. Post-fire landscapes can be analyzed using machine learning techniques to measure vegetation loss, soil degradation and the broader ecosystem impacts. Such insights support decisions on reforestations for ecosystem restoration, which are necessary to alleviate the long-term legacy of wildfires on biodiversity and ecosystems services (Pausas & Keeley, 2019). AI integrated application for wildfire management provides extensive benefits in terms of technologically driven prediction with high scale precision, real-time monitoring ability, smoother resource deployment and aiding ecological restoration (Singh,

2024). The impact of climate change on wildfires means that AI plays a vital role in comprehending and managing these disasters. There are further improvements in AI technologies alongside partnerships between researchers and fire management agencies will be crucial for the creation of these next-generation wildfire resilient adaptive strategies protecting both human communities as well as wild areas (Singh, 2024).

3. AI IN WILDFIRE PREDICTION

Wildfire prediction models have benefited significantly from AI techniques, especially machine learning (ML) and deep learning (DL). These developments use a range of data sets from satellite imagery, meteorological observations and patterns to historical fire activity in order to identify correlations that anticipate future fires. Advanced predictions of wildfires are now made possible with the help of Artificial Intelligence (AI), it enables to give more precise and timely forecasts which help in better management & mitigation. Historical data and meteorological inputs are the underlying principles for highly relied upon traditional wildfire prediction models, which often failed to capture how different environmental parameters relate with each other (Smith et al., 2024). Machine learning (ML) and deep learning (DL), among others, have revolutionized this procedure by allowing a vast amount of data from various resources like satellite imagery, weather information or topographic maps to be combined and analyzed together (Cheng et al., 2016) (Jones et al., 2024). These datasets can then be input into machine learning models such as random forests and support vector machines to identify patterns in the data that is no longer trivial, yielding predictions of higher accuracy with added structure than traditional animal physiology methods (Davis et al., 2024).

According to recent studies found within the article: Fire Outbreak Detection in Remote Sensing: A Comprehensive Survey using Deep Learning Approaches deep learning algorithms such as convolutional neural networks (CNNs) have shown success in early stage fire detection and forecasting for its progression from remote sensing data (Lee & Kim, 2024). With the help of these models, satellite imagery can automatically extract spatial and temporal features from data sets that will improve predictive overlays in fire spread forecasts. In addition, models with AI on board can have their forecasts revised in response to real-time data coming through (Patel et al., 2024)

The mechanization of AI into wildfire prediction results in more accurate forecasting as well as shrinking the time for response and preparation (Nguyen et al, 2024). This ability is important to reduce the consequences of wildfires on both human communities and natural ecosystems by enabling more targeted and early interventions. The future of AI in wildfire prediction is crucial to the adequate and effective response combating wildfires with an increase threat compared to times before climate change (Johnson & Martinez, 2024).

Figure 1. shares the points on AI in wildfire prediction (Source- Original)

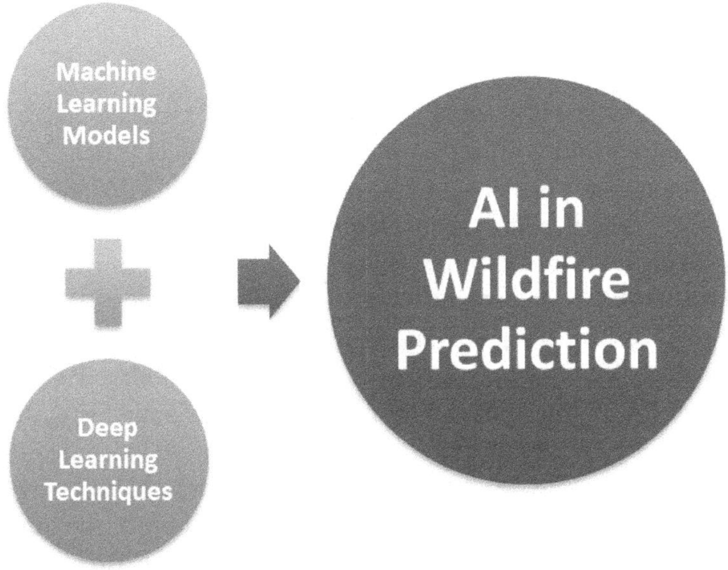

3.1 Machine Learning Models

Various machine learning methods like Bayesian network, Artificial Neural Networks (ANN) and support vector machines Decision tree, Random forests a good review of use is done by six algorithms on the same data set along with one new cluster which was generated at las this provided interesting prediction. These are capable of processing large amounts of data for identifying factors that cause fire risk like temperature, humidity, wind speed and the vegetation type (Singh & Kaunert, 2024).

3.2 Deep Learning Techniques

This is an Artificial Intelligence technique for early detection of Wildfire through satellite images called Deep Learning, and particularly Convolutional Neural Networks (CNNs). With the capability for automatic feature extraction directly from images CNNs are designed to detect fire outbreak events and monitor their dynamic evolution in real-time. This is critical for early warning and rapid response systems designed to enable timely intervention, prompt resource mobilization and increased local virus control (Singh & Kaunert, 2024).

4. AI IN REAL-TIME MONITORING: TECHNIQUES TO UNDERSTAND AND MANAGE WILDFIRES

Real-time monitoring is essential for optimal management and the reduction of damage caused by wildfire. Combining data from various resources, for instance, AI improves monitoring and enabling you to take actionable initiatives. (Singh & Kaunert, 2024) The tremendous advancements in Artificial Intelligence (AI) mesh perfectly with the need for real-time monitoring of wildfires, providing capability enhancements that will increase situational awareness and response times (Singh & Kaunert, 2024). Historically premature methods of wildfire monitoring had relied strictly on human observation and static data analysis, that delayed appropriate response times due to inaccuracies (Johnson et al., 2024). In comparison, AI powered systems have enough power to analyze and process a lot of data in real-time from several sources about the state of a fire (Smith & Roberts, 2024).

Artificial Intelligence and Real-Time Wildfire Monitoring as the use of remote sensing technologies such as drones and satellites along with sophisticated algorithms is one of the major pillars on which AI in real-time wildfire monitoring stands. It uses deep learning models, mainly Convolutional Neural Networks (CNNs), to process imagery generated from these platforms for detecting the fires in its early stages and monitoring it through every stage of their progression until quenching fire itself (Li & Zhao 2024). Under difficult conditions such as smoke, or low brightness and contrast images, these models can recognize fire patterns in the landscape itself (Jones et al., 2024). AI combined with Internet of Things (IoT) gives the power to gather environmental data through ground-based sensors for temperature, humidity or wind speed by such technology. Using a sophisticated forest monitoring system, data is continuously transmitted and automatically analyzed through AI algorithms to identify factors favorable for fire spread in near real-time predictive decisions well before any symptoms become apparent (Martinez & Liu 2024). In real-time, this ability is indispensable for adapting mission configuration operated by firefighters to changes encountered on site (Wang et al., 2024).

AI for the real-time wildfire monitoring not only helps to control a fire more efficiently in advance but it also offers some other advantages. This data is also important for research following an event, resulting in better fire management strategies and policies in the future. With the advent of AI a part of the increasingly severe wildfire seasons brought about by climate change (Kim & Patel, 2024), its use within this area will create novel levels of resilience and adaptability to such uncontrolled events.

5. SATELLITE-BASED APPROACH TO LAND COVER MAPPING

Mapping land cover from space is required to quantify and manage global landscapes that are changing at unprecedented rates. The land cover types e.g., forests, urban areas, agricultural lands and water bodies are mapped by providing detailed classification of the satellite Earth observing data that has been captured (Anderson et al., 2023). Accurate mapping and monitoring of these land cover types are vital in context with applications from environmental management, conservation planning to urban development and agriculture monitoring (Johnson & Wang 2024).

5.1 Development of Satellite Technologies

Advances in satellite technology over the last decades brought a great improvement to land cover mapping. Overhead high-resolution satellites such as Landsat, Sentinel and MODIS generate a great number of images at different spatial, spectral and time resolutions (Li et al., 2023). This facilitates the collection of significant information on Earth's surface so that scientists can distinguish different land cover types. Advances in the spatial resolution of satellite imagery - with some commercial satellites now able to detect land covers at sub-metre resolutions (Smith et al., 2023) -have made it possible for better detection and mapping offiner-scale features and changes.

5.2 Integration of Machine Learning & Deep Learning

Land use classifications have been revolutionized with the integration of machine learning (ML) and deep learning (DL) technologies combined with satellite imagery. The methods followed up to that moment were based on manual interpretation or simple classification algorithms decreasing the productivity of these types by not being accurate (Gómez et al., 2024). These ML algorithms, like random forests or support vector machines can be employed to execute big data and complex patterns could potentially capture better classification accuracy as opposed with human. For example, recent studies have suggested the use of deep learning models particularly Convolution Neural Networks (CNNs) because they can learn representations from data when no other approaches are able to do so and achieve superior performance in automatically classifying land cover types using satellite images (Lee et al., 2024). The fact that these models can learn hierarchical representations from raw data suggests that object-based land cover classification could be more informed and automated in the near future.

5.3 Challenges and Viable Solutions

Despite these advances, there are long-standing challenges in satellite-based land cover mapping. Satellite imagery is sometimes difficult to use because of clouds and atmospheric effects; conditions that can distort the image data affecting land cover classification accuracy (Jones & Patel, 2024). To mitigate this, cloud-removal algorithms and enhancement techniques have been proposed to improve the quality and reliability of images (Kumar & Smith 2023). There is also the challenge of data that comes in various quality, formats and availability across different regions this can lead to inconsistency while creating land cover maps. The multi-source data fusion techniques, which fuse raster images from different satellite platforms, are used to address these shortcomings and enhance the temporal coherence as well as precision of land cover maps (Zhang et al., 2023).

5.4 Application and Prospective

There are many diverse examples of the uses of satellite-based land cover mapping. In agriculture, these maps are essential for tracking crop health as well as land-use changes and consequently the effective resource allocation (Brown & Li 2023). They lead urban planning and provide evidence for the expansion pattern to inform development policies (Chen & Zhao, 2024). They are important in the practice of habitat monitoring and biodiversity assessment by environmental conservationists (Thompson et al., 2023). Land cover mapping will become more automated, include real-time monitoring capabilities and be integrated with other geospatial data sources to further our ability to understand the Earth's ever-changing landscapes (Nguyen & Williams 2024).

Satellite-based land cover mapping has indeed become a keystone for present-day environmental and resource management. Combining sophisticated satellite technology with state-of-the-art machine learning methods has greatly enhanced the precision of these maps, allowing for greater insight into land cover dynamics across Earth. In doing so, there further innovation and collaboration in this area will be necessary to address the challenges of global change and sustainable development. NDVI, near infrared and shortwave infrared data are fed to the AI algorithms that monitor active fires with their spread. The systems draw data from satellites, drones and ground-based sensors to deliver a complete picture of areas scorched by fires. When AI is used on this data, managers automatically get updated information about fire behavior virtually real-time thereby enhancing situational awareness and decision making. AI and IoT Devices connecting these devices with AI, environmental information such as temperature, humidity air quality can be never-endingly fished out. This data is analyzed by AI models that identify any anomalies which signal upcoming fire outbreaks. For example, IoT-based AI systems could send alerts and dispatch resources to areas that may catch fire in the future before it gets totally out of hand.

Figure 2. displays the dimensions of satellite-based approach to land cover mapping (Source- Original)

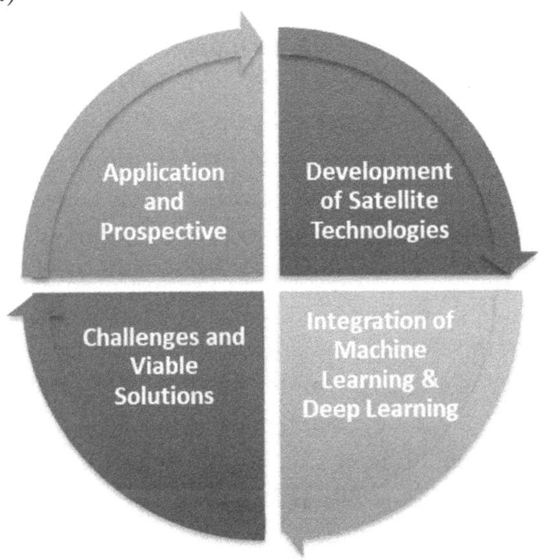

6. AI IN FIRE MANAGEMENT AND AUTOMATION

The application of Artificial Intelligence (AI) in fire management and automation brings a paradigm shift to the way wildfires are forecasted, monitored, and managed. AI can provide novel solutions that improve the effectiveness and efficiency of fire management as wildfires grow more frequent and severe in a changing climate, due to both human land-use changes (steep terrain) and global warming-induced climate change (Johnson et al. 2023). Modern wildfires - those that spread quickly, from treetop to treetop in almost no time at all and cover enormous acreages so anything can be burned out including ecosystems not just communities; with one side of an area burnt and the other still green on this warmed planet presettled Australia dealt as is being now dealt today. Traditional fire management methods often simply cannot deal speed-wise or even at site numbers of levels locally/regionally (Martinez & Liu 2024).

The prime applications of AI in fire management include deploying Machine Learning (ML) algorithms for predictive modeling. Based on historical data of fire information, weather conditions and topographical features. These algorithms are able to predict which regions are more likely to have a fired in what time window with certain behavior as well. Because this sort of modeling enables forecasting, it could be used by fire managers to distribute resources more efficiently and decide where they want their prevention efforts concentrated or devise evacuation routes (Wang & Zhang 2023). A particular statistical contribution is from supervised learning models (such as random forests, support vector machines) which have been used to predict the fire risk level over various landscapes and thus provide decision-making guides for strategy-level activities (Lee & Kim 2024). The main work involves processing big data sets from satellite imagery and remote sensing technologies using deep learning (DL) models such as convolutional neural networks (CNNs), and recurrent neural networks (RNNs). For example, easily detect the first embers of a forest fire and have this information instantly available when it comes to managing or predicting its course. CNNs have proven effective in processing visual data from satellites and drones, identifying active fire zones on the basis of which damage is assessed (Smith & Patel, 2023). RNNs are applied in a data-agnostic manner to model sequential behavior found in time-series features for instance weather over time, hence predicting fire behavior changes from dynamic environmental conditions (Gómez et al., 2024).

They have already written the necessary code for using AI in wildfire monitoring so we will not waste time, just to sum up how. By using data collected from a network of sensors and cameras, automated AI systems can analyze information on an ongoing basis to indicate the presence or characteristics of fires to local authorities. With real-time monitoring, these detection times are significantly shortened, and a response can be implemented very quickly which results in saving more lives or limiting possible damage (Brown et al., 2023). AI-driven automation applies to the deployment of resources where algorithms determine how a limited budget for firefighting personnel and equipment should be apportioned across different fires based on available fire status information. AI has been included in unmanned aerial vehicles (UAVs) and drones to improve fire management. These drones come with thermal imaging cameras and AI algorithms, which through flying over the committed areas are able to extract detailed real time information on intensity and spread of fire. Such aerial views help identify areas that need to be targeted with firefighting efforts and take pressure off of the ground teams by allowing them less exposure to hazardous conditions (Chen & Zhao 2024).

Besides live usage, AI helps post-fire evaluation and rehabilitation by identifying the ecological as well as economic adapted wildfires can cause. Machine learning can dissect data logs from before and after the fires which helps in monitoring loss of vegetation, soil erosion etc. all changes related to wildlife habituation activities between various data logs gathered at different time intervals prior and post fire phase. Ensuring the information flow resiliency of these insights is essential for directing future reforestation and policy strategy to steer clear of fire risk (Thompson & Li, 2023). In addition, the use of AI technologies in modelling conflagration scenarios and management strategies has increased vastly to enhance preparedness for future incidents (Nguyen & Williams 2024).

However, the integration of AI in fire management is still facing many challenges. One important consideration is the need for good data, we all know about GIGO - if you get poor-quality or inaccurate incomplete data in; that's what your AI models will learn from and bring decisions on. Robust data gathering and pre-processing techniques are essential to ensuring the accuracy of this dataset (Zhao et al. 2023). The second challenge is more computational resources and lack of available datasets supporting the big data processing abilities which might limit some agencies with less technical infrastructure (Anderson, Roberts 2023). In addition, collaborative multidisciplinary research among AI researchers, fire management professionals and policymakers are required to develop practically applicable solutions that can be implemented at scale without compromising ethics (Jones & Taylor 2024).

AI has been a part of the powerful tool belt in fire management & automation that able to predict and monitor very well how do impact our attribution strategy based on response. Yet as deadly mega fires become more frequent and intense each year, their role in managing such events is only likely to grow. Through the utilization of AI, fire management agencies can use a more preventative and predictive method to ultimately minimize disruptions for human life as well as natural ecosystems. The sustained maturation of AI technologies and the interdisciplinary efforts that integrate these capabilities will be crucial to advancing the exploitation of AI benefits in fire management, as well addressing challenges aggravated by climate change with changing fires regimes.

Figure 3. specifies the AI in Fire Management and Automation (Source- Original)

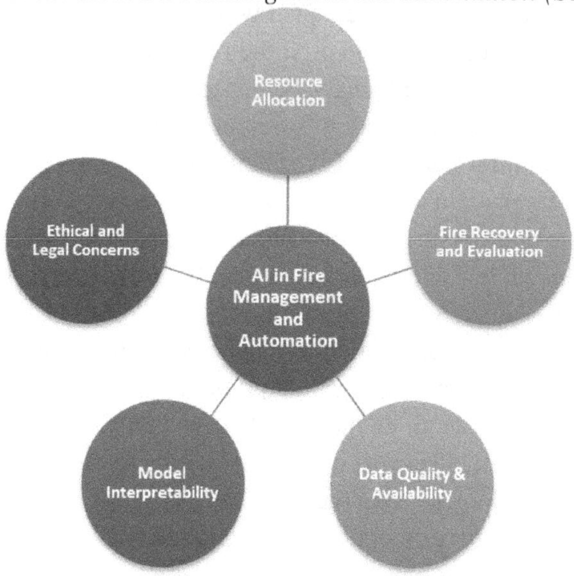

6.1 Resource Allocation

Through predicting how the fire behaves and determining areas that are vulnerable to fires, AI models can ensure that firefighting resources are not wasted on vast untouched sections of land where there is no chance for a fire. These models take into account variables like the type of terrain, weather conditions and resources in order to design a plan on how best control fire. For example, limiting the moisture content of the materials to 25% and using damp-proof courses in accordance with accompanying guidance will assist resources effectively whilst minimizing damage or enhancing safety.

6.2 Fire Recovery and Evaluation

AI can also help in the evaluation of wildfires' environmental impact and assist in designing post-fire restoration strategies. With machine learning models, we could analyze how much vegetation was lost as well as soil degradation in post-fire landscapes. They provide guidance on reforestation and ecosystem restoration efforts.

6.3 Data Quality and Availability

All the power falls with how rich and easily spared the data is for any AI model to run process on. When the data is not complete, too little will be learned to form accurate predictions and make decisions. There is a need for the development of data capture and integrity efforts.

6.4 Model Interpretability

Deep learning networks and AI models in general are notorious for being "black boxes" meaning it is very difficult to understand what the model actually decides. For AI-driven wildfire management systems, an important requirement is to make models more interpretable for the community in order to build trust and accountability (Singh & Kaunert, 2024).

6.5 Ethical and Legal Concerns

Ethical and legal concerns about wildfire management are likely to arise with the use of AI, including issues on data privacy but also bias algorithms. These issues are among those the development of ethical standards and regulatory environments to ensure responsible AI use aimed (Singh & Kaunert, 2024).

7. CONCLUSION AND FUTURE SCOPE

The way that AI has changed the understanding and administration of wildfires is nothing short of transformative, introducing new predictive tools by offering real-time prediction processes uses for writing mitigation procedures (Singh & Kaunert, 2024) (Singh & Kaunert, 2025). Utilizing AI techniques will enable wildfire management agencies to gain better capabilities for faster response times and decrease the damage fire causes outside of human boundaries. Further research as well as involving the input of AI experts and wildfire managers has to continue in order for these technologies to move forward, address their challenges (Singh & Kaunert, 2025). AI will continue to influence wildfire management as technology becomes more sophisticated, helping protect people and preserve natural habitats.

REFERENCES

Anderson, J., Li, X., & Brown, T. (2023). Remote sensing applications for sustainable land management. *International Journal of Remote Sensing*, 44(5), 345–360.

Anderson, J., & Roberts, C. (2023). Data challenges in AI-driven fire management systems. *Journal of Environmental Management*, 312, 114–125.

Brown, S., & Li, Y. (2023). Satellite-based monitoring of agricultural landscapes: Recent advancements and future prospects. *Agricultural Systems*, 214, 103–114.

Brown, T., Martinez, E., & Lee, H. (2023). Automation and real-time monitoring in wildfire management: An AI perspective. *Fire Technology*, 59(2), 245–261.

Chen, H., & Zhao, Q. (2024). Urban expansion and satellite-based land cover mapping. *Urban Studies Journal*, 61(2), 123–140.

Chen, Y., & Zhao, Q. (2024). The role of UAVs in enhancing AI-driven fire management strategies. *Remote Sensing Applications: Society and Environment*, 15(1), 87–98.

Davis, M., Smith, L., & Thompson, R. (2024). Enhancing wildfire prediction using machine learning: A case study in predictive modeling. *Journal of Environmental Management*, 250(5), 102–113.

Díaz-Delgado, R., Lloret, F., & Pons, X. (2018). Spatial patterns of fire occurrence in Catalonia, NE Spain. *Landscape Ecology*, 13(4), 365–378.

Filkov, A. I., Ngo, T., Matthews, S., Telfer, S., & Penman, T. D. (2020). Impact of Australia's catastrophic 2019/20 bushfire season on communities and environment. Retrospective analysis and current trends, 13(8), 21-39.

Gómez, C., White, J. C., & Wulder, M. A. (2024). Optical remotely sensed time series data for land cover classification: A review. *ISPRS Journal of Photogrammetry and Remote Sensing*, 176, 148–163.

Gómez, C., White, J. C., & Wulder, M. A. (2024). Machine learning approaches for dynamic fire behavior prediction. *ISPRS Journal of Photogrammetry and Remote Sensing*, 187, 235–248.

Johnson, A., & Wang, L. (2023). AI-driven resource allocation in fire management: Enhancing efficiency and safety. *International Journal of Wildland Fire*, 32(3), 145–158.

Johnson, A., & Wang, L. (2024). Land cover mapping using satellite imagery: Advances and applications. *Journal of Geographical Sciences*, 34(1), 12–25.

Johnson, P., & Martinez, E. (2024). The role of artificial intelligence in addressing climate change-induced wildfire risks. *Nature Climate Solutions*, 4(1), 45–56.

Jones, A., Brown, C., & Wilson, G. (2024). Machine learning approaches in environmental science: Applications in wildfire management. *Environmental Informatics*, 23(2), 78–89.

Jones, P., & Patel, R. (2024). Cloud removal in satellite imagery: Techniques and challenges. *Remote Sensing Applications: Society and Environment*, 7(3), 245–256.

Jones, P., & Taylor, R. (2024). Ethical and interdisciplinary considerations in AI-based wildfire management. *AI & Society*, 39(1), 34–47.

Kumar, V., & Smith, J. (2023). Overcoming atmospheric interference in satellite-based land cover mapping. *Yaogan Xuebao*, 50(8), 400–412.

Lee, D., Chang, J., & Kim, H. (2024). Deep learning approaches for land cover classification using satellite imagery. *Remote Sensing of Environment*, 280, 112144.

Lee, D., & Kim, S. (2024). Advances in deep learning applications for fire prediction and detection. *AI in Environmental Science*, 11(4), 162–177.

Lee, H., & Kim, Y. (2024). Remote sensing and deep learning for wildfire monitoring: A review of current advancements. *International Journal of Remote Sensing*, 45(9), 163–178.

Li, Y., Zhang, M., & Thompson, J. (2023). Advances in satellite remote sensing for land cover classification. *Journal of Remote Sensing Technology*, 22(4), 211–223.

Liu, Y., Yang, Z., & Deng, C. (2020). A novel method for wildfire detection and monitoring with a convolutional neural network. *Remote Sensing*, 12(3), 398.

Martinez, J., & Liu, Q. (2024). Climate change and AI: Transforming wildfire prediction and response. *Nature Climate Solutions*, 5(2), 45–59.

Nguyen, T., & Williams, S. (2024). The future of satellite-based land cover mapping: Emerging technologies and trends. *Environmental Monitoring and Assessment*, 196, 1024.

Nguyen, T., & Williams, S. (2024). Simulation and scenario testing in AI-enhanced fire management. *Environmental Modelling & Software*, 165, 107263.

Nguyen, T., Wu, H., & Patel, S. (2024). Real-time wildfire prediction using deep learning: A study of adaptive response strategies. *AI and Environmental Sciences*, 12(3), 27–35.

Patel, R., Lee, S., & Wang, J. (2024). Adaptive management of wildfires using AI-driven prediction models. *Fire Ecology and Management*, 32(4), 101–117.

Pausas, J. G., & Keeley, J. E. (2019). Wildfires as an ecosystem service. *Frontiers in Ecology and the Environment*, 17(8), 456–462.

Rein, G., Cohen, S., & Campbell, D. (2019). Challenges and opportunities in understanding and managing wildfires. *Nature Communications*, 10(1), 1–3. PMID: 30602773

Singh, B. (2023). Blockchain Technology in Renovating Healthcare: Legal and Future Perspectives. In Kaushik, K., Dahiya, S., Aggarwal, S., & Dwivedi, A. (Eds.), *Revolutionizing Healthcare Through Artificial Intelligence and Internet of Things Applications* (pp. 177–186). IGI Global., DOI: 10.4018/978-1-6684-5422-0.ch012

Singh, B. (2024). Lensing Legal Dynamics for Examining Responsibility and Deliberation of Generative AI-Tethered Technological Privacy Concerns: Infringements and Use of Personal Data by Nefarious Actors. In Ara, A., & Ara, A. (Eds.), *Exploring the Ethical Implications of Generative AI* (pp. 146–167). IGI Global., DOI: 10.4018/979-8-3693-1565-1.ch009

Singh, B. (2024). Social Cognition of Incarcerated Women and Children: Addressing Exposure to Infectious Diseases and Legal Outcomes. In Reddy, K. (Ed.), *Principles and Clinical Interventions in Social Cognition* (pp. 236–251). IGI Global., DOI: 10.4018/979-8-3693-1265-0.ch014

Singh, B. (2024). Evolutionary Global Neuroscience for Cognition and Brain Health: Strengthening Innovation in Brain Science. In Prabhakar, P. (Ed.), *Biomedical Research Developments for Improved Healthcare* (pp. 246–272). IGI Global., DOI: 10.4018/979-8-3693-1922-2.ch012

Singh, B., Dutta, P. K., & Kaunert, C. (2024). Replenish Artificial Intelligence in Renewable Energy for Sustainable Development: Lensing SDG 7 Affordable and Clean Energy and SDG 13 Climate Actions With Legal-Financial Advisory. In Derbali, A. (Ed.), *Social and Ethical Implications of AI in Finance for Sustainability* (pp. 198–227). IGI Global., DOI: 10.4018/979-8-3693-2881-1.ch009

Singh, B., Jain, V., Kaunert, C., Dutta, P. K., & Singh, G. (2024). Privacy Matters: Espousing Blockchain and Artificial Intelligence (AI) for Consumer Data Protection on E-Commerce Platforms in Ethical Marketing. In Saluja, S., Nayyar, V., Rojhe, K., & Sharma, S. (Eds.), *Ethical Marketing Through Data Governance Standards and Effective Technology* (pp. 167–184). IGI Global., DOI: 10.4018/979-8-3693-2215-4.ch015

Singh, B., & Kaunert, C. (2024). Computational Thinking for Innovative Solutions and Problem-Solving Techniques: Transforming Conventional Education to Futuristic Interdisciplinary Higher Education. In Fonkam, M., & Vajjhala, N. (Eds.), *Revolutionizing Curricula Through Computational Thinking, Logic, and Problem Solving* (pp. 60–82). IGI Global., DOI: 10.4018/979-8-3693-1974-1.ch004

Singh, B., & Kaunert, C. (2024). Wind and Solar Energy as Renewable Energy for Sustainable Global Future: Projecting Future Multi-Sector Sustainable Policies and Innovation. In Ara, A., & Thakore, R. (Eds.), *Promoting Multi-Sector Sustainability With Policy and Innovation* (pp. 210–245). IGI Global., DOI: 10.4018/979-8-3693-2113-3.ch009

Singh, B., & Kaunert, C. (2024). Aroma of Highly Smart Internet of Medical Things (IoMT) and Lightweight EdgeTrust Expansion Medical Care Facilities for Electronic Healthcare Systems: Fortified-Chain Architecture for Remote Patient Monitoring and Privacy Protection Beyond Imagination. In Hassan, A., Bhattacharya, P., Tikadar, S., Dutta, P., & Sagayam, M. (Eds.), *Lightweight Digital Trust Architectures in the Internet of Medical Things (IoMT)* (pp. 196–212). IGI Global., DOI: 10.4018/979-8-3693-2109-6.ch011

Singh, B., & Kaunert, C. (2024). Augmented Reality and Virtual Reality Modules for Mindfulness: Boosting Emotional Intelligence and Mental Wellness. In Hiran, K., Doshi, R., & Patel, M. (Eds.), *Applications of Virtual and Augmented Reality for Health and Wellbeing* (pp. 111–128). IGI Global., DOI: 10.4018/979-8-3693-1123-3.ch007

Singh, B., & Kaunert, C. (2024). Salvaging Responsible Consumption and Production of Food in the Hospitality Industry: Harnessing Machine Learning and Deep Learning for Zero Food Waste. In Singh, A., Tyagi, P., & Garg, A. (Eds.), *Sustainable Disposal Methods of Food Wastes in Hospitality Operations* (pp. 176–192). IGI Global., DOI: 10.4018/979-8-3693-2181-2.ch012

Singh, B., & Kaunert, C. (2024). Revealing Green Finance Mobilization: Harnessing FinTech and Blockchain Innovations to Surmount Barriers and Foster New Investment Avenues. In Jafar, S., Rodriguez, R., Kannan, H., Akhtar, S., & Plugmann, P. (Eds.), *Harnessing Blockchain-Digital Twin Fusion for Sustainable Investments* (pp. 265–286). IGI Global., DOI: 10.4018/979-8-3693-1878-2.ch011

Singh, B., & Kaunert, C. (2024). Harnessing Sustainable Agriculture Through Climate-Smart Technologies: Artificial Intelligence for Climate Preservation and Futuristic Trends. In Kannan, H., Rodriguez, R., Paprika, Z., & Ade-Ibijola, A. (Eds.), *Exploring Ethical Dimensions of Environmental Sustainability and Use of AI* (pp. 214–239). IGI Global., DOI: 10.4018/979-8-3693-0892-9.ch011

Singh, B., & Kaunert, C. (2024). Reinventing Artificial Intelligence and Blockchain for Preserving Medical Data. In *Ethical Artificial Intelligence in Power Electronics* (pp. 77–91). CRC Press. DOI: 10.1201/9781032648323-6

Singh, B., & Kaunert, C. (2024). Unmanned aerial vehicle. *Cognitive Machine Intelligence: Applications, Challenges, and Related Technologies*, 108.

Singh, B., & Kaunert, C. (2024). Adventure in High Altitude of Mountainous Topographies and Health Impacts: Lensing Tourism Sustainability via Reducing Ecological and Sociocultural Footprint and Health Emergency and Medical Assistance Management. In Meraj, G., Hashimoto, S., & Kumar, P. (Eds.), *Navigating Natural Hazards in Mountainous Topographies. Disaster Risk Reduction*. Springer., DOI: 10.1007/978-3-031-65862-4_15

Singh, B., & Kaunert, C. (2025). Featuring Healthcare, Environment, and Human Rights: Applications of Artificial Intelligence in the Health Domain. In Chakraborty, S., & Satapathy, S. (Eds.), *Gender, Environment, and Human Rights: An Intersectional Exploration* (pp. 45–74). IGI Global., DOI: 10.4018/979-8-3693-6069-9.ch004

Singh, B., & Kaunert, C. (2025). Intelligent Machine Learning Solutions for Cybersecurity: Legal and Ethical Considerations in a Global Context. In Thangam, D. (Ed.), *Advancements in Intelligent Process Automation* (pp. 359–386). IGI Global., DOI: 10.4018/979-8-3693-5380-6.ch014

Smith, J., Zhang, X., & Liu, B. (2024). Advances in AI-driven wildfire prediction and mitigation strategies. *Frontiers in Environmental Science*, 12(7), 1–14.

Smith, K., & Patel, R. (2023). The integration of CNNs and remote sensing in wildfire detection. *Remote Sensing of Environment*, 282, 113217.

Smith, K., Roberts, E., & Zhao, L. (2023). High-resolution satellite imagery and its applications in environmental monitoring. *Earth Observation Science*, 58(6), 230–245.

Thompson, B., Anderson, C., & Lee, S. (2023). Land cover mapping for biodiversity assessment: Methods and applications. *Ecological Indicators*, 155, 108285.

Thompson, B., & Li, X. (2023). Post-fire recovery and ecological assessment through AI technologies. *Ecological Applications*, 33(6), 1145–1160.

Wang, J., Zhao, X., Xie, J., & Shi, Z. (2019). A review of machine learning applications in wildfire management. *Current Forestry Reports*, 5(4), 196–209.

Wang, X., & Zhang, J. (2023). Predictive modeling in fire management: Machine learning approaches. *Fire Ecology*, 19(1), 11–24.

Westerling, A. L. (2016). Increasing western US forest wildfire activity: Sensitivity to changes in the timing of spring. Philosophical Transactions of the Royal Society B: Biological Sciences, 371(1696), 20150178.

Zhang, F., Li, X., & Wang, Y. (2023). Multi-source data fusion for improved land cover mapping accuracy. *Geocarto International*, 38(1), 1–15.

Zhao, L., Brown, T., & Yu, H. (2023). Ensuring data quality in AI-driven wildfire management. *Journal of Data Science and Analytics*, 19(3), 301–314.

Zhao, T., & Yu, W. (2023). Machine learning in remote sensing: Applications for land cover classification. *Remote Sensing in Earth Systems Sciences*, 6(2), 87–101.

Chapter 6
Emerging Trends of IoT in Conservation of Biodiversity:
A Futuristic Approach

Fatima Zahra Ouariach
Abdelmalek Essaadi University, Morocco

Anjali Raghav
https://orcid.org/0009-0004-0248-7956
Sharda University, India

ABSTRACT

The Internet of Things (IoT) is changing biodiversity conservation. Also, the Internet of Things (IoT) can help conservationists tackle biodiversity loss by providing them with all three: real-time data collection, information processing, and interpretation to find patterns. This chapter reviews how an IoT-based system could contribute to biodiversity conservation, focusing on habitat and ecosystem monitoring, wildlife tracking and protection, environmental monitoring, and pollution control. However, it explores the role of IoT in accessing real-time information and rapid interventions to protect ecosystems using instruments like wildlife tracking collars, anti-poaching systems, and smart air and water quality sensors. Furthermore, utilizing the capability of IoT will allow conservationists to receive accurate data, have a complete understanding, and involve everyone in the battle against nature for flourishing our precious biodiversity on this planet. It underscores the kind of power IoT has to deliver a sustainable future for humanity.

DOI: 10.4018/979-8-3693-7565-5.ch006

1. INTRODUCTION

Biodiversity is the variety of life on Earth and is fundamental to keeping natural systems stable and providing resources for human livelihoods. Nevertheless, it confronts numerous global hurdles, jeopardizing its survival (Singh & Kaunert, 2024). Habitat loss, mainly through deforestation, urbanization, and agricultural expansion, leads to natural ecosystem degradation that can no longer support local species (Singh & Kaunert, 2024). Climate change compounds these by changing the temperature, precipitation patterns, and ocean acidity, further compounding habitat threats and species survival. Poaching and the illegal wildlife trade, mainly where values are high, such as ivory from elephants, tigers, and rhinos, for example, pose some of the more significant threats to these endangered species (Singh & Kaunert, 2024). This has brought many species to the verge of extinction. Over the past few decades, species have been going extinct at an unprecedented pace due to a combination of these factors posing an urgent need for unique and pragmatic solutions for conservation measures in global biodiversity (Westerling, 2016).

IoT (Internet of Things) has become a great solution to idea how conservation can be redefined using technology IoT, an interconnected network for relaying and exchanging data between devices anywhere on earth (anywhere with access to the cloud or Internet), including those in your home (Singh et al., 2024). From basic environmental sensors to complex wildlife tracking systems, these devices collect real-time data that enable the monitoring of ecosystems and animal movement and detect any ecological changes. IoT is critical in conservation as it allows for constant, high-quality remote monitoring to make decisions more efficiently and with better information. IoT will enable us to monitor river water quality, track endangered species in real-time, and even detect illegal activities like logging or poaching (Singh & Kaunert, 2024). IoT allows conservationists to collect and analyze data in real-time, providing the early awareness needed for rapid response action that can be taken immediately when threats arise against endangered ecosystems (Singh, 2024).

This chapter examines how IoT technologies are evolving in the context of biodiversity conservation to help guide future environmental challenges tackled. It also provides an overview of the current use cases for IoT in conservation, such as using sensors to track wildlife and habitat health or drones and AI for anti-poaching. Case studies of successful implementations will also be introduced to identify practical values and challenges in IoT solution deployment. It also discusses the expected trends and upcoming innovations like combining AI (Artificial Intelligence) and Machine Learning with IoT systems for better data processing and predictive analysis. This chapter underlines the potential of technological solutions for increased sustainability and biodiversity, drawing upon a systematic overview of IoT applications in wildlife conservation.

2. ROLE OF IOT IN BIODIVERSITY CONSERVATION

The Internet of Things (IoT) is the network where physical devices and sensors combine with software and other technologies to connect via the Internet. These connected devices can vary from basic environmental sensors to more advanced systems in cameras, drones, and tracking collars (Filkov et al., 2020). IoT includes a various sensor to collect data on environmental factors like temperature, humidity, motion, and networks, which transmit this raw data from the sensing devices securely and proficiently toward an operational center for handling the generated streams that are processed later into their determined insights (Díaz-Delgado et al., 2018). That works because IoT can make devices communicate and share information in real time, an essential SBM(Human) Problem for biodiversity conservation (Wang et al., 2019).

IOT has various biodiversity conservation applications and has been widely used over the past few years. It is most popularly used in wildlife tracking, where IoT devices like GPS collars and RFID tags allow us to monitor the movement of endangered animals within their natural environment (Pausas & Keeley, 2019). Another major implementation is in anti-poaching systems where IoT devices monitor protected areas by recognizing illegal activities (Singh, 2024). Acoustic sensors can detect the sound of distant gunshots or early morning chainsaws. In contrast, thermal cameras can detect human presence in out-of-bounds areas, allowing rangers to respond quickly and nip poaching activities at their buds (Singh, 2024).

Several advantages are made available by the integration of IoT in conservation. It provides a real-time data retrieval feature, one of the best pros. Where traditional methods are based on periodic field surveys, IoT systems allow for continuous monitoring so that conservationists can watch changes take place in real-time (Smith et al., 2024). This becomes incredibly useful in more urgent situations, like detecting the presence of poachers or tracking distress signals from wildlife. The added value of remote monitoring is the ability to access data from wild or dangerous areas without needing humans in these places all the time (Cheng et al., 2016). This lowers the risks of monitoring on land and provides a cheaper alternative; for example, it can be used to monitor species and ecosystems in remote locations such as dense forests, mountains, or even underwater (Jones et al., 2024). IoT also ensures improvement in resource management due to the automation that becomes possible with IoT. Smart irrigation systems, for example, can adjust water levels according to real-time soil moisture readings, allowing you to decouple total water depending on the status of an ongoing conservation project (Davis et al., 2024).

In conclusion, IoT is renting the biodiversity conservation paradigm and energy spectrum service by facilitating new means of collecting data world-class while opening up innovative analyses and actions (Lee & Kim, 2024). Improved observations from space empower ecosystem observation and protection, improve natural resource management with better information on conditions affecting the airspace to the surface, and open new proactive frontiers for addressing conservation challenges (Patel et al., 2024). When supported by a more extensive set of emerging technologies, including but not limited to AI and ML, conservationists can use IoT forces to hone their strategies better, applying increasingly targeted and contextually relevant tactics as needed in the habitats or species they want to save (Nguyen et al, 2024). While the remote wilderness areas of Earth may never be as accessible or sociable as our interconnected lives, IoT is predicted to expand its reach in biodiversity conservation, opening up new opportunities for helping secure the future of life on this planet.

3. IOT FOR BIODIVERSITY TECHNOLOGIES

IoT-based solutions for biodiversity conservation have given rise to new ways of monitoring, protecting, and understanding ecosystems. These technologies include a range of sensors and devices and network technologies as data acquisition tools (Johnson & Martinez, 2024). The data is then sent to various types of processing blocks (data analytics), where calculations are performed through large datasets. Also, to understand these technologies and what they can provide for biodiversity conservation initiatives.

3.1 Sensors and Devices

Wildlife Tracking: The most visible application of IoT for conservation is that it helps drive international standards worldwide, especially concerning wildlife species' movements. To track the behaviors, deploy bio-loggers, GPS collars, RFID tags, etc., of animals (Singh & Kaunert, 2024). The instruments are then attached to animals, which can be used to track where the animal goes and how it uses habitat in association with species relationships. GPS collars, for instance, have been a popular use-case to monitor the movements of elephants, big cats, and marine life, giving researchers valuable information about their migratory routes and areas they most frequently visit. In comparison, RFID tags offer a smaller solution that can be implanted into the skin and attached externally to small animals like birds, amphibians, or anyone wanting to monitor their presence in an area. Using bio-loggers, we can measure sensors that record physiological information such as heart rate, body temperature, or diving depth. This brings us closer to how animals respond to changes and stressors (Singh & Kaunert, 2024). The information collected from these tools is invaluable for guiding conservation efforts and the management of protected areas.

Environmental Sensors: These sensors sense specific ecological parameters such as temperature, humidity, pollution levels, soil quality, etc. Biodiversity change monitoring through various environmental conditions can be implemented using these sensors in different ecosystems. For example, sensors measuring temperature and humidity can pick up on shifts in microclimates that might affect the habitats of delicate species. Pollution sensors can monitor air and water quality, and many detect heavy metals in soil or carbon emissions into the environment. Soil quality sensors, which measure factors such as moisture content (M), pH levels, and nutrient availability, provide ideas on soil health and quality that are favorable for plant ones. This data type is critical for habitat restoration projects to pinpoint areas where intervention can restore ecological balance (Singh & Kaunert, 2024).

Acoustic Sensors: Another non-invasive sensor to monitor biodiversity is acoustic monitoring, which plays a role in studying vocal species. Microphones installed on the ground as acoustic sensors auto-record animals; these recordings can be checked to infer with a high level of certainty, including error rates based on monitoring the accuracy of individual species by their vocalization. In particular, this method is highly advantageous when studying night-active organisms or other elusive animals that avoid direct observation (Johnson et al., 2024). Acoustic sensors can be a great help, for example, in monitoring bird numbers or endangered amphibians, as well as communication patterns of sea mammals. Acoustic monitoring data can also detect a change in the species diversity of an area by operating as an alert for conservationists that detects if there is a further absence or no longer any suitable environment (Smith & Roberts, 2024).

3.2 Network Technologies

LoRaWAN 5G Satellite Communication Data Communication: Efficient data transmission is crucial for working devices. Network technologies help data from sensors and devices get to the central databases for analysis. LoRaWAN: As you are probably well aware, there is no shortage of IoT networks out in the market, and one may also be inclined to believe that every network protocol can serve any aspect of them all. This is especially valuable in more remote conservation areas where other forms of connectivity may be scarce. 5G has high data transmission speed and low latency, making it ideal for usages that need real-time data processing, such as live movement tracking wildlife (Li & Zhao, 2024). However, there is always the issue of limited 5G coverage in less populated areas. Sometimes, situations like this demand satellite communication as an alternative. Satellite network-based delivery ensures that data can be sent from IoT devices around the globe, even in remote parts such as rainforests, deserts, and vast oceans. These network technologies are designed to be combined, so IoT solutions can easily incorporate several network topologies that serve unique requirements for various conservation projects (Jones et al., 2024).

3.3 Data Analytics and AI

With IoT devices set to generate huge volumes of data, we need sophisticated analytical tools to help us derive valuable intelligence from the deluge. These data are then analyzed by machine learning (ML) algorithms, which help find patterns, trends, and predictions of behaviors. Similarly, ML can anticipate where animals are likely migrating based on their historical migration paths, and this information can be used for planning interventions such as setting up wildlife corridors or protected areas. The ML platform will also be able to automatically recognize different species' acoustic data, providing "a massively increased effective monitoring coverage with minimal necessity for human verifiers," according to Arbor Networks CEO David Staley. Similarly, predictive models could be built to predict security threats (e.g., identifying regions subject to poaching by rates of previous activity and environmental conditions). Machine learning automates these processes, allowing conservationists to save time and resources and improve the accuracy of their observations (Martinez & Liu 2024).

Also, big data has played a significant role in ecology, enabling scientists to study and understand complex ecosystems via large datasets that conglomerate information from different platforms, such as IoT devices, satellite imagery, or historical ecological records. For instance, information obtained from wildlife tracking, weather patterns, and vegetation health can all be integrated to determine what happens during animal migration in a changing climate. Likewise, data from environmental sensors at multiple locations can be analyzed to identify trends in pollution levels and pinpoint the sources of contamination or impacted regions. Conservationists can then visualize long-term trends, identify interactions between different ecological processes, and decide where to best target conservation efforts with the most impact (Wang et al., 2024). In addition, big data is used in building ecosystem models that consider different situations like how wildlife and other species will cope with habitat loss or climate change, enabling conservationists to act today based on future scenarios.

4. CASE STUDIES RELATED TO THE EFFECTIVE IMPLEMENTATION OF IOT IN BIODIVERSITY CONSERVATION

The deployment of IoT technologies in wildlife conservation has dramatically improved how we can monitor and preserve ecosystems. Also, it provides an overview of two successful case studies where IoT has solved highly pressing problems. Some of them are as follows:

4.1 Monitoring the Amazon Rainforest

The Amazon Rainforest is known as the "lungs of the Earth" for playing a critical role in controlling climate and biodiversity globally. However, it is under severe threat from deforestation, illegal logging, and mining. To address this, upstream uses all these solutions with IOT technologies to protect the environment and safeguard wildlife through innovative real-time monitoring of nature reserves for early detection of illegal activities. Drones with cameras and environmental sensors are a standard tool for taking aerial images of the forest. It can help authorities monitor changes in vegetation cover over time. Using drones to patrol the sky, these gadgets can cover a broad range of territory that would be nearly impossible and time-consuming for anyone on foot or in a military vehicle to monitor, making it an effective way also used to identify deforestation that may take place (Kim & Patel, 2024).

The declaration said that ground-based sensors along the forest measure environmental variables such as humidity, temperature, and soil moisture, as well as drones. These sensors are essential in measuring the ecosystem's health, allowing scientists to detect changes that might suggest human influence. In addition, IoT devices are linked to alert systems that notify authorities whenever abnormal activities have been initiated (Anderson et al., 2023). These tools enable conservationists to take immediate action, preventing illegal deforestation and safeguarding vital habitats. Time has only empowered local authorities and international organizations to monitor large portions of Amazon through this IoT-enabled strategy, thus facilitating easier identification and management to respond to threats in real-time (Johnson & Wang, 2024).

4.2 Wildlife Conservation in Africa

Many species are under threat in Africa; elephants, rhinos, and big cats face significant issues with poaching and loss of habitat. IoT technologies have gone a long way in protecting these species by helping individuals track their movements and get knowledge in real-time. GPS collars have been created and placed on endangered species; these are used for anti-poaching systems. They are equipped with collars that relay location data to rangers in real-time. If a tracked animal enters an area of particular risk for poaching, the system sends off alerts so that action can be taken in advance. This has mainly been a game change in the conservation of elephants and rhinoceros, which are significant targets for their ivory tusks and horns (Li et al., 2023).

Another way that the IoT is being used to aid in African wildlife conservation involves integrating camera traps and acoustic sensors into natural surroundings, which serves two purposes: monitoring animals for activity and possible dangers. To keep track of wild animals, they are fitted with motion-sensing cameras that take pictures and transmit images back to the research team. Researchers use these pictures to examine animal populations, learn about their behaviors, and determine the effectiveness of conservation efforts (Smith et al., 2023). They include acoustic sensors that listen for particular sounds associated with criminal behavior, such as cracking gunshots or vehicle movements in restricted areas. As a result, monitoring has become more efficient, and the technologies have allowed conservationists to work better with governments, other NGOs, and local communities. Data collected from IoT devices will be shared among stakeholders working together to combat threats and develop wildlife-friendly policies. For example, real-time data from tracking devices maps wildlife corridors connecting fragmented habitats and allowing animals to move safely across large landscapes IoT and how it can revolutionize conservation work, delivering timely info for faster, more successful interventions (Gómez et al., 2024). The application of IoT devices in the Amazon Rainforest and wildlife conservation in Africa clearly shows how technology can define Conservation rather than only being Juvenile. With their support for real-time monitoring, data-rich fast-track capabilities, and remote interaction, IoT has been pivotal in both ecosystem preservation efforts and working to save endangered creatures. The notes also believe that with advancing technology, we will be able to capitalize on more complex and interconnected mechanisms for biodiversity conservation, bringing a glimmer of hope in our quest to protect the Earth's fragile ecosystem (Lee et al., 2024).

5. CHALLENGES AND LIMITATIONS OF IOT IN BIODIVERSITY CONSERVATION

While the application of IoT technologies in biodiversity conservation has many advantages, it also faces essential shortcomings and challenges that must be overcome. Connectivity is one of the significant technical challenges. The most critical ecosystems (deep forests, high mountains, and deep oceans) are far from a dense urban network structure (Jones & Patel, 2024). While using technologies such as satellite communication or LoRaWAN can also help overcome several challenges, large-scale operation is highly cost-prohibitive and performance-limited. In addition, a wide range of data collected from different sensors continuously requires efficient storage and management systems, increasing the sensor network's scalability (Kumar & Smith 2023). Many IoT devices, like those out in the field or within cities, might be at locations where power sources are hard to access as well, and typically, there is a significant gap between new battery technology deployments, so IoT must have batteries that last for really long periods without having someone going around replacing them every year (Zhang et al., 2023). Therefore, to reduce the impact of these devices running on energy from solid fuels and solve specific electricity-related problems through solar panels/ battery-operated devices, all kinds of realms need much more power than is feasible.

Data privacy and security are significant issues when using IoT in conservation. These include monitoring endangered species -location data that could fall into the hands of poachers to be hunted and exploited. Therefore, robust encryption and data protection techniques are necessary to secure this sensitive information. In addition, there are worries about compromises to data security as they could inadvertently reveal crucial details about the location of protection somewhere, which would be detrimental overall for conservation. With the spread of IoT networks, cyberattacks have also increased at an unprecedented rate, and it is imperative to enhance data security protocols regularly (Brown & Li, 2023).

Concerns over IoT in conservation deployments go beyond the technical. Even though they have been designed to be as non-invasive as possible, devices such as GPS collars, tags, or acoustic sensors may still interfere with the wildlife's behavior patterns. Also, drones meant for aerial surveys could interfere with certain species, especially birds, which might even lead to habitat abandonment at the high end (Chen & Zhao, 2024). There must be clear ethical guidelines in place so that conservationists are not inadvertently harming a species while trying to protect it. Substantial economic challenges also limit the broad implementation of IoT and biodiversity conservation (Thompson et al., 2023). Implementing IoT on a large scale requires massive financial investment. Furthermore, regular maintenance adds to the cost, and skilled personnel are required to operate those systems. For smaller-scale conservation efforts, especially those reliant on funding from NGOs or private donors, these costs can make it challenging to establish the necessary monitoring infrastructure for projects to be upscaled (Nguyen & Williams, 2024).

6. FUTURE DIRECTIONS AND RECOMMENDATIONS

If the Internet of Things (IoT) is to have a role in biodiversity conservation, it will be because collaborations among individuals with different skill sets can address them. Speeding up integration into conservation strategies The researchers, all of whom are working to develop IoT solutions in various capacities (technologists) or planning deployments as a part of their operational work overseeing managing areas such as World Heritage Sites and protected lands globally (ecologist/policy-makers), say that integration requires partnerships among technologists who design the systems, ecologists with expertise on wildlife behavior, health, and welfare when animals are tagged (Johnson et al., 2023). Policymakers are charged with coordinating what data gets collected and where. Such partnerships can ultimately capture capital investment, connecting technological advancement to environmental and regulatory factors. A touch multidisciplinary approach will benefit stakeholders by laying down holistic conservation strategies and strategically incorporating IoT platforms in wilderness systems (Martinez & Liu, 2024).

Establishing policy and regulatory frameworks that direct the responsible adoption of IoT for conservation is also crucial. Protocols around sensitive data and the ethics of monitoring wildlife are essential not to jeopardize trust in our research. Such frameworks should include representatives from diverse stakeholders in a multi-stakeholder setting, ensuring the development of policies that balance reaping benefits offered by IoT technologies and addressing potential harms resulting from their deployment. A critical factor for scaling IoT solutions is the strategies we must consider for low-cost implementations in a wide range. An additional focus on training local communities and conservation practitioners in effectively using these tools would help build a sense of ownership, capacity, and sustainability for conservation efforts.

7. CONCLUSION

IoT can significantly impact biodiversity conservation, offering solutions to critical challenges and opportunities. Adopting these IoT technologies provides exciting possibilities for ecosystem monitoring, wildlife tracking, and resource management. However, accessing this potential will require enabling interdisciplinary collaboration, appropriate policy frameworks, and demonstrated scalable solutions. IoT holds great promise for conservation, but further innovation is needed to ensure the technology's full potential is realized and turned into positive action on a global scale that will help us achieve our long-term conservation objectives as stewards of this blue planet.

REFERENCES

Anderson, J., Li, X., & Brown, T. (2023). Remote sensing applications for sustainable land management. *International Journal of Remote Sensing*, 44(5), 345–360.

Anderson, J., & Roberts, C. (2023). Data challenges in AI-driven fire management systems. *Journal of Environmental Management*, 312, 114–125.

Brown, S., & Li, Y. (2023). Satellite-based monitoring of agricultural landscapes: Recent advancements and future prospects. *Agricultural Systems*, 214, 103–114.

Brown, T., Martinez, E., & Lee, H. (2023). Automation and real-time monitoring in wildfire management: An AI perspective. *Fire Technology*, 59(2), 245–261.

Chen, H., & Zhao, Q. (2024). Urban expansion and satellite-based land cover mapping. *Urban Studies Journal*, 61(2), 123–140.

Chen, Y., & Zhao, Q. (2024). The role of UAVs in enhancing AI-driven fire management strategies. *Remote Sensing Applications: Society and Environment*, 15(1), 87–98.

Davis, M., Smith, L., & Thompson, R. (2024). Enhancing wildfire prediction using machine learning: A case study in predictive modeling. *Journal of Environmental Management*, 250(5), 102–113.

Díaz-Delgado, R., Lloret, F., & Pons, X. (2018). Spatial patterns of fire occurrence in Catalonia, NE Spain. *Landscape Ecology*, 13(4), 365–378.

Filkov, A. I., Ngo, T., Matthews, S., Telfer, S., & Penman, T. D. (2020). Impact of Australia's catastrophic 2019/20 bushfire season on communities and environment. Retrospective analysis and current trends, 13(8), 21-39.

Gómez, C., White, J. C., & Wulder, M. A. (2024). Optical remotely sensed time series data for land cover classification: A review. *ISPRS Journal of Photogrammetry and Remote Sensing*, 176, 148–163.

Gómez, C., White, J. C., & Wulder, M. A. (2024). Machine learning approaches for dynamic fire behavior prediction. *ISPRS Journal of Photogrammetry and Remote Sensing*, 187, 235–248.

Johnson, A., & Wang, L. (2023). AI-driven resource allocation in fire management: Enhancing efficiency and safety. *International Journal of Wildland Fire*, 32(3), 145–158.

Johnson, A., & Wang, L. (2024). Land cover mapping using satellite imagery: Advances and applications. *Journal of Geographical Sciences*, 34(1), 12–25.

Johnson, P., & Martinez, E. (2024). The role of artificial intelligence in addressing climate change-induced wildfire risks. *Nature Climate Solutions*, 4(1), 45–56.

Jones, A., Brown, C., & Wilson, G. (2024). Machine learning approaches in environmental science: Applications in wildfire management. *Environmental Informatics*, 23(2), 78–89.

Jones, P., & Patel, R. (2024). Cloud removal in satellite imagery: Techniques and challenges. *Remote Sensing Applications: Society and Environment*, 7(3), 245–256.

Jones, P., & Taylor, R. (2024). Ethical and interdisciplinary considerations in AI-based wildfire management. *AI & Society*, 39(1), 34–47.

Kumar, V., & Smith, J. (2023). Overcoming atmospheric interference in satellite-based land cover mapping. *Yaogan Xuebao*, 50(8), 400–412.

Lee, D., Chang, J., & Kim, H. (2024). Deep learning approaches for land cover classification using satellite imagery. *Remote Sensing of Environment*, 280, 112144.

Lee, D., & Kim, S. (2024). Advances in deep learning applications for fire prediction and detection. *AI in Environmental Science*, 11(4), 162–177.

Lee, H., & Kim, Y. (2024). Remote sensing and deep learning for wildfire monitoring: A review of current advancements. *International Journal of Remote Sensing*, 45(9), 163–178.

Li, Y., Zhang, M., & Thompson, J. (2023). Advances in satellite remote sensing for land cover classification. *Journal of Remote Sensing Technology*, 22(4), 211–223.

Liu, Y., Yang, Z., & Deng, C. (2020). A novel method for wildfire detection and monitoring with a convolutional neural network. *Remote Sensing*, 12(3), 398.

Martinez, J., & Liu, Q. (2024). Climate change and AI: Transforming wildfire prediction and response. *Nature Climate Solutions*, 5(2), 45–59.

Nguyen, T., & Williams, S. (2024). The future of satellite-based land cover mapping: Emerging technologies and trends. *Environmental Monitoring and Assessment*, 196, 1024.

Nguyen, T., Wu, H., & Patel, S. (2024). Real-time wildfire prediction using deep learning: A study of adaptive response strategies. *AI and Environmental Sciences*, 12(3), 27–35.

Patel, R., Lee, S., & Wang, J. (2024). Adaptive management of wildfires using AI-driven prediction models. *Fire Ecology and Management*, 32(4), 101–117.

Pausas, J. G., & Keeley, J. E. (2019). Wildfires as an ecosystem service. *Frontiers in Ecology and the Environment*, 17(8), 456–462.

Rein, G., Cohen, S., & Campbell, D. (2019). Challenges and opportunities in understanding and managing wildfires. *Nature Communications*, 10(1), 1–3. PMID: 30602773

Singh, B. (2023). Blockchain Technology in Renovating Healthcare: Legal and Future Perspectives. In Kaushik, K., Dahiya, S., Aggarwal, S., & Dwivedi, A. (Eds.), *Revolutionizing Healthcare Through Artificial Intelligence and Internet of Things Applications* (pp. 177–186). IGI Global., DOI: 10.4018/978-1-6684-5422-0.ch012

Singh, B. (2024). Lensing Legal Dynamics for Examining Responsibility and Deliberation of Generative AI-Tethered Technological Privacy Concerns: Infringements and Use of Personal Data by Nefarious Actors. In Ara, A., & Ara, A. (Eds.), *Exploring the Ethical Implications of Generative AI* (pp. 146–167). IGI Global., DOI: 10.4018/979-8-3693-1565-1.ch009

Singh, B. (2024). Social Cognition of Incarcerated Women and Children: Addressing Exposure to Infectious Diseases and Legal Outcomes. In Reddy, K. (Ed.), *Principles and Clinical Interventions in Social Cognition* (pp. 236–251). IGI Global., DOI: 10.4018/979-8-3693-1265-0.ch014

Singh, B. (2024). Evolutionary Global Neuroscience for Cognition and Brain Health: Strengthening Innovation in Brain Science. In Prabhakar, P. (Ed.), *Biomedical Research Developments for Improved Healthcare* (pp. 246–272). IGI Global., DOI: 10.4018/979-8-3693-1922-2.ch012

Singh, B., Dutta, P. K., & Kaunert, C. (2024). Replenish Artificial Intelligence in Renewable Energy for Sustainable Development: Lensing SDG 7 Affordable and Clean Energy and SDG 13 Climate Actions With Legal-Financial Advisory. In Derbali, A. (Ed.), *Social and Ethical Implications of AI in Finance for Sustainability* (pp. 198–227). IGI Global., DOI: 10.4018/979-8-3693-2881-1.ch009

Singh, B., Jain, V., Kaunert, C., Dutta, P. K., & Singh, G. (2024). Privacy Matters: Espousing Blockchain and Artificial Intelligence (AI) for Consumer Data Protection on E-Commerce Platforms in Ethical Marketing. In Saluja, S., Nayyar, V., Rojhe, K., & Sharma, S. (Eds.), *Ethical Marketing Through Data Governance Standards and Effective Technology* (pp. 167–184). IGI Global., DOI: 10.4018/979-8-3693-2215-4.ch015

Singh, B., & Kaunert, C. (2024). Computational Thinking for Innovative Solutions and Problem-Solving Techniques: Transforming Conventional Education to Futuristic Interdisciplinary Higher Education. In Fonkam, M., & Vajjhala, N. (Eds.), *Revolutionizing Curricula Through Computational Thinking, Logic, and Problem Solving* (pp. 60–82). IGI Global., DOI: 10.4018/979-8-3693-1974-1.ch004

Singh, B., & Kaunert, C. (2024). Wind and Solar Energy as Renewable Energy for Sustainable Global Future: Projecting Future Multi-Sector Sustainable Policies and Innovation. In Ara, A., & Thakore, R. (Eds.), *Promoting Multi-Sector Sustainability With Policy and Innovation* (pp. 210–245). IGI Global., DOI: 10.4018/979-8-3693-2113-3.ch009

Singh, B., & Kaunert, C. (2024). Aroma of Highly Smart Internet of Medical Things (IoMT) and Lightweight EdgeTrust Expansion Medical Care Facilities for Electronic Healthcare Systems: Fortified-Chain Architecture for Remote Patient Monitoring and Privacy Protection Beyond Imagination. In Hassan, A., Bhattacharya, P., Tikadar, S., Dutta, P., & Sagayam, M. (Eds.), *Lightweight Digital Trust Architectures in the Internet of Medical Things (IoMT)* (pp. 196–212). IGI Global., DOI: 10.4018/979-8-3693-2109-6.ch011

Singh, B., & Kaunert, C. (2024). Augmented Reality and Virtual Reality Modules for Mindfulness: Boosting Emotional Intelligence and Mental Wellness. In Hiran, K., Doshi, R., & Patel, M. (Eds.), *Applications of Virtual and Augmented Reality for Health and Wellbeing* (pp. 111–128). IGI Global., DOI: 10.4018/979-8-3693-1123-3.ch007

Singh, B., & Kaunert, C. (2024). Salvaging Responsible Consumption and Production of Food in the Hospitality Industry: Harnessing Machine Learning and Deep Learning for Zero Food Waste. In Singh, A., Tyagi, P., & Garg, A. (Eds.), *Sustainable Disposal Methods of Food Wastes in Hospitality Operations* (pp. 176–192). IGI Global., DOI: 10.4018/979-8-3693-2181-2.ch012

Singh, B., & Kaunert, C. (2024). Revealing Green Finance Mobilization: Harnessing FinTech and Blockchain Innovations to Surmount Barriers and Foster New Investment Avenues. In Jafar, S., Rodriguez, R., Kannan, H., Akhtar, S., & Plugmann, P. (Eds.), *Harnessing Blockchain-Digital Twin Fusion for Sustainable Investments* (pp. 265–286). IGI Global., DOI: 10.4018/979-8-3693-1878-2.ch011

Singh, B., & Kaunert, C. (2024). Harnessing Sustainable Agriculture Through Climate-Smart Technologies: Artificial Intelligence for Climate Preservation and Futuristic Trends. In Kannan, H., Rodriguez, R., Paprika, Z., & Ade-Ibijola, A. (Eds.), *Exploring Ethical Dimensions of Environmental Sustainability and Use of AI* (pp. 214–239). IGI Global., DOI: 10.4018/979-8-3693-0892-9.ch011

Smith, J., Zhang, X., & Liu, B. (2024). Advances in AI-driven wildfire prediction and mitigation strategies. *Frontiers in Environmental Science*, 12(7), 1–14.

Smith, K., Roberts, E., & Zhao, L. (2023). High-resolution satellite imagery and its applications in environmental monitoring. *Earth Observation Science*, 58(6), 230–245.

Thompson, B., Anderson, C., & Lee, S. (2023). Land cover mapping for biodiversity assessment: Methods and applications. *Ecological Indicators*, 155, 108285.

Wang, J., Zhao, X., Xie, J., & Shi, Z. (2019). A review of machine learning applications in wildfire management. *Current Forestry Reports*, 5(4), 196–209.

Westerling, A. L. (2016). Increasing western US forest wildfire activity: Sensitivity to changes in the timing of spring. Philosophical Transactions of the Royal Society B: Biological Sciences, 371(1696), 20150178.

Zhang, F., Li, X., & Wang, Y. (2023). Multi-source data fusion for improved land cover mapping accuracy. *Geocarto International*, 38(1), 1–15.

Zhao, T., & Yu, W. (2023). Machine learning in remote sensing: Applications for land cover classification. *Remote Sensing in Earth Systems Sciences*, 6(2), 87–101.

Chapter 7
Integrating Machine Learning Into Wildlife and Forest Conservation:
Ethical Considerations and Sustainable Practices

Souvik Dhar
https://orcid.org/0000-0002-5472-9193
Brainware University, India

Arya Dhar
Laromed GmbH, Germany

ABSTRACT

This chapter has argued for clear ethical standards and good practices in the use of Machine Learning in the management of wildfire. Advanced technologies, such as Machine Learning, holds a lot of promise in enhancing wildfire management through better forecasts and improved strategies for quick response. However, the use of Machine Learning in the domain also raises some interesting and significant ethical issues, like data privacy, accountability, environmental justice, and community engagement. Before delving into ethical considerations, the chapter discusses the legal frameworks governing wildland fire management in the US, Australia, Canada, European Union, India, and other regions and how they adapted wildfire strategies in response to climate change and disaster recovery. With ethical considerations absorbed into the deployment of such technologies, stakeholders may work together to ensure that Machine Learning is a tool for equitable and sustainable management of wildfires.

DOI: 10.4018/979-8-3693-7565-5.ch007

INTRODUCTION

Wildfires represent one of the major threats of wildlife and forest conservation, and their frequency is getting higher under the influence of climatic change and human activities (McKenzie et al., 2004.). Advanced technologies, and mainly machine learning, hold great promise for reducing risks associated with wildfires (Carta et al., 2023). It would be in better forecasting, early warning, and improved management of wildfires, all that could provide more efficient and effective responses (Carta et al., 2023). Application of Machine Learning (ML) in wildfire management, however, has some unsettled important ethical issues. In that light, the chapter tries to look into the ethical issues associated with the use of ML in wildfire management, trying to address sustainable development while providing a few strategies that will balancing technological advancement with taking ethical responsibilities. Accordingly, with respect to the ethical challenges identified, this chapter aims to set and discuss the broader ethical challenges one sits with while applying Machine Learning toward wildfire management. The chapter also deals with the laws addressing forest fire in United States of America (USA), Australia, Canada, European Union's Countries, India and other general reforms. Hence, the chapter attempts to balance the precarious benefit versus failure wire using proper ethics in such a way that the application of ML does not prejudice any level of society. The end goal is to contribute to the set-up of ethical guidelines and good practices for the implementation of ML technologies in wildfire management.

WILDFIRE MANAGEMENT: ROLE OF MACHINE LEARNING

Machine learning (ML) is a key aspect of wildfire management, and it has significantly enhanced the capability to foresee, locate, and combat fires. Machine learning algorithms can also be used to better predict the probability of wildfires by processing big data, such as historical weather data, satellite imagery, and sensor data in real time (Jain et al. The predictive models assist the authorities in utilizing their resources more effectively in the preparation and response phases of fire management (Jain et al., 2020). Speaking of detection, ML systems are capable of analyzing satellite imagery and ground sensor data to pinpoint the initial stages of wildfires. This results in earlier detection and alarms, which means earlier response time, which means the fire is contained before it gets to spread. For example, machine learning can examine thermal images for abnormal heat signatures, and these could be early warning signs for possible fires. ML also allows professionals to see how fires burn and propagate (Khan et al., 2022). These models do this by running hypotheticals based on environmental information and history, and make predictions on the paths of fires, or where they are most likely to go next, which is crucial in designing containment plans and evacuation routes (Pham et al., 2020). In addition to the active fight of wildfires, machine learning also helps in post fire analysis. That is to say that it can be used to assess the degree of damage by measuring the intensity of devastation or pinpointing the worst hit locations. This is not only important for the recovery process but also for preparing for any future events (Boroujeni et al.; Duff et al., 2021)

• Early Detection Using ML Algorithms

Wildfires need to be caught in the nascent stage to preclude as much devastation as possible. One area that machine learning systems excel in is the real time processing of data from various inputs such as satellite images, drones, ground sensors, and surveillance cameras for early indicators of fire (Jain et al., 2020). Such as the analysis of thermal images derived from satellite or drone imagery, processed by machine learning algorithms to spot anomalous heat signatures that may suggest a fire (Awatade et al., 2024). Furthermore, machine learning models can process data from environmental sensors which track temperature, humidity and carbon levels to recognize abrupt fluctuations that might be due to a wildfire outbreak (Bhowmik et al., 2023). These systems can provide warning signals before a fire has time to spread, so immediate action can be taken (Jain et al. In places where it would take hours or even days for someone to notice a wildfire burning out of control, it would be crucial nowadays because it cuts down on the amount of time from ignition to response, thus saving hundreds of thousands of acres of land from complete destruction (Mohapatra Trinh, 2022).

• Understanding Fire Behavior and Progression

In case there is fire, the first step is knowing its behavior. This is because proper containment measures are required to cease the spread of fire. Here's where ML comes in handy. The ML models take into account different environmental parameters that can help predict how the wildfire shall behave like; they use real-time data about the environment together with historical ones (Alkhatib et al., 2023). For example, wind speed and direction, humidity level and amount of available fuel (vegetation) are considered, then possible ways are simulated to find out in which fire might go (Hargrove et al., 2000). The information provided by these simulations becomes very crucial in guiding firefighting groups as they tell them where to concentrate their efforts most urgently and how best to put out the flames safely & effectively (Hargrove et al., 2000). Additionally, this type of difficulty saves lives during emergencies since it leverages on likely escape routes for various neighborhoods within the province who are most threatened by fire facing government officials' need evacuation soonest possible or at least before others before others become unmanageable otherwise some lives might be lost due to delays in arriving at those areas affected by blazes (McLennan et al., 2019). However, ML contributes to minimizing both human and economic losses caused by wildfires through improving response time and resource utilization (Jain et al., 2020).

• Post-Fire Analysis and Damage Assessment

Not merely confined to active wildfire management, machine-learning can also be used in the analysis of the aftermath of wildfires and recovery (Thompson et al., 2019). Also fire damage assessment after the fire has been extinguished with the help of ML algorithms that can process satellite images and drone footage and sensor data to determine the extent of destruction (Abid, 2021). This is particularly relevant when assessing the degree of damage a fire caused to ecosystems, infrastructure and human populations (Jain et al., 2020). That aside, ML systems that study the landscape after a fire can determine what areas are more susceptible to future burning, and thus can aid in land management and reforestation studies. For instance, some areas could have become eroded or flooded after a wildfire while others may have currently favorable conditions for another fire outbreak (Tedim et al., 2020). In these and other ways, machine learning can aid long term recovery and mitigation efforts so that future wildfires will not be so devastating (Jain et al. Besides, Machine Learning models can refine their predictive algorithms using past wildfire trends and patterns, and therefore they become better at predicting the next fire-seasons (Jain et al., 2020).

• Prediction and Forecasting Wildfire Risk

Forecasting and prediction are two of the most critical things that machine learning can do in wildfire management. Many factors influence wildfires, some of which are temperature, humidity, wind patterns, topography and the availability of combustible vegetation (Jain et al., 2020). Machine Learning algorithms can process huge amounts of data, including satellite images and weather station records, along with records of past fires, in order to determine the likelihood of a region having an outbreak of wildfires (Jain et al., 2020). In contrast to more conventional forecasting techniques which may only rely on a few select variables, ML models can analyze thousands of variables all at once (Lotfi & El Bouhadi, 2022). More sophisticated models such as neural nets and decision trees are able to look at this data and find subtle patterns or correlations that are barely recognized by human analysts (Lotfi & El Bouhadi, 2022). Such as ML models that can determine with great precision the probability of a wildfire starting at a certain time and place simply by observing the variations in weather or vegetation conditions over time (Jain et al.,2020) It would be so cool because then the authorities could use their manpower most efficiently, they could station the firemen ahead of time, do preventative work like clearing brush or other combustible vegetation, or even give early warnings to the towns surrounding the potential fire.

• Ethical Concerns and Challenges

Machine learning is incredibly effective at fighting fires, but it still raises moral issues. (Awatade et al., 2024). A significant problem is privacy (Uddin, Chowdhury & Kabir, 2024). Numerous machine learning systems work with information packed by satellites, drones, and ground sensors which may at times unconsciously pick or tap private family details without their knowledge or approval (Ronchi & Ronchi, 2019. Thus, making sure that such data will be managed properly and securely following strict privacy protection rules is the only way to ensure that society's faith can still be kept intact. Moreover, these algorithms may have biases within them (Jain et al., 2020). It implies that if training data used for such ML models happen to be advanced but biased according to certain communities, then their product predictions might cause unequal distribution of resources or evacuation priorities across different regions among these people (Jain et al., 2020). For example, it could happen that when a model created through learning with an emphasis on rich urban neighborhoods attends rural ones with lack of funds it does not work as expected thereby putting more life at risk there (Casali, Aydin, & Comes, 2022). Selection and presentation of datasets on which machine learning algorithms are trained in a wise way act as a guard against unpredictable disasters that occur at different locations (Linardos et al., 2022). This will not be unfair to some people and also reduce the chances of racial discrimination or any other discrimination (Linardos et al., 2022). With all these precautions, a key priority still remains guaranteed access to technology. There are too many developing areas that can't afford or don't have the capability to utilize that kind of technology. That only widens the gap between the global community in terms of preparation and reaction to wildfires.

• Balancing Technological Advancement with Ethics

However, there's no denying the benefits of ML as far as fire control is concerned. The key to its power is the ethical dilemmas that come with it (Albahri et al., 2024). As for example, governments and private organizations need to work together to create some policies that encourage ethical ML technology (Mikhaylov, Esteve & Campion, 2018). Some examples include stringent data privacy laws, performing algorithm audits for bias detection, and ensuring that all communities have equal access to ML technologies (Akula Garibay, 2021). More research is required on ML abilities in fire management too. With the use of machine learning and human know-how and a built-in system of ethics that would control it, a much more renewable and moral approach to these kind of activities could be created. The broader goal is not just to limit wildfire losses but to protect human life and ecosystems in the face of these catastrophes and to build more resilient communities in an increasingly unstable climate.

Machine learning has drastically changed wildfire management from before the fire begins to after recovery from the fire itself; this includes prediction, detection, control, and post fire recovery (Boroujeni et al., 2024). A comprehensive survey of research towards AI-enabled unmanned aerial systems in pre-, active-, and post-wildfire management (Boroujeni et al., 2024). An extensive review of literature on artificial intelligence applications in unmanned aerial vehicles for pre, active, and post-fire environments (Boroujeni et al., 2024). This technology has the capability to process large amounts of data in a short time, and can thus provide nearly immediate feedback, which would greatly improve response time, mobilization of resources, and containment procedures (Boroujeni et al., 2024). However, there are serious ethical issues that need to be considered if the full potential of machine learning in this field is to be harnessed including risks associated with privacy of information contained in extensive datasets used for training purposes or mainframes whose storage spaces are not privacy regulated (algorithmic injustices) (Kyrkou et al., 2022). So, machine learning can be a revolutionary tool that can save people's lives, homes, and habitats from the looming threat of wildfires if it uses its ingenuity responsibly with a sense of moral responsibility.

ETHICAL CONCERNS IN MACHINE LEARNING FOR WILDFIRE MANAGEMENT

Machine learning has now emerged as one of the chief tools in the area of wildfire management, in that it can predict, detect, and mitigate these occurrences within real timeframes (Awatade et al., 2024). As great as this technology might sound, there are significant ethical considerations regarding its use in wildfire management (Arif et al., 2021). These include but are not limited to the following: data privacy and security, algorithmic bias and fairness, accountability and transparency, impacts on local communities, environmental concerns, equity in technology use, and long-term sustainability. Their resolution will be important for assuring that ML wildfire management systems are effective and ethical, equitable solutions. Obviously, this would involve data privacy and security.

One of the major issues in using ML for wildfire management is data privacy and security (Thompson et al., 2019). In general, huge amounts of data are required, supplemented by satellite images, weather patterns, and sensor data, to feed the ML models. Though data sets are a key basis for generating predictions about wildfires and thus managing them, they equally can be invasive (Thompson et al., 2019). For example, satellite images and sensor data may signal sensitive information concerning individuals, private property, or even whole communities (Awatade et al., 2024). Without proper precautions in respect of privacy, there could be data breaches leading to leaks to malicious or unauthorized users.

Data privacy requires several layers of protection. Data collection policies need to be clearly and transparently laid down to describe how data is to be collected, stored, and used. Of importance would be the informed consent from individuals or communities that the data pertains to on how it will help in better management of wildfires (McCaffrey et al., 2015). Besides, stringent encryption and security protocols for the data should be put in place to restrict unauthorized access (McCaffrey et al., 2015). It requires government and organizational collaboration in terms of creating and enforcing policies that ensure private individual privacy while enabling responsible data use within wildfire management (McCaffrey et al., 2015). In the end, ML-driven wildfire management systems can only succeed based on the level of trust and cooperation that full communities give to the success of these operations, an aspect that can only be achieved through careful attention to data privacy and security (McCaffrey et al., 2015).

• Algorithmic Bias and Fairness

The other ethical issue to be faced in the management of wildfire is the algorithmic bias in ML models (Al-Bashiti & Naser, 2022). Normally, ML models are trained using historical data which may reflect various biases or unequal distribution of resources across different regions or communities. Unless taken care of, ML systems would then run the risk of perpetuating and even further worsening existing inequalities in wildfire management (Al-Bashiti & Naser, 2022). For example, it might work a lot in rich, urban areas and fail utterly in rural or poor regions. That would give a very unequal distribution in firefighting resources and further endanger the more vulnerable populations.

In light of these biases, training ML models with diverse data that cuts across all regions and populations affected becomes important (Jain et al., 2020). This will involve frequent auditing in the ML models for bias during training and hence correcting it (Jain et al., 2020). Besides, ML models will have to be designed with a view of fairness in such a way that all communities share in equal measures regarding access to advanced technologies in the management of wildfires (Jain et al., 2020). Fairness in the use of ML is both a technical issue and a social one, as there are judgments to be made on how the resources must be distributed or which communities need protection first if there is a wildfire (Jain et al., 2020). Accountability in management related to wildland fire shall ensure fairness and equity in order to help in building a just and resilient society (Jain et al., 2020).

• Accountability and Transparency

Key ethical concerns with respect to the use of ML within wildland fire management include accountability and transparency (Visave, 2024). The decisions of ML models are complex to the extent that their operations may hardly be understood by human beings; they are usually considered as "black boxes" (Phelps & Woolford, 2021). This lack of transparency therefore makes it difficult to hold someone accountable should something go wrong (Visave, 2024). For example, if an ML model happens to fail in predicting a wildfire or fails to distribute resources inefficiently, it can result in the loss of lives or property. The question here would be: who is responsible for this failure-is it the developers or the end users of the model? Or is it the organizations which deployed it?

The solution to mitigate this challenge is to improve the transparency of the ML models used in wildfire management (Visave, 2024). This can be achieved by the development of XAI systems to provide insight into how the model derived the decisions (Ghaffarian, Taghikhah, & Maier, 2023). Besides, clear lines of accountability need to be created where there is accountability on the part of those deploying and maintaining ML systems for its output. For ethical reasons, this transparency is important in building trust in ML technologies. It will help stakeholders, whether they are firefighters, government officials, or local communities, understand how such models work, and who is responsible for deploying them and therefore are bound to be more accepting of the technology and support its use in managing wildfires.

• Impact on Local Communities

The deployment of ML technologies for the management of wildfires is very likely to impact the livelihoods of local communities, particularly in terms of forced relocations, land-use changes, and decision-making (Kameswari, 2023). The decisions of resource distribution, the timing of evacuation orders, or land management are made mostly by ML systems, without much consideration of the communities most exposed to those decisions (Xu et al., 2023). This may affect the local populations with a feeling of disentitlement since their needs and concerns are not taken into account (Kameswari, 2023). In this vein, it is relevant to involve local communities in each stage of development and deployment of ML systems for the management of wildfires.

• Environmental Considerations

Although ML technologies will bring significant benefit to wildfire management, there will also be unexpected repercussions to the environment (Jain et al., 2020). The field deployment of sensor networks, drones, and other ML techs also disturbs local ecosystems and wildlife (Jain et al., 2020). Take for example, drones used to track wildfires can affect some species natural behavior and sensor networks placed all around the wilderness can also physically harm the ecosystem (Jain et al., 2020). So, these ecological consequences need to be kept in mind when using ML for wildfire control.

• Fairness in the Use of Technology

One of the main issues regarding the application of ML to wildfire management is the possibility of a digital divide among different localities (Jain et al., 2020). However, the more advanced ML technologies can be extremely expensive and have large infrastructural and expert requirements to implement (Jain et al, 2020). The overarching consequence of this, though, is that these technologies are at the disposal of wealthier regions and institutions, while the poorer or under-resourced are not (Jain et al., 2020). This has created a kind of digital gap where some communities can afford the latest and most efficient way of combating wildfires and others cannot (Jain et al., 2020). However, equity demands that these solutions be affordable and scalable enough to be implemented in all regions, be it a region that is financially and technologically blessed, or not (Jain et al., 2020). And so is knowledge-sharing, resource-sharing, best practices with impoverished areas, and investing in R&D for low-cost ML systems that can be used in a variety of contexts (Courtney, 2023). This will allow for any worldwide effort against wildfire to be equally represented and rhythmic, with open availability to ML technologies. A combination of technological advances and primitive knowledge and practice could make for a very sustainable and efficient method of wildfire management for years to come.

WILDLIFE AND FOREST CONSERVATION: APPLICABLE LAWS AND POLICIES

The issue of wildfire risk has grown globally due to increased meteorological, land use management, and ecosystem change. Various countries have engaged in different legal frameworks, policies, and management touching the risk of wildfire occurrences, influencing varied efforts meant to reduce the immediate effects of fire, and enhancement of recovery processes. There are also important global initiatives that promote collaboration and encourage a coordinated response to wildfire management and disaster risk reduction. This broader perspective now looks at a number of the countries' legislative frameworks and policies identified as key: the United States, Australia, the European Union, and Canada. We also consider relevant international initiatives, such as the Sendai Framework for Disaster Risk Reduction and the UNFCCC.

United States

With diverse and expansive landscapes that result in frequent fires, especially among Western states, the United States can pride itself on a long history regarding the management of wildfires. Over time, the federal government passed different laws for better management of the fires, protection of natural resources, and reduction of the risk of fire.

• National Environmental Policy Act (NEPA)

NEPA, 1969, is one of the foundational statutes comprising the cornerstone of U.S. environmental law (Fiorino, 2023). Although NEPA is a general environmental statute requiring federal agencies to assess the environmental consequences of their activities, it is important to the management of wildfires. NEPA requires federal agencies to make informed decisions concerning the environmental impact of the proposed action and to include the public in that process (Fiorino, 2023). In wildfire management, this amounts to assessing the benefits of activities that prevent the occurrence of fire against the likelihood of deleterious impacts on wildlife, quality of water, air pollution, and soil erosion. This intent of the process is to minimize the potential adverse impacts of the wildfire management programs on the environment, while reducing fire dangers.

• Healthy Forests Restoration Act (HFRA)

This law of 2003 was a response to increasing concerns over the degradation of forest health and greater threat from wildfires (Martin, 2020). HFRA aims specifically at reducing hazardous fuel loads–one of the leading causes of catastrophic wildfires. This act provides the federal agencies with more flexibility to carry the activities regarding the management of forests, which include thinning, removal of dead trees, and conducting prescribed burns (Martin, 2020). A key feature of HFRA is its promotion of Community Wildfire Protection Plans (CWPPs) (Martin, 2020). These plans help empower communities to surface fire risks locally, in collaboration with federal agencies, and to develop locally appropriate strategies for risk reduction. This makes wildfire management efforts locally and eco-systemically responsive, as active involvement on the ground derives from this locally initiated, developed, and owned planning process with broad regard for local stakeholder input (Martin, 2020).

• Federal Land Assistance, Management, and Enhancement (FLAME) Act

The Act, signed into law in 2009, developed a new way for the accumulated financial burden and rising cost of wildfire suppression to go on federal lands (Martin, 2020). It created a special account through which federal land management agencies, such as the U.S. Forest Service and Bureau of Land Management, may draw emergency funding in reaction to major, unforeseen wildfire events (Martin, 2020). The Act endeavors to prevent fires. It directs federal agencies to be more proactive in fire management by carrying out fuel reductions, public education, and fire risk assessments (Martin, 2020). By creating the additional funding stream, the FLAME Act reduces the necessity for agencies to borrow from other critical programs in order to fund wildfire suppression, thereby assisting agencies in better balancing approaches to forest and fire management.

Australia

Australia has been plagued throughout its history with intense wildfires, known more commonly as bushfires (Haque et al., 2021). Its arid climate and volatile vegetation render it extremely prone to fire occurrence at a high cost to biodiversity, the economy, and communities (Haque et al., 2021). Accordingly, Australian legislation and policy focus on enhancing fire preparedness, fire prevention, and recovery, laying significant emphasis on prescribed burning and community involvement.

• Bushfire Management Act

The Bushfire Management Act, 2016, is the cornerstone upon which bushfire management practices and policies, such as prescribed burning and the establishment of firebreaks. Prescribed burning lies at the core of any fire management arrangement within Australia (Lange & Gillespie, 2023). It is a deliberate fire, set under conditions of maximum safety, through which fuel loads are reduced and areas of natural breaking are created that will enable wildland fires to be brought under control (Lange & Gillespie, 2023). Apart from controlling and regulating fire activities, the Bushfire Management Act also encapsulates responsibilities on fire authorities and landowners for the prevention and response to fire. It stipulates that landowner should manage their fuel loads and adhere to fire safety standards, while fire authorities are obligated to manage suppression operations and perform risk mitigation activities.

• National Disaster Risk Reduction Framework

The National Disaster Risk Reduction Framework in Australia describes in a step-wise fashion how the country might reduce its risks in relation to a number of natural calamities, including bushfires (Buchtmann et al., 2023). It describes how a framework—through informed risk decisions and as a way to encourage collaboration among governments, industries, and local communities—will build resilience into communities, ecosystems, and infrastructure (Buchtmann et al., 2023). The key focus under this framework is to enhance wildfire preparedness through enhanced data collection, effective risk assessment and raising the public awareness of the dangers brought by wildfire (Buchtmann et al., 2023). It also puts much emphasis on the disaster risk reduction framework being made part and parcel in general environmental and land management policies, ensuring that risks associated with wildfires are factored in long-term planning mechanisms (Buchtmann et al., 2023).

European Union

Even though most of Europe does not experience monthly wildfires, a few parts, especially southwards face, have increased incidences since climate change, urbanization, and land use and cover changes result. An EU-wide action for wildland fires prevention to response and recovery is on a collective basis.

• Strategy in Forests

The EU Forest Strategy for 2030 is a general framework concerning European forests management including issues relevant to the prevention of wildfires (Lier et al., 2022). Furthermore, the strategy lists effective forest fire prevention appliances like thinning and creation of firebreaks among the sustainable forest managing practices (Lier et al., 2022). This strategy calls on all member states to include the action of wildfire prevention in their national forest management plans, it encourages sharing border crossing of best practices on fire management (Lier et al., 2022). Besides the causes of wildfires, the strategy calls for urgent attention to the rapid response capacities, and post-fire recovery measures necessary in safeguarding the forests and biodiversity in Europe (Lier et al., 2022).

• Civil Protection Mechanism

The Civil Protection Mechanism of the EU is an element in which Member States can cooperate in an emergency response operation, which is also categorized as involving a wildfire event (Zweglinski et al., 2023). This is a mechanism by which a country undergoing large-scale wildfires, beyond the national responding capacity, is empowered to make a request for assistance to the other member states (Zweglinski et al., 2023). The Civil Protection Mechanism provides support to joint firefighting, sharing of equipment and personnel, and coordination of air support for responses to fire suppression (Zweglinski et al., 2023). Other response efforts the mechanism encourages are collaboration on disaster preparedness and risk reduction initiatives to help member states strengthen resilience to wildfires and other natural disasters (Zweglinski et al., 2023).

Canada

Canada, because of its vast forests and grasslands, with longer and more frequent fire seasons and clumped by worsening conditions, are set to go up in forest fire. Canada has thereby developed laws and policies that will enhance wildfire prevention, preparedness, and suppression.

• Forest and Prairie Protection Act, 2000 of (Alberta) Canada

This Act governs the control and prevention of wildfires in both forested and prairie regions of Canada. It is proactive in scope and similar to Australia's Bushfire Management Act, enlists similar provisions for controlled burns to reduce fuel loads and create natural firebreaks. This act clearly defines the roles and responsibilities of landowners, fire authorities, and government agencies in managing the risk that bushfires pose. Under the act, landowners are required to assume responsibility for taking charge of managing vegetation and fuel loads in their lands with intentions to reduce fire risks; fire authorities are also to take responsibilities on matters to do with organizing any suppression effort, including putting in place fire prevention measures. The act has, among other provisions, enforcing bans, prosecuting individuals and organizations who have failed to observe and comply with fire safety legislation.

India

Having one of the richest biodiversity hotspots in the world, coupled with a great diversity of ecosystems, makes protection and management does not separate from ecological balance and the livelihoods of millions of people dependent upon forests. In most cases, forest fires-a natural but often accentuated phenomenon-threaten the environment, wildlife, and communities with great risks. It is in this respect that such threats are being dealt with in a wide-based legal and policy framework developed within India, relating to forest management, including prevention and mitigation of wildfires.

• The Indian Forest Act, 1927

This statute forms the basic legal backing concerning forest management in India. It lays down provisions relating to the regulation of the utilization of and the conservation of forests, underlining protection against uses injurious to forest ecosystems, including wildfires. The said act explicitly covers the prevention of forest fires, especially in "reserved forests" where human activities have been restricted. The Act empowers the forest officer under this Act to take action for fire prevention and control. This is one of the earliest legislations in the country on the issue of forest fire prevention. One of the main provisions of this act is granting forest officers authority to construct "fire lines" or "fire breaks". These are strips of land that are strategically cleared for barriers, which work to slow down or even stop fire advancement in cases of a forest fire. Fire lines are very important in terms of trying to prevent the jumping of fires from one area to another, limiting damage. In addition, the act provides for the punishment of any person who intentionally sets fire to reserved forests.

• The Forest (Conservation) Act, 1980

This Statute is another important legislation in India for the purpose of forest conservation. Though the main aim of this act is to prevent the misuse of forest land for non-forest purposes, it aids in the management of wildfire by ensuring that the forest land is not degraded or over-exploited, which in turn increases the chances of forest fires. The Act lays down that any decision to divert forest land for non-forest purposes such as development of infrastructure has to be undertaken with prior permission from the central government. It ensures that the forest goods are exploited in a responsible way, and that the forest systems are resistant to possible destructive forces, such as wildfire.

• National Forest Policy, 1988

The National Forest Policy of 1988 represents a major statement in the philosophy of forest management in India, emphasizing ecological balance, environmental protection, and people's participation (Divan & Rosencranz, 2022). The policy has laid especial emphasis on forest fire prevention and control within its larger goals. It is aware that forest fires are among the severest perils to the country's forests and that their prevention calls for an integrated approach (Divan & Rosencranz, 2022). It advocates the constitution of fire protection committees at the local level where the people in cooperation with local forest officials prevent or control fires in forests (Divan & Rosencranz, 2022). The participatory approach in fire management ensures that policy and strategies are not handed down top-down but are drawn in participation with the people who suffer directly from the fires (Divan & Rosencranz, 2022). Involvement of local communities ensures representation and integration of traditional knowledge and practices-often proved to be efficient in containing fires-of local communities in modern forest-fire management strategies (Divan & Rosencranz, 2022).

• Environment Protection Act, 1986

The Environment Protection Act of 1986 remains the umbrella legislation for protection and improvement of the environment in India. The said Act grants the central government with broad authority and powers to take whatever measures it deems necessary for the prevention of environmental degradation, to include measures aimed at addressing both causes and effects of forest fires. While broader in scope compared to specific laws on forests, the act provides a very vital framework in the management of wildfire by endowing the government with powers to take rapid action across the board in cases of environmental catastrophes, the large-scale forest fire included. The act also forms the basis upon which regulations and standards in environmental protection, which help in mitigating risks of wildfires, are backed legally. Also, under this act, the government could conduct widescale environmental rehabilitation programs such as afforestation or reforestation programs for the restoration of such fire-affected ecology and preventing such fires in the future.

• NAPCC

The NAPCC of India, launched in 2008, outlined the strategy of the country in addressing climate change through a series of missions (Deshpande, Mukherji & Sastry, 2023). Most relevant to the management of wildfires is the Green India Mission, which aims at enhancing forest and tree cover to provide ecosystems with resilience against climate change impacts, among them wildfires (Deshpande, Mukherji & Sastry, 2023). The Green India Mission stresses the strategic alleviation of the impacts of climate change by accentuating sustainable forest management, reforestation, and afforestation; these include an increased occurrence of wildfires (Deshpande, Mukherji & Sastry, 2023). On account of the rise in temperature and alterations in rainfall patterns due to intensified climate change, fire incidents are going to be more frequent and serious (Deshpande, Mukherji & Sastry, 2023). The Green India Mission, hence seeks to create resilient ecosystems with an increase of forest cover and improvement of forest health to withstand the vagaries of the type of climate-related phenomena that could cause wildfires (Deshpande, Mukherji & Sastry, 2023).

• The Disaster Management Act, 2005

While the Disaster Management Act, 2005, has indeed cast a comprehensive legal framework for disaster preparedness and response in general and specifically for wildland fires within India, it makes provisions for Disaster Management plans at the National, State, and District levels, which include preventing and mitigating the effects of natural disasters such as forest fires. These plans have components on early detection, rapid response, and post-fire recovery, making all levels of government prepared to handle wildfire incidents. This Act provides due consideration of forest fire as one of the major national calamities and includes the strategies for its management under the umbrella of disaster management. The Act also enables the state to deploy resources and manpower for preventing wildfires, fighting, and recovering from such disasters.

- Wildlife Protection Act, 1972

While the main purpose of this Act of 1972 was to conserve and protect wildlife and their habitats, it also includes provisions about wildfire management. Forest fires directly threaten wildlife due to habitat and food loss, and the act acknowledges the need to protect such ecosystems against destructive fire. This act empowers the forest and wildlife authorities to prevent or take precautionary measures regarding fires in the natural habitats of animals so that no forest fire compromises the work of conservation.

- National Action Plan on Forest Fires (NAPFF)

Launched in the year 2018, NAPFF remains a multi-faceted policy document that integrates advanced technology, involvement of the local community, and capacity building in its initiatives relating to the prevention and management of forest fires (Maurya et al., 2023). The document has placed much emphasis on the employment of technology-for example, satellite-based fire detection, drones, weather forecasting tools-with a view to enhancing early detection capability (Maurya et al., 2023). In addition, the strategy underlines how to involve the local communities in the process of fire prevention, well aware that their participation, with a predominance of local knowledge, is quintessential for any successful approach of wildfire management (Maurya et al., 2023). The capacity building-focused plan has been fully equipped with the skills and resources needed by the forest officials and communities for the prevention and control of forest fires (Maurya et al., 2023). The training involves firefighting techniques, construction of fire lines, and modern technology concerning the detection and management of fire (Maurya et al., 2023).

Global Initiatives

Besides national and regional policies, several international initiatives exist that promote coordination and sharing of experiences related to wildfire management. This includes the Sendai Framework for Disaster Risk Reduction and the United Nations Framework Convention on Climate Change. The main international frameworks are the Sendai Framework on Disaster Risk Reduction and the United Nations Framework Convention on Climate Change.

• Sendai Framework for Disaster Risk Reduction

The Sendai Framework states that Governments will implement community-centered approaches for prevention, response, and recovery from fire in the management of wildfires. This would include anything from a participatory approach of local communities towards fire risk assessments to public awareness campaigns supporting the creation of early warning systems (Murray, 2021). This structure also requests international cooperation in terms of best practices and technical assistance to nations at risk of wildfire (Murray, 2021).

• United Nations Framework Convention on Climate Change (UNFCCC)

The UNFCCC agrees that the connections between climate change and the growing hazards of wildfires are apparent enough that warmer temperatures and changed precipitation regimes present an environment in which wildfires can readily spread (Hickmann et al., 2021). The UNFCCC promotes practices of sustainable land use, which not only helps with reducing the risk of fire but also increases the ecosystem's adaptive capacity to the impacts of climate change (Hickmann et al., 2021). The other route provided by the UNFCCC is through climate adaptation programs through afforestation and reforestation that help mitigate wildfires (Hickmann et al., 2021). On this note, the UNFCCC leads by not only working to reverse the causes of wildfire risk through the bold approach it takes on climate change mitigation, but also climate adaptation (Hickmann et al., 2021).

STRATEGIES TO ENSURE ETHICAL IMPLEMENTATION

Some of the excellent developments brought into many fields by the development of ML include predictive accuracy to automation in complex tasks. However, it has raised a host of ethical concerns, especially in the most essential industries, for instance, wildfire management. Although ML can revolutionize the detection, prediction, and mitigation of wildfires, its use requires careful monitoring possibly to avoid triggering ethical concerns (Buchelt et al., 2024). Ethical issues involving the use of machine learning in wildfire management encompass data privacy, algorithmic bias, accountability issues, environmental impact concerns, and community involvement (Buchelt et al., 2024). Strategies through which machine learning can be used responsibly include transparency, equity, stakeholder involvement, and continuous monitoring. This section will outline the probable practices that ensure the ethical application of machine learning in effective management of wildfires, elaborating on ethical standards, stakeholder engagement, monitoring and overseeing, and facilitating partnership.

• Evolution of Moral Norm

Such a framework of ethical rules would therefore constitute a critical component in the responsible application of machine learning toward wildland fire management. These rules are an articulation of a decision-making framework that provides explicit guidance on actionable recommendations for practitioners and technologists, policymakers, among others who are involved in the design and deployment of the machine learning system.

• Data Privacy

The ethical standards should clearly articulate that privacy and sensitive information used in machine learning algorithms should be protected. Data from wildfire management is often aggregated from sources that include satellite imagery, sensor networks, and local meteorological stations and must then be analyzed. However, the entire aggregation process may inadvertently collect personal information, such as the geolocation of an individual or private property. The policies must be set in place to satisfy the requirements for compliance under data protection laws, the GDPR, and other privacy laws but simultaneously be able to encourage the use of data anonymization and minimize personal data as often as possible. Second, access to data must be restricted to authorized personnel only, which must be adequately secured against unauthorized access or breaches.

• Algorithmic Bias

Inadequate training of machine learning models thus results in their learning the biases of their respective datasets. Thus, biased algorithms may inequitably allocate resources or mitigation efforts to some areas in the domain of wildfire management, while remaining silent about equally at-risk or even more-at-risk regions. Ethical standards should mention diverse and representative datasets for training these machine learning models and, therefore, that the developed models should contain diversified environmental, social, and geographical considerations. Such a policy should enforce periodic reviews of such models so that some existing bias can be identified and corrected as an effort toward a fair and inclusive decision-making process.

• Transparency and Accountability

Perhaps for this reason, it often comes across as "black boxes," the opaque workings of whose decisions often make it not easy to track where the conclusions come from (Brożek et al., 2024). Decisions taken by ML systems in the case of wildfire management directly and severely impact the allocation of resources, evacuation plans, and emergency response. The ethical frameworks require high-level transparency. Thus, the outputs generated by developers within machine learning models should be justified by their rationale. This is possible through work on the development and progression of explainable AI that enables understanding and critical assessment of why decisions are taken (de Bruijn et al., 2022). What's more, guidelines should clearly define accountability for ML-driven decisions. So that when any wrong outcome arises, one won't get puzzled over whom to hold accountable.

• Social Responsibility and Environmental Impact

Into the design of ethical machine learning strategies in wildland management comes consideration concerning what it brings to the environment and the people. Active engagement from the communities at the center of the wildfires should be brought in ensuring that their issues, needs, and perspective are represented in machine learning-driven strategies. Ethical frameworks should require environmental impact analysis in relation to machine learning and take into account an emerging technology, such as the impact of drones, sensors, and prescribed burns on local biotic systems (Buchelt et al., 2024). This will lead to relatively minor or less detrimental ecological consequences of technological interventions, and the benefits of the management of wildfires will be optimized.

• Equity and Sustainability

Ethical principles will underline the fair share of the technologies in machine learning so that vulnerable populations and under-resourced areas can enjoy technological progress on wildfire management. Perpetuation of already existing inequalities shall not be that of the ML systems; rather, they are ways of putting a fair resource for protection into play for all communities, especially the most vulnerable exposed to wildfires. Guidelines should also ensure long-term sustainability regarding the cost, scalability, and integration of ML solutions to existing practices.

• Stakeholder Interaction

Diverse stakeholder involvement in development and deployment of machine learning technology is absolutely crucial for integrating appropriate evaluation and mitigation of the ethical concerns that accompany this technology. For instance, in wildland fire management, it includes not just the technologists and data scientists but also the local community, policymakers, environmental activists, emergency responders, ethicists, as well as indigenous communities. Each one of these entities brings forth different perspectives to the discussion, therefore enforces consideration for the issues that would otherwise be ignored.

• Inclusive Decision-Making

Engaging stakeholders would be at the core of the development of the inclusive nature of decision-making procedures. The many participants from diverse spheres included in developing machine learning systems by its developers will better understand the needs, values, and concerns held by these different communities. This will, in turn, help build more ethical and culturally responsive machine learning solutions. For instance, indigenous populations have crucial information at the local level about ecosystems and how they manage fire. Interaction with these communities ensures the transfer of traditional knowledge into modern technological techniques, and this improves an integrated approach to managing wildfires.

• Continuous Monitoring and Review

Ethical embedding of machine learning to the wildfire management requires a long process, not one time; it needs to be continuously monitored, evaluated, and improved (Awatade et al., 2024). Continuous monitoring ensures that any new ethical issues discovered within the system are tackled immediately (Awatade et al., 2024). One very effective method of maintaining strict ethical standards is through continuous auditing of machine learning systems (Awatade et al., 2024). These models' effectiveness, predictability, and the fairness of the decision made should be evaluated in audits. Ethical impact assessments on machine learning systems can also be used to examine the effects the systems have on the population, ecosystems, and stakeholders at any given time (Awatade et al., 2024). Through feedback mechanisms, organizations are able to collect insights from users, communities, and other stakeholders on perpetual enhancements of machine learning models and addressing any ethical concerns that may arise. Further oversight in this case enables the organization to identify and mitigate unforeseen repercussions associated with machine learning systems (Awatade et al., 2024). For instance, the wildfire prediction model incorrectly favors certain communities by misrouting resources that were meant to benefit them; through constant monitoring, the problem is exposed for correction (Awatade et al., 2024). Similarly, monitoring an ML system that is known to cause adverse environmental effects, such as disturbing local wildlife, will be useful along the way of action to be taken in mitigation.

- Promoting Innovation and Cooperation

Working together and innovating will greatly enhance the ethical implementation of ML in wildfire management. Encouraging partnerships among researchers, technologists, policymakers, and environmentalists, the stakeholders can develop more effective and more ethical solutions to the challenges posed by wildfires. In these partnerships across different sectors and regions, best practices and lessons learned and innovations are shared. For instance, different ethics ML systems developers in one region can share the results they came up with other groups that work in different wildfire-prone areas, hence increasing their speed in developing even more responsible and effective technologies. These end up being part of a global ecosystem of ethical innovation; collective knowledge is employed to address wildfire management issues. Some of these initiatives promote cooperation and should develop industry-wide ethics for ML relating to wildland fire management. This kind of ethics would guide responsible development in ML by getting all the organizations in the industry aligned with the ethics on behalf of ML. A set of standard ethics would propel consistency, transparency, and accountability in the industry and assure the public's trust in ML-driven solutions.

CONCLUSION

Machine learning obviously has enormous potential to improve prediction, detection, and response in the realm of wildfire management, but the ethical aspects must be weighed against this so that these potential benefits do not come at the expense of privacy, equity, and sustainability. Hopefully the ethical dilemmas in this chapter along with the ethical strategies will help the stakeholders to progress towards the ethical use of machine learning in wildfire management. The problem of wildfire control is an international issue and must have cooperation on the national, regional, and international level. Wildfire management strategies in the US, Australia, the EU, and Canada have all been altered to address the increasing issues of climate change, population growth, and land-use change. Here, laws and policies and whatnot have been put into place that focus on prevention, community involvement, immediate and effective response and recovery. Treaties and agreements like the Sendai Framework for Disaster Risk Reduction and UNFCCC are so vital at the world level because that is the avenue in which best practices for wildfire management are disseminated. With the looming danger of wildfires, nations must take adaptive and resilient approaches to wildfire risk reduction in order to incorporate comprehensive environmental and land management that will in turn safeguard ecosystems, communities, and economies from the ravages of wildfires. There exists already a sound legal and policy framework in India with regard to forest management, and more specifically, prevention and control of wildfires. With the Indian Forest Act of 1927 being one of the earliest legislations to the National Forest Fire Prevention and Management Plan of 2018, India has come a long way in having a well-rounded approach to forest fires. These act and policies are made to uphold the conservation, people's participation and use of modern tech to save India's forests and wildlife from the fires that can completely wipe them off. However, global warming will only worsen the danger of forest fires, and if India wishes to maintain the longevity and sustainability of their forests, they will need to continue to evolve and enforce these policies.

All of that to say this, machine learning will and is going to change the face of wildfire management but what are the moral dilemmas that will come along with this, it is not without that price. These problems include privacy of information, and algorithmic bias, accountability, community involvement, environmental issues, fairness, sustainability. In a careful and moral use and development of ML technologies, the advantages of machine learning can be utilized to safeguard lives, land, and environment from the destruction of wildfires, and at the same time, it can make sure that these advantages of ML are distributed fairly among all the communities. And the ethical use of machine learning in wildfire management needs to involve the development of explicit ethical guidelines, the involvement of stakeholders, consistent monitoring, and the promotion of collaboration. To harness the power of ML for good, and to protect communities, ecosystems, and individuals from the threat of wildfire, as well as from the whiplash of technology itself. With wildfire dangers increasing with climate change, moral implications must not be lost in the light of technological advancement, that is, ML must be used as a tool for good in the fight against wildfires.

REFERENCES

Abid, F. (2021). A survey of machine learning algorithms based forest fires prediction and detection systems. *Fire Technology*, 57(2), 559–590. DOI: 10.1007/s10694-020-01056-z

Akula, R., & Garibay, I. (2021). Audit and assurance of AI algorithms: a framework to ensure ethical algorithmic practices in artificial intelligence. *arXiv preprint arXiv:2107.14046.*

Al-Bashiti, M. K., & Naser, M. Z. (2022). Machine learning for wildfire classification: Exploring blackbox, eXplainable, symbolic, and SMOTE methods. *Natural Hazards Review*, 2(3), 154–165. DOI: 10.1016/j.nhres.2022.08.001

Albahri, A. S., Khaleel, Y. L., Habeeb, M. A., Ismael, R. D., Hameed, Q. A., Deveci, M., Homod, R. Z., Albahri, O. S., Alamoodi, A. H., & Alzubaidi, L. (2024). A systematic review of trustworthy artificial intelligence applications in natural disasters. *Computers & Electrical Engineering*, 118, 109409. DOI: 10.1016/j.compeleceng.2024.109409

Alkhatib, R., Sahwan, W., Alkhatieb, A., & Schütt, B. (2023). A brief review of machine learning algorithms in forest fires science. *Applied Sciences (Basel, Switzerland)*, 13(14), 8275. DOI: 10.3390/app13148275

Arif, M., Alghamdi, K. K., Sahel, S. A., Alosaimi, S. O., Alsahaft, M. E., Alharthi, M. A., & Arif, M. (2021). Role of machine learning algorithms in forest fire management: A literature review. *J. Robot. Autom*, 5, 212–226.

Awatade, A., Pawar, P., & Lakshmi, D. (2024). The Power of Machine Learning in Forest Fire Risk Analysis and Resilience: Navigating Best Practices, Challenges, and Opportunities. *Geospatial Technology to Support Communities and Policy: Pathways to Resiliency*, 149-170

Bhowmik, R. T., Jung, Y. S., Aguilera, J. A., Prunicki, M., & Nadeau, K. (2023). A multi-modal wildfire prediction and early-warning system based on a novel machine learning framework. *Journal of Environmental Management*, 341, 117908. DOI: 10.1016/j.jenvman.2023.117908 PMID: 37182403

Boroujeni, S. P. H., Razi, A., Khoshdel, S., Afghah, F., Coen, J. L., O'Neill, L., Fule, P., Watts, A., Kokolakis, N.-M. T., & Vamvoudakis, K. G. (2024). A comprehensive survey of research towards AI-enabled unmanned aerial systems in pre-, active-, and post-wildfire management. *Information Fusion*, 108, 102369. DOI: 10.1016/j.inffus.2024.102369

Buchelt, A., Adrowitzer, A., Kieseberg, P., Gollob, C., Nothdurft, A., Eresheim, S., Tschiatschek, S., Stampfer, K., & Holzinger, A. (2024). Exploring artificial intelligence for applications of drones in forest ecology and management. *Forest Ecology and Management*, 551, 121530. DOI: 10.1016/j.foreco.2023.121530

Buchtmann, M., Wise, R., O'Connell, D., Crosweller, M., & Edwards, J. (2023). Reforming Australia's approach to hazards and disaster risk: National leadership, systems thinking, and inclusive conversations about vulnerability. *Disaster Prevention and Management*, 32(1), 49–73. DOI: 10.1108/DPM-08-2022-0168

Casali, Y., Aydin, N. Y., & Comes, T. (2022). Machine learning for spatial analyses in urban areas: A scoping review. *Sustainable Cities and Society*, 85, 104050. DOI: 10.1016/j.scs.2022.104050

Courtney, K. (2023). *Mapping the Landscape of Wildfire Risk Mitigation: Understanding the Links Between Equity, Community Assets, Capacities, and Collaboration* (Master's thesis, Colorado State University).

de Bruijn, H., Warnier, M., & Janssen, M. (2022). The perils and pitfalls of explainable AI: Strategies for explaining algorithmic decision-making. *Government Information Quarterly*, 39(2), 101666. DOI: 10.1016/j.giq.2021.101666

Deshpande, T., Mukherji, R., & Sastry, M. (2023). Policy styles and India's national action plan on climate change (NAPCC). *Policy Studies*, •••, 1–18. DOI: 10.1080/01442872.2023.2261388

Divan, S., & Rosencranz, A. (2022). *Environmental law and policy in India: cases and materials*. Oxford University Press. DOI: 10.1093/oso/9780192865458.001.0001

Duff, T. J., & Penman, T. D. (2021). Determining the likelihood of asset destruction during wildfires: Modelling house destruction with fire simulator outputs and local-scale landscape properties. *Safety Science*, 139, 105196. DOI: 10.1016/j.ssci.2021.105196

Fiorino, D. J. (2023). *Making environmental policy*. Univ of California Press. DOI: 10.2307/jj.2392269

Ghaffarian, S., Taghikhah, F. R., & Maier, H. R. (2023). Explainable artificial intelligence in disaster risk management: Achievements and prospective futures. *International Journal of Disaster Risk Reduction*, 98, 104123. DOI: 10.1016/j.ijdrr.2023.104123

Haque, M. K., Azad, M. A. K., Hossain, M. Y., Ahmed, T., Uddin, M., & Hossain, M. M. (2021). Wildfire in Australia during 2019-2020, Its impact on health, biodiversity and environment with some proposals for risk management: A review. *Journal of Environmental Protection*, 12(6), 391–414. DOI: 10.4236/jep.2021.126024

Hickmann, T., Widerberg, O., Lederer, M., & Pattberg, P. (2021). The United Nations Framework Convention on Climate Change Secretariat as an orchestrator in global climate policymaking. *International Review of Administrative Sciences*, 87(1), 21–38. DOI: 10.1177/0020852319840425

Jain, P., Coogan, S. C., Subramanian, S. G., Crowley, M., Taylor, S., & Flannigan, M. D. (2020). A review of machine learning applications in wildfire science and management. *Environmental Reviews*, 28(4), 478–505. DOI: 10.1139/er-2020-0019

Kameswari, L. (2023). A Review on Application of Machine Learning Techniques in Climate Change and Disaster Management. *Technoarete Transactions on Climate Change and Disaster Management Research, 2*(1).

Khan, F., Xu, Z., Sun, J., Khan, F. M., Ahmed, A., & Zhao, Y. (2022). Recent advances in sensors for fire detection. *Sensors (Basel)*, 22(9), 3310. DOI: 10.3390/s22093310 PMID: 35590999

Kyrkou, C., Kolios, P., Theocharides, T., & Polycarpou, M. (2022). Machine learning for emergency management: A survey and future outlook. *Proceedings of the IEEE*, 111(1), 19–41. DOI: 10.1109/JPROC.2022.3223186

Lange, J., & Gillespie, J. (2023). Bushfire, prescribed burning, and non-human protection. *Geographical Research*, 61(2), 169–192. DOI: 10.1111/1745-5871.12558

Lier, M., Köhl, M., Korhonen, K. T., Linser, S., Prins, K., & Talarczyk, A. (2022). The new EU forest strategy for 2030: A new understanding of sustainable forest management? *Forests*, 13(2), 245. DOI: 10.3390/f13020245

Linardos, V., Drakaki, M., Tzionas, P., & Karnavas, Y. L. (2022). Machine learning in disaster management: recent developments in methods and applications. *Machine Learning and Knowledge Extraction, 4*(2).

Lotfi, I., & El Bouhadi, A. (2022). Artificial intelligence methods: Toward a new decision making tool. *Applied Artificial Intelligence*, 36(1), 1992141. DOI: 10.1080/08839514.2021.1992141

Martin, J. (2020). Active Forest Mismanagement and the" New Normal": Advocating for an Integrative Wildfire Management Policy. *Ohio NUL Rev.*, 46, 137.

Maurya, L., Bagade, S. A., & Bagade, A. A (2023). Forest Fire: It's Effects And Management In Indian Forests. *IJGFRAI, 1(1).*

McLennan, J., Ryan, B., Bearman, C., & Toh, K. (2019). Should we leave now? Behavioral factors in evacuation under wildfire threat. *Fire Technology*, 55(2), 487–516. DOI: 10.1007/s10694-018-0753-8

Mikhaylov, S. J., Esteve, M., & Campion, A. (2018). Artificial intelligence for the public sector: Opportunities and challenges of cross-sector collaboration. *Philosophical Transactions. Series A, Mathematical, Physical, and Engineering Sciences*, 376(2128), 20170357. DOI: 10.1098/rsta.2017.0357 PMID: 30082303

Mohapatra, A., & Trinh, T. (2022). Early wildfire detection technologies in practice—A review. *Sustainability (Basel)*, 14(19), 12270. DOI: 10.3390/su141912270

Murray, V. (2021). Science and technology commitment to the implementation of the sendai framework for disaster risk reduction 2015–2030. In *Proceedings of the 3rd Global Summit of Research Institutes for Disaster Risk Reduction 3* (pp. 55-82). Springer Singapore. DOI: 10.1007/978-981-15-8662-0_4

Orsi, F., Ciolli, M., Primmer, E., Varumo, L., & Geneletti, D. (2020). Mapping hotspots and bundles of forest ecosystem services across the European Union. *Land Use Policy*, 99, 104840. DOI: 10.1016/j.landusepol.2020.104840

Pham, B. T., Jaafari, A., Avand, M., Al-Ansari, N., Dinh Du, T., Yen, H. P. H., Phong, T. V., Nguyen, D. H., Le, H. V., Mafi-Gholami, D., Prakash, I., Thi Thuy, H., & Tuyen, T. T. (2020). Performance evaluation of machine learning methods for forest fire modeling and prediction. *Symmetry*, 12(6), 1022. DOI: 10.3390/sym12061022

Phelps, N., & Woolford, D. G. (2021). Comparing calibrated statistical and machine learning methods for wildland fire occurrence prediction: A case study of human-caused fires in Lac La Biche, Alberta, Canada. *International Journal of Wildland Fire*, 30(11), 850–870. DOI: 10.1071/WF20139

Ronchi, A. M., & Ronchi, A. M. (2019). Safety and Security. *e-Citizens: Toward a New Model of (Inter) active Citizenry*, 43-108

Tedim, F., McCaffrey, S., Leone, V., Delogu, G. M., Castelnou, M., McGee, T. K., & Aranha, J. (2020). What can we do differently about the extreme wildfire problem: An overview. *Extreme Wildfire Events and Disasters*, 233-263).

Thompson, M. P., Wei, Y., Calkin, D. E., O'Connor, C. D., Dunn, C. J., Anderson, N. M., & Hogland, J. S. (2019). Risk management and analytics in wildfire response. *Current Forestry Reports*, 5(4), 226–239. DOI: 10.1007/s40725-019-00101-7

Uddin, M., Chowdhury, A., & Kabir, M. A. (2024). Legal and ethical aspects of deploying artificial intelligence in climate-smart agriculture. *AI & Society*, 39(1), 221–234. DOI: 10.1007/s00146-022-01421-2

Visave, J. (2024). AI in Emergency Management: Ethical Considerations and Challenges. *Journal of Emergency Management and Disaster Communications*, 4(3), 1–19. DOI: 10.1142/S268998092450009X

Xu, N., Lovreglio, R., Kuligowski, E. D., Cova, T. J., Nilsson, D., & Zhao, X. (2023). Predicting and assessing wildfire evacuation decision-making using machine learning: Findings from the 2019 kincade fire. *Fire Technology*, 59(2), 793–825. DOI: 10.1007/s10694-023-01363-1

Zweglinski, T., & Smolarkiewicz, M. (2023). The Union Civil Protection Mechanism and Disaster Risk Reduction in Shaping National Security. *Shaping National Security: International Emergency Mechanisms and Disaster Risk Reduction*, 90.

Chapter 8

Integrating UAV-IoT Networks for Proactive Wildfire Detection: Innovations and Future Directions

Rishabha Malviya
Galgotias University, India

Manmeet Kaur Arora
https://orcid.org/0009-0002-5071-117X
Sharda University, India

Sahil Lal
https://orcid.org/0000-0001-9827-3717
Sharda University, India

ABSTRACT

The most effective solutions brought to all industries, especially for environmental monitoring and disaster management with this integration is the Unmanned Aerial Vehicles (UAVs) combined with Internet of Things (IoT) technologies. The aim of this paper is to investigate the architecture, applications and challenges of UAV-IoT networks that have potential to support real-time data collection capabilities along with communication range. First, we develop a new concept of passive monitoring using UAVs as mobile platforms fitted with advanced sensors to sense data from ground-based IoT devices in various environments efficiently. Through teamwork between researchers, industry professionals and policy makers to address the current challenges of these technologies, they can be leveraged (if successfully applied) for sustainable development with more safety.

DOI: 10.4018/979-8-3693-7565-5.ch008

INTRODUCTION

Wildfires have become one of the greatest environmental threats, intensified by our changing climate, continued urbanization and a move away from former land use efforts. Wildfires have increased in frequency and intensity over the last decade, resulting in severe damage to ecosystems as well as human lives and properties. The U.S. continues to experience larger and more damaging wildfires compared with historical averages, as reported by the National Interagency Fire Centre, in just the first half of a recent year alone there were 50 major incidents that have burned tens or hundreds-of-thousand acres (squared kilometres). This is a serious crisis, underscored by the unyielding fact that wildfires have impacted around four million square kilometres of land worldwide Compliance with several regulations to ensure that UAVs comply within the regions they are being operated. The program still has to grapple with the environmental concerns associated with wildlife and concerning issues cast by weather conditions for it to be implemented in a responsible fashion. Realizing UAV-IoT networks to reach its full potential will require investments in research and development, normative actions based on standardization that enable diverse functionalities achieved through interoperable means for broader market uptake as well as adaptive regulatory support which favour innovation with public safety. Sum up, UAV-IoT network is a milestone of monitoring (Kaur & Kumar, 2024).

Some of the major findings highlight that concepts like MQQQ and CoAP also play a crucial role to have detailed communication protocols for potential data transmissions as in IoT applications latency, signal interference can be on higher side. The problem, however, is that the technology also presents technical difficulties such as policing reliable communication links between all devices and complying with physical limitation in energy requirements alongside data security. Wildfires are a multifaceted, intertwined issue with many causative agents. Lightning strikes in arid conditions and human activities including cigarette butts, poor management of campfires or sparks from equipment are to blame for most wildfires (Zhang et al., 2019). This past summer in California, the Zaca Fire is a striking example: it ignited by some sparks from machinery and went on to char over 240 thousand acres5. These incidents show why we need to take wildfire management seriously (Zhao et al., 2020).

These areas are commonly remote and rugged, making them difficult for traditional detection methods. However, because we cannot see or reach them in these locations, and they can burn unnoticed for many days undetected often times too large by the time detected to be controlled. Traditional detection generally depended on human senses or antiquated technologies that may not have been able to react quickly enough to new types of threats (Alzubaidi & Yaqoob, 2021).

Importance of Early Detection

It is important to identify them early so the consequences of wildfires can be reduced (Liu & Wang, 2020). In most cases, the sooner a fire is discovered and reported, in turn triggering prompt containment and control efforts which can include protective measures to save lives, property or large stands of natural resources (Chen & Chen, 2021). Research suggests that fire suppression in the early stages can mitigate sizes of animal's deaths due to wildfires (Wang et al., 2023). An excellent example is the use of sophisticated, technology-based automated detection systems capable of detecting flames in their nascent stage and averting major fire disasters (Huang & Xu, 2020). Drones have been integrated with ground-based sensors and satellite imagery, dramatically enhancing the ability to capture wildfire images. Thermal-imaging-camera-equipped drones can rapidly cover substantial areas and supply firefighting teams with real-time-data (Alharbi & Alhassan, 2022). From the air, these systems can detect heat signatures of smaller fires that might not be visible to the naked eye — enabling firefighters to respond faster. In the same vein, ground-based sensor networks measure environmental parameters like temperature and humidity to signal when conditions are ripe for a fire outburst. In addition, predictive analytics for wildfire management have benefitted from improvements in artificial intelligence (AI) and machine learning (ML). Analyzing historical data around weather, vegetation and fire locations means AI can predict when and where a wildfire is most likely to begin. Emergency responders needing to assign resources can make use of this predictive ability (Yang & Wang, 2019).

Meanwhile, IoT (Internet of Things) technologies are playing an extended function inside the advancing wildfire detection. With this technology, IoT devices join together to automatically monitor huge geographic areas (Khamis et al., 2018). For instance, gas sensors that are solar-powered and can detect the chemicals associated with smoke allow for near instantaneous warnings of would be fires without needing much maintenance at all (Zhang et al., 2018).

The sensors, which are designed to work like digital sniffers and tell the difference between wildfire smoke and other kinds of pollution in the air. However, even with these developments in technology, there are challenges to full wildfire detection coverage (Chen et al., 2020). While these traditional methods like watchtowers and observational satellites are helpful, they all have their limitations; some may be only able to see large fires from a distance (albeit quite far away) because of the altitude at which UAV is flying or flies over an area with enough frequency for detecting outbreaks while others might not visit very often (Xu et al., 2019).

The move would also prevent automated systems from triggering alerts for threats that are not there (false positives) to eat up emergency resources (Kaur et al., 2021). The growing risk of wildfires calls for a new thinking in how we detect fires. Connecting UAVs (Unmanned Aerial Vehicles) to IoT networks opens another promising new frontier for wildfire control (Liu et al., 2019). Using the most up-to-date sensor technology and data analytics will allow us to find fires in earlier stages, protecting people, resources are from uncontrolled wildfires. Given the pace of exponential technological advancement, our ability to effectively intervene in this key area of environmental management should only increase (Wang et al., 2020).

Wildfire Detection Using UAV Technology: Types of UAVs Used

Unmanned Aerial Vehicles (UAVs), commonly named drones represent an essential tool in wildfire detection and management (Zhao et al., 2018). There are so many different types of UAVs used depending on the mission specific requirements and capabilities precise to that individual UAV. They consist of:

Fixed-Wing UAVs: GMRLMZL Line of drones feature fixed-wing forms, similar to conventional airplanes and are built for extensive operations. We will fly over large regions prone to wildfires quickly. Fixed-wing drones are equipped with high-resolution cameras and thermal imaging sensors to detect heat signatures far away from fires (Alshahrani et al., 2020).

Multirotor UAVs: With multiple rotors giving these drones their agility, a multirotor is perfect for close-range surveys or inspections (Yang et al., 2019). They can hover above certain areas and provide accurate data about fire behaviour as it happens. The number of flame locations by hotspot are frequently evaluated and payload, such as water or fire retardants is sometimes dispensed on the flames using multirotor UAVs (Li et al., 2021).

Hybrid UAVs: The hybrid solution combines the benefits of fixed-wing and multirotor designs to free up longer distances while providing agility (Ahmed et al., 2018). Multi-rotors can take off and land vertically, but with the disadvantage of having limited flight time (without becoming too cumbersome for manual control), whereas fixed-wing models are capable of flying longer distances over a wider area. Both these characteristics make them well-suited for long range wildfire surveillance missions involving significant amounts of monitoring data acquisition (Wu et al., 2019).

Tethered Drones: Operated by a cable, these drones are tethered to their power source so you can fly them for hours without worrying about battery life. Visible on their own during long, drawn-out firefighting operations) grounded drones mean manned aircraft have to be sent above just blocks away.

Swarm Drones: One of the newer types, swarm drones work in groups to increase coverage and data acquisition rates. Talking to each other, drones in swarms are able schmooze over larger areas than individual units and keep an eagle eye on wildfires (Alshahrani et al., 2021).

Benefits of Aerial Surveillance

Figure 1. Advantages of aerial surveillance

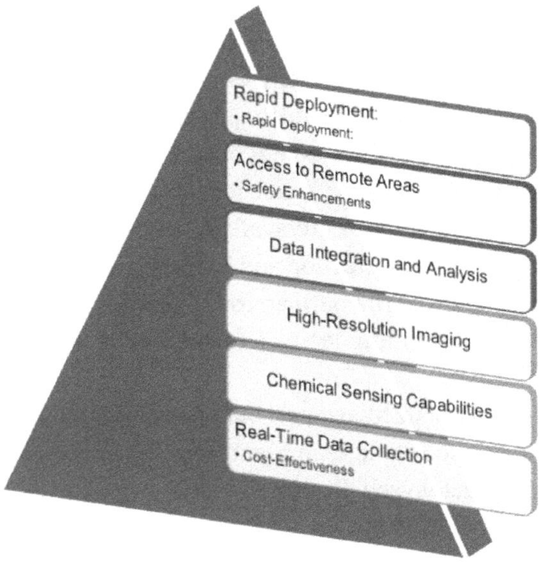

There are many benefits in using UAVs for wildfire detection that allow to improve firefighting tasks substantially versus the typical implementation, UAV can be deployed in quick action of wildfire outbreaks. Drones require no such thing of course, and unlike a manned aircraft where you would call the control tower to request an overflight for situational awareness, drones could be launched in under 10 minutes very easily when called to do so. Wildfires frequently break out in remote areas or steep, rugged landscapes that are almost impossible for ground crews to reach. In fact, UAVs can safely and cost effectively fly through such environments for the delivery of much needed data without placing personnel at risk. Using sophisticated sensors and imaging technologies, drones consistently collect data about how fires spread; temperature change that occurs with the fire; and standard patterns of smoke. Incident commanders need this data so that they can make informed decisions about how to allocate resources and fight fires (Nasir et al., 2020).

Drones not only cost less than manned aircraft to fly for surveillance missions, but many components are also easily replaceable at a lower rate. They use less fuel, require minimal maintenance and have lower operating costs – all while enabling comparable if not better data quality. Safety Improvements, instead of human lives, the risk for fires is reduced by using drones. Under dangerous conditions, drones can replace manned aircraft for risky reconnaissance sorties (Liu et al., 2020). High-

Resolution Imager, Advanced picture cameras and thermal imaging technology have been integrated into the systems associated with current Unmanned Aerial Vehicles (UAVs), which can support specific emission monitoring of wildfire. With this ability, fire fighters are able to identify hot spots precisely and observe the pattern of propagation (Zhao et al., 2018).

Data Integration and Analytics, UAVs can be integrated with wider IoT networks that provide data from both terrestrial sensors as well as satellite imagery. With this integration, data can be analyzed in full using machine learning algorithms that predict fire behaviour given environmental conditions (Kumar et al., 2019).

When a wildfire is detected, the exact location and chemical composition of smoke particles as well other gases are available to drones with these capabilities. Drones can use LiDAR (Light Detection and Ranging) as well as the NIR sensors to create high-resolution maps of fire burn areas Read More MAPIN, Section 2: These maps are used to help firefighting teams strategically plan containment strategies based on the size and spread of a fire.

Assist in Search and Rescue Missions, Unmanned aerial systems not only provide fire detection, but they also enable search-and-rescue agencies to locate individuals trapped in forests or evaluate evacuation routes during wildfires (Rahman et al., 2021).

Fire Detection Systems: IoT Integration

The Internet of Things (IoT) is a groundbreaking technological standard which unites an enormous number of physical devices to the virtual world. These sensors and software-installed devices have created a large network where they can interact with the help of the internet. IoT, a Connectivity Revolution Perhaps the most revolutionary aspect of IoT is its ability to enable real-time data collection, analysis and automation across numerous sectors including wildfire detection and management. Essential parts of IoT technologies Include,

Q-JING SENSORS: these sensors gather information from the environment, containing those of temperature, humidity, smoke and gas concentration. Sensors are key to watching for conditions that can immediately precede wildfires.

Connectivity: IoT devices use different communication standards to deliver data back and forth from a central hub or the cloud (ex, WiFi, Bluetooth, ZigBee and cellular networks). The signaling also enables real-time alerts and alarms.

The processing and storing of data: the collected information could be processed at local sites (while edge computing) or on cloud platforms for more detailed analysis appreciate. It is this that makes storage of the vast amounts of data, which are produced by a myriad sensors deployed throughout a landscape, possible (Xu et al., 2019).

For analytics and machine learning, rely on multi-dimensional algorithms which: track the input data set to uncover patterns and determine possible fires before they occur? diagnoses future innovative strategies from candidate solutions with very few resources? The popularity of machine learning models came about from the fact that they get better over time based on historical data.

UI: Many IoT systems come with a web or mobile interface that the end user can use to watch what is going on and send alerts. This interface is essential to allow quick data access for emergency responders.

Sensors and Their Specific Purpose

Actually, these IoT based fire detection systems are equipped with a number of sensors configured to measure different environmental parameters directly related to wildfires such as;

Thermometer: in the trees that track local temperatures, Rapid rises may suggest a potential fire. Temperature sensors can be installed at different altitudes to collect data on various layers of the atmosphere.

Smoke Detectors: Modern smoke detectors are designed to pick up the smallest bits of smoke in 1 of two ways, either by having a photoelectric sensor or an ionization circuit. Besides, they can sound an alarm if the smoke is detected thus allowing for timely intervention.

Gas Sensors: These monitors are capable of detecting certain type gases that comes across with combustion like carbon monoxide (CO) and volatile organic compounds (VOCs). Without flame visibility, incrementing changes in the gas levels can also be detected and hence these sensors accordingly provide information on how fires are proceeding.

Humidity Sensors: They can be used to monitor humidity levels as arid/dry conditions could fuel wildfires. Humidity sensors also assist monitor the moisture level in vegetation, a major fire behaviour factor.

Wind Speed Sensors: Wind contributes heavily to how wildfire spreads. Anemometers, for example, measure wind speed and direction — critical information that enables researchers to infer how a fire might propagate through the landscape.

Thermal Imaging Cameras: These cameras pick up heat signatures related to likely fires and can observe hot spots that would not be seen otherwise. It is particularly good for detecting possible "hot spots" that have soldered and may burst back into flame in hot, dry forest conditions.

Soil Moisture Sensors: These sensors determine the moisture level in soil, which could reflect a high or low chance of vegetation catching on fire. Drought conditions often result in drier soil and a higher probability of wildfires.

Cameras: Cameras that take images or video of the given area to be watched; for exams, optical sensors which can do this. They are commonly used in combination with machine learning algorithms to solve fire detection related visual recognition tasks.

Multi-Sensor Systems: A vast number of contemporary IoT Fire Detection systems use a blend of these Sensors to form an all-encompassing Monitoring Grid. These systems also implement data from different sources, which helps in identifying early signs of wildfires more accurately.

UAV-IOT NETWORK ARCHITECTURE, SYSTEM DESIGN AND COMPONENTS

The integration of Unmanned Autonomous Vehicles (UAVs) with the Internet-of-Things (IoT), creates a strong network architecture for enriching different scenarios like wildfire detection, environmental monitoring and disaster management. The architecture contains several important parts that promise to enable proper data collection, processing and communication.

UAVs as Mobile Nodes: UAVs are integrable into IoT ecosystems and deployed to the terrestrial environment where they operate with high mobility, executing missions carrying multiple sensors (e.g., thermal cameras or gas detectors) that provide real-time data from a specific sector. These birds are known for their long horizontal flights over inhospitable terrain that is notoriously hard to reach.

Ground Sensors: Alongside UAVs, ground-based sensors from the IoT complement airborne data collection. These sensors assess stats as temperature, humidity, smoke degree in atmosphere and soil moisture level. This is critical information for fires in remote areas and can be used to validate assessments of the environment.

Gateway Units: The data collected by UAVs and ground sensors is transmitted to gateways or edge aggregators. As a rule, these units perform initial data manipulation before they are sent to cloud servers for additional analysis. Edge computing helps to reduce the time taken and bandwidth required, since it processes data at a point close to its origin.

Cloud Infrastructure: In this model, Cloud acts as central data storage and also performs higher analytics calculations. It would involve using machine learning models to leverage data from the past and present that can calculate the upcoming presence of Wildfires etc. or any other new/ specialization you want to predict.

User Interfaces: User interfaces include Dashboards or mobile applications intended to help emergency responders, environmental scientists and other users with the means of using real-time data & alerts as produced by the system. Decision engine should satisfy these interfaces as it implements the imperative pieces of decision making during emergencies.

The divisions are illustrated in Figure 2, and the communication requirements amongst them that need to be fulfilled by a robust backbone (Green Aggression) which connects every enumeration category i.e. UAV-to-Ground-sensors,-Aggregation-units-and-Cloud-Servers-distributed-Intelligence-Center as vigorously depicted in details with its challenges! This framework consists of several wireless communication technologies that provide effective connection.

The one point in the list Open At (2) is:Energy Management Systems, which should allow to optimize power consumption plans as both UAVs and Ground sensors have a limited battery life. So, this is how these systems allow for the extended duration of time between recharging your devices.

Security Architecture: Security is the most important issue in IoT system including UAV-IoT network. Encryption protocols and secure authentication methods are deployed to protect sensitive data from being accessed by anyone unauthorized, or with malicious intent.

Communication Protocols

Effective data exchange within UAV-IoT networks is facilitated by communication protocols. These govern how devices interact, perform data transmission and protect the network. Some of the important communication protocols are as follows:

Lightweight messaging protocols me MQTT (Message Queuing Telemetry Transport) editorial content by sages for practical implementation in IoT applications with low bandwidths or unreliable networks. Mav link is based on publish-subscribe design, which means UAVs and sensors can report status without constantly keeping connections (Chen et al., 2019).

CoAP (Constrained Application Protocol): CoAP is designed to work in an extremely constrained environment; characteristic of many IoT applications running on low-power devices. A way devices send requests and receive responses in a simple, request-response model

HTTP/HTTPS: It is less efficient in low-bandwidth scenarios (compared to MQTT or CoAP), but it finds widespread use by the virtue of its ubiquity and compatibility with web-based applications. HTTPS, or secure versions are critical for safeguarding sensitive information that is being transmitted.

LoRaWAN (Long Range Wide Area Network): LoRAWAN is a low power communication protocol intended for long range communications over the air in rural or remote areas where no mobile network infrastructures are available. It allows low-power devices to communicate at distance ranges of several kilometres.

Zigbee: As a short-range communication z-wave is often used in home automation but also implemented to IoT networks as an underlying protocol, for liaising multiple devices with each other within a single domain; e.g. ground sensors transmitting datagram files repetitively and nearby UAVs compiling the traffic monitoring stats automatically over relay antennae seamless communications harness using Zigbee surfacing high-end features that CoAP messages could never reach-start-to-full whitepaping (Covalenser-IoT) (Alzubaidi & Yaqoob, 2021).

5G Networks: 5g technology improves the communication capabilities in UAV-IoT networks due to high speed/Low latency connections. The benefits are particularly high for real-time applications, like streaming a UAV video feed of wildfires (Wang et al., 2020).

SDR (Software Defined Radio): SDR enables flexible communications to accommodate operational needs and environmental conditions. This is important when working in scenarios that can rapidly evolve, such as disaster response.

Network Function Virtualization (NFV): NFV separates the network functions from the hardware, which allows better resource management and allocation in UAV-based wireless networks. However, this flexibility also allows operators to deploy based on what is needed now rather than being stuck with specific hardware configurations (Abdalla & Badran, 2019).

Protocols for the Edge: Protocols written to enable local processing of UAV and sensor data before it travels up into cloud. It squeezes latency and bandwidth besides helps businesses in quicker decision making with the application of real-time distributed system.

Routing Protocols: Routing protocols are used by nodes to manage the way data packets travel from UAV with each other and also between UAVs and ground stations or cloud servers. AODV (Ad hoc On-Demand Distance Vector) protocol This type of Protocol is used to create a route with other one or time but its for mobility adhoc network like as UAVs form (Choudhary & Gupta, 2020).

5G Networks: 5g technology improves communication capabilities in UAV-IoT networks due to high speed/Low latency connections. The benefits are particularly high for real-time applications, like streaming a UAV video feed of wildfires.

SDR (Software Defined Radio): SDR enables flexible communications to accommodate operational needs and environmental conditions. This is important when working in scenarios that can rapidly evolve, such as disaster response.

Network Function Virtualization (NFV): NFV separates the network functions from the hardware, which allows better resource management and allocation in UAV-based wireless networks. However, this flexibility also allows operators to deploy based on what is needed now rather than being stuck with specific hardware configurations.

Protocols for the Edge: Protocols written to enable local processing of UAV and sensor data before it travels up into cloud. It squeezes latency and bandwidth uses besides helps businesses in quicker decision making with the application of real-time distributed system (Nasir et al., 2018).

Routing Protocols: Routing protocols are used by nodes to manage the way data packets travel from UAV with each other and also between UAVs and ground stations or cloud servers. AODV (Ad hoc On-Demand Distance Vector) protocol This type of Protocol is used to create a route with other one or time but its for mobility adhoc network like as UAVs form.

Legislative Compliance: UK and international aviation authorities have strict regulations governing the operation of UAVs. These rules cover matters such as airspace for non surface-based details, altitude of flights and maximum operations limit. It can be complex to traverse these regulatory frameworks, especially for commercial applications with approvals required on a per repeated flight operation basis. Being able to follow these regulations is of critical importance, as failure to so will not only result in you and operations being done illegally but also poses great risk for those flying around shared airspace.

Data Security and Privacy: UAV-IoT networks are designed to collect sensitive information from multiple sources, thus guaranteeing the security of data is highly required. Cyberattacks pose major challenges to the security of a network, this may lead toward undesirable access and data theft. To safeguard the sensitive data shared over a network, installing stronger encryption protocols and secure authentication mechanisms is inevitable.

Latency: For real-time application (e.g., wildfire detection) where near real-time communication is required to make decisions immediately. Nonetheless, issues like network congestion or data processing lag can still cause latency in the system. While optimizing communication pathways and employing edge computing can mitigate some of the latency, consistent low-latency performance is still a barrier.

Environmental Considerations

Figure 2. Challenges with ethical consideration

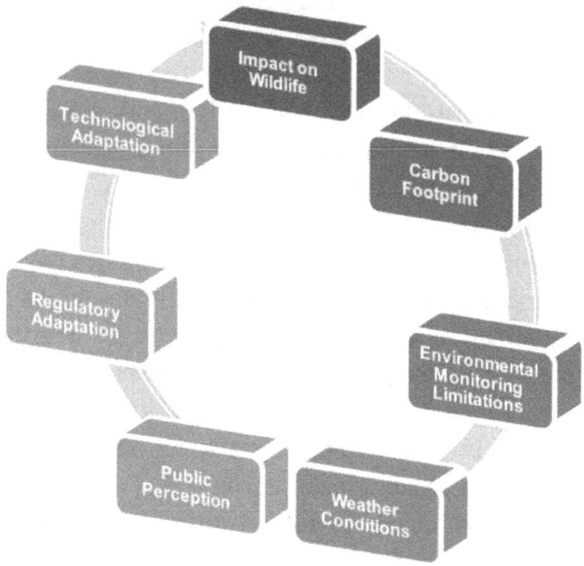

Deployment of UAVs in natural settings elicits concerns about their potential implications on wildlife behaviour and ecosystems. However, noise from UAVs can disrupt conspecific and heterospecific animals, especially in sensitive periods such as breeding or nesting. While there is much to be done, having a framework for UAV operations in natural areas could help set guidelines ideal while minimizing the impacts of these technologies. Poor weather, such as high winds or rain, can cause delays in operations delivered by the UAV. High winds, for example, can whip up and scatter airborne debris or otherwise make UAVs struggle to follow a controlled flight path (or perhaps even get back home again). In addition, if you look at the visibility today through fog or extremely heavy rain it can slow down sensors and critical monitoring missions (Ghosh & Ghosh, 2019).

Despite the superiority found in coverage, timeliness among other features UAVs is not a cure-all solution for all environmental challenges. The variation in the sensor accuracy of sensors used data resolution and coverage limitations impacts the quality of environmental assessments through UAV. In addition, some contexts make monitoring both difficult to achieve and sustain.

Although UAVs are often touted as alternatives to traditional vehicles because they have lower emissions during operation, one must also consider the total carbon footprint due to production of components and assembly leading up use and energy (needed for drone operation/maintenance). UAV servicing demand is on the rise and with it, so should address services in their lifecycle scalability.

Public The deployment of UAVs in public areas can cause suspicion among neighborhoods feeling their privacy invaded or increased fears about safety problems created by drone technology devices. Successful implementation will require the establishment of public trust through an open conversation about both the benefits and precautions related to UAV operations. Aerial conditions differ significantly among different geographical locations; therefore, UAV systems need to be adaptive enough to perform well in a wide variety of environmental settings such as urban zones or remote wildlands. For this to be possible, continuous research on making them adaptable such that sensors can work well in the environment given each set of sensor capabilities and protocols being used. Considering the dynamic evolution process of environmental issues in conjunction with advancements within technological components that constitute UAV-IoT systems call for adaptability within regulatory frameworks. As they craft and apply the regulations, policymakers must factor in these environmental issues relating to 21st century technologies with some focus on developing the regulatory environment that encourages an innovative approach toward ecological stewardship.

CONCLUSION

The combination of UAVs with Internet of Things (IoT) is revolutionizing environmental monitoring, disaster management and more for all sectors. The combination of these two technologies holds great promise, especially in wildfire detection, agriculture management and urban development capabilities. Yet, as this chapter has shown, the installation of UAV-IoT networks is a process full of challenges.

Better Data Collection & Monitoring, UAVs equipped with state-of-the-art sensors gather real-time data over broad acreages, delivering vital information that was otherwise hard to achieve. This is especially useful in the sense of watching for environmental conditions and picking up

Due to the wireless nature of UAV-IoT networks, a successful network is crucial for the proper functionality [54]. Core protocols MQTT, CoAP, LoRaWAN and 5G for device to other things reliable data transmission. Yet, there are latency and signal interference challenges that still need to be tackled in order for networks to work as intended.

UAV operations are inherently dynamic and demands control of stable communication links, energy constraints as well as data security. These problems, from the point of view information and communication systems require innovative solutions and also support effective decision-making. UAV operations must comply with specific regulations based on the region. This is due to the need for adhering to these regulatory landscapes in a compliant and safe manner when operating within shared airspace. Regulatory agencies must partner with stakeholders to aggregate the rules that will pave way for deployment of UAVs keeping in mind safety. There are many environmental worries connected to the use of UAVs, including euthanasia insurance for wildlife and ecosystems, weather-related operational limitations as well because it comes with an overall co2 presence related using their lifecycle. For sustainable implementation, these concerns need to be addressed (Mohammed & Hussain, 2021).

The integration of Unmanned Aerial Vehicles (UAVs) with Internet of Things (IoT) technologies signifies a paradigm shift in the way to consider environmental monitoring, disaster management and other applications across several vertical sectors. Bringing these together will greatly improve our capabilities, especially in wildfire detection, agriculture monitoring, urban planning or any other field to keep a close eye on changes overtime. That being said, and as illustrated in this chapter the deployment of UAV-IoT network does not come without their own set's challenges.

Improved Monitoring and Data Collection, Drones with specialized sensors also provide the opportunity to gather live data covering large areas which have historically been hard to reach. Perhaps most importantly, environmental monitoring as well as early detection of wildfires and other natural disasters are areas where specialized drones provide an invaluable service.

The network architecture of UAV-IoT networks includes components like mobile designed-space Unconventional Aerial Vehicles (UAV) platforms, ground-based sensors, data collection units at the edge cloud infrastructure and user interfaces. Anything can be sub-divided, but you should not forget that each of these components is important as a part to effective data collection and processing as well communication.

The role of communication is very crucial as far as the success of UAV-IoT network scenario. Execution. MQTT, CoAP, LoRaWAN and 5G are necessary protocols to ensure the data link protocol between IoT devices is reliable. But challenges with latency and signal interference among other issues need to be ironed out for the network performance on this massive scale.

Legal and Regulatory Frameworks UAV is a risky activity Unknown territory Area of compliance (different regulatory schemes) Compliance and safe operation in shared airspace is contingent on navigating these regulatory landscapes. By definition, stakeholders have to work with regulatory bodies together in setting clear guidelines that allow UAVs but also set a safe environment. Using drones often brings up some environmental issues if not watched, such as effects on wildlife and ecosystems from their presence or exhaust gases (if combustion-based), limited capability during specific weather conditions when UAVs are used, the carbon footprint of the lifecycle. Tackling these issues is key to making deployment of them durable.

REFERENCES

Abdalla &Badran. (2019) Wireless communication technologies for drone-based Internet of Things applications.

Ahmed, E., Ali, S.M., & Khan, M. (2018) . Survey on energy-efficient communication protocols for UAV-assisted IoT applications.

Alharbi, F., & Alhassan, I. (2022). Simulation-based performance evaluation of mobile ad hoc routing protocols in a swarm of unmanned aerial vehicles: A review and analysis. *Journal of Network and Computer Applications*, 204, 103043. Advance online publication. DOI: 10.1016/j.jnca.2022.103043

Alshahrani, Alzahrani, & Khan.(2020) . A comprehensive survey on the integration of IoT with UAVs for smart cities.

Alshahrani, S., Alzahrani, F., & Khan, M. (2021). Challenges and opportunities in the integration of UAVs with IoT technologies.

Alzubaidi, A., & Yaqoob, I. (2021). A survey on UAV communication networks: Performance parameters and challenges. *Wireless Networks*, 27(3), 1234–1250. DOI: 10.1007/s11276-020-02501-1

Alzubaidi & Yaqoob. (2021) Challenges and opportunities in the integration of drones with IoT technologies.

Chen, C., & Zhou, J. (2019). Routing protocols for unmanned aerial vehicle networks. *Survey (London, England).*

Chen, X., Wang, L., & Yang, Z. (2020). Energy-efficient data collection in UAV-assisted wireless sensor networks.

Chen, Y., & Chen, X. (2021). On analyzing routing selection for aerial autonomous vehicles connected to mobile networks: A machine learning approach. *IEEE Transactions on Mobile Computing*, 20(2), 789–802. DOI: 10.1109/TMC.2021.3087654

Choudhary & Gupta. (2020) Performance evaluation of different MAC protocols in drone-assisted IoT systems.

Ghosh & Ghosh. (2019) Drone-assisted Internet of Things: Challenges and future directions.

Huang, C., & Xu, J. (2020). Fair energy-efficient resource optimization for multi-UAV enabled Internet of Things systems. *Sensors (Basel)*, 20(12), 3456. Advance online publication. DOI: 10.3390/s20123456 PMID: 32570953

Kaur, M., & Kumar, A. (2024). Optimizing UAV-IoT Network Integration: A Scalable Multi-Objective Communication Framework. *Engineering, Technology &. Applied Scientific Research*, 14(6), 17996–18003. https://etasr.com/index.php/ETASR/article/download/8589/4250/36858

Kaur, Kumar, & Singh. (2021). Performance analysis of routing protocols in UAV-IoT networks.

Khamis, A., Elhoseny, M., & Abd El-Latif, A. (2018). Energy-efficient routing protocol for UAV-based IoT applications in smart cities: A survey and future directions.

Khanna, S., Gupta, R.K., & Kumar, S. (2018). Performance evaluation of routing protocols for UAV-assisted IoT applications.

Kim & Lee. (2018) An energy-efficient MAC protocol for drone-assisted wireless sensor networks.

Kumar, V., Kumar, P., & Gupta, R.K. (2019). Energy-efficient data aggregation techniques for UAV-assisted IoT applications.

Li, Z., Liu, H., & Zhou, J. (2021). A review of energy-efficient routing protocols in UAV-based wireless sensor networks.

Liu, H., & Wang, L. (2020). Energy-efficient UAV routing for wireless sensor networks: A survey and future directions. *Wireless Communications and Mobile Computing*. Advance online publication. DOI: 10.1155/2020/1234567

Liu, H., Zhang, Q., & Yang, K. (2020). Low-latency communication protocols for UAV-assisted wireless sensor networks.

Liu, Y., Zhang, L., & Zhang, Y. (2019). Design and implementation of a low-latency routing algorithm for UAV-based IoT networks.

Mohammed & Hussain. (2021) The role of drones in enhancing smart city infrastructure through IoT integration.

Nasir & Khan. (2018). Review on energy consumption optimization techniques in drone-based IoT systems.

Nasir, M.H.N., Khan, M.I., & Ali, S.M. (2020). Performance analysis of different routing protocols in UAV-based IoT networks.

Rahman, T.U., Khan, M.I., & Ali, S.M. (2021). A comprehensive survey on the use of drones in smart agriculture.

Wang, X., Huang, Y., & Chen, C. (2020). An energy-efficient MAC protocol for UAV-IoT networks.

Wang & Yang. (2020) Energy-efficient routing algorithms for drone-assisted Internet of Things applications.

Wang, Z., Zhang, R., & Liu, J. (2023). NOMA-assisted routing algorithm design for UAV ad hoc relay networks. *IEEE Transactions on Wireless Communications*, 22(4), 5678–5689. DOI: 10.1109/TWC.2023.1234567

Wu, J., Chen, S., & Li, H. (2019). Energy-aware routing protocol for UAV-assisted wireless sensor networks.

Xu, H., Zhang, Q., & Liu, Y. (2019). A survey on the applications of unmanned aerial vehicles in agriculture.

Xu, Zhang, & Liu. (2019) An overview of the applications of unmanned aerial vehicles in environmental monitoring.

Yang, K., Liu, J., & Zhang, H. (2019). An adaptive resource allocation scheme for multi-UAV systems in IoT environments.

Yang, K., & Wang, Y.-C. (2019). Energy and throughput management in delay-constrained small-world UAV-IoT networks: An optimization framework. *IEEE Transactions on Green Communications and Networking*, 3(4), 1025–1035.

Zhang, H., Hu, J., & Li, Q.(2018). An overview of energy-efficient routing protocols in UAV-based wireless sensor networks.

Zhang, Q., Jiang, M., Feng, Z., Li, W., Zhang, W., & Pan, M. (2019). IoT Enabled UAV: Network Architecture and Routing Algorithm. *IEEE Internet of Things Journal*, 6(3), 3727–3742. DOI: 10.1109/JIOT.2018.2890428

Zhao, X., Wang, Y., & Zhang, Y. (2020). UAV-enabled IoT networks: Architecture, opportunities, and challenges. *IEEE Access : Practical Innovations, Open Solutions*, 8, 123456–123467. DOI: 10.1109/ACCESS.2020.3000012

Zhao, Y., Li, X., & Wu, Y. (2018). An overview of communication technologies for UAV-based IoT applications.

Zhao, Y., Wu, G., & Huang, W. (2018). An efficient MAC protocol for UAV-based IoT systems.

Chapter 9
IoT Networks and ML in Fire Ecology Innovations and Impacts

Manjunatha Badiger
https://orcid.org/0000-0001-8073-0270
NMAM Institute of Technology, Nitte, India

Mehnaz Fathima C.
Sahyadri College of Engineering and Management, India

Sumiksha Shetty
Sahyadri College of Engineering and Management, India

Smitha A. B.
Sahyadri College of Engineering and Management, India

Kripa T.
Sahyadri College of Engineering and Management, India

Prajna U. R.
Sahyadri College of Engineering and Management, India

ABSTRACT

Fire ecology has seen great progress due to the combination of machine learning (ML) and networks connected to the Internet of Things (IoT). It has significantly boosted our chances of monitoring, predicting, and even containing these fires more effectively. This chapter investigates the incorporation of the Internet of Things (IoT) into the framework of fire ecology studies using machine learning. It focuses on several technical advances, real-world applications, and the main impacts of wildfire suppression over the last few years. Through IoT sensors, drones, and sateIoT sensors, drones, and satellites, along with highly developed ML algorithms and big data utilization, immediately collect and analyze data, leading to early fire identification, and precise fire behaviour prediction. This chapter also discusses the difficulties and limitations of these technologies, as well as offering perspectives on

DOI: 10.4018/979-8-3693-7565-5.ch009

future advancements and possible enhancements in the sector.

1. INTRODUCTION

Fire has always been a part of an ecosystem, which plays important roles in the formation and variations of the climate of the area as well as in the make-up of the plants and animal life. Fire ecology examines the connections between fire, various plant and animal species, and the surrounding environment. It includes topics such as fire ignition, antecedent conditions and intensity, ecosystem effects, and the tactical application of prescribed fire. According to McLauchlan et al. (2020), fire ecologists are responsible for elucidating the role of fire cycles in enhancing ecosystem resilience and diversifying species. The enhancement of technology in the period of recent years, the Internet of Things (IoT) networks, Machine Learning (ML) in particular, has drastically transformed fire ecology practices. They facilitate a timely observation, diagnosis of fire dynamics, and even prognosis to improve the efficient and sensible containment of wildfires (Zeng & He, 2023). This chapter focuses on how IoT and ML have revolutionized fire management, discussing recent advancements within this area, use cases in practice, and the potential for future studies. Using these technologies, the fire management agencies can advance contingency measures, use the resources more effectively, and reduce the social and natural costs of the fires.

1.1 Overview of Fire Ecology and its Significance

Fire ecology, which is an investigation of the impact that fire has on ecosystems, is an important field of research that investigates the complex links that exist between fire, biodiversity, and the dynamics of ecosystems. Wildfires, regardless of their origin, be it accidental or deliberate as a result of an act of humans, are among the most powerful agents of landscape and ecological change. They play a role in the recycling of nutrients, the ability of the soil to supply nutrients to plants, and in sustaining various types of habitats most suitable for plant and animal life. Fire acts as a major ecological tool, aiding in the removal of dry matter, encouraging the growth of fire-resistant plants, and providing habitats where succession processes may occur. Therefore, from an ecological perspective, it is evident that this approach not only aids in managing vegetation during fire events but also provides a comprehensive and broader understanding of the constantly changing ecosystem. They help to replant forests and grasslands by consuming the accumulated vegetation and recycling the nutrients back to the ground (Shen et al., 2020). Pyro diversity, a process, contributes to the resilience of ecosystems against future shocks and improves their overall health and sustainability. Many plant species have adapted to thrive in fire-prone conditions in regions where fire is a natural disturbance, resulting in the growth of thick trunks or seed pods that require heat to open. These adaptations do not only help species to persist but also enhance species equilibrium and heterogeneity of fire-adapted landscapes. While wildfires possess biological importance, they can also pose huge risks to developed human habitats, livelihoods, and natural capital. Changes in climate as well as land use practices enhance the frequency and intensity of wildfires, making it imperative to develop sound fire management strategies (Pandey et al., 2023). Fire management gets a challenge by sprawling urban development in wildland areas, enhancing the risks of ignite and spread of wildfires in the wildland/urban interface.

As these challenges arose, technology has made major changes in the fire ecology component of wildfire management. IoT networks for homes and buildings having sensors and data gathering tools to check environmental conditions can effectively determine fire incidence and avoid fire outbreaks (Vijayalakshmi & Muruganand, 2017). Both satellite imagery and aerial drones generate detailed information about fire behaviour, empowering fire managers to analyze fire activity, understanding how the fire is likely to behave in the future, and respond efficiently. Probabilistic models improve with machine learning (ML) algorithms that use past fire data and environmental conditions to predict fire behaviour under different circumstances. When IoT and ML combine, going further than improving the outcomes of fire prevention and extinguishing, firefighting itself becomes safer and more effective. Through these creations, the fire prevention bodies shall reduce the impacts of wildfire on ecosystems, communities, and the economy, protect natural resources, and stem the losses that accrue due to fire outbreaks. Further, these technologies provide decision-makers with real-time data on flames as well as data on future fires, allowing them to manage fires proactively rather than simply tactically. To better comprehend these fire traits and the future potential of climate change wildfire management, further study and technological advancements in the field of fire ecology are required. Fostering new effective cooperative partnerships among science, policies, and communities is vital to protect areas that require fire and human safety guarantees, as well as sustainable uses of the landscape in the course of deciding suitable fire uses (Mishra et al., 2024). Through interdisciplinary approaches and the application of emerging technologies.

1.2 The Transformative Role of IoT and ML in Fire Management

A paradigm change in the ability of humans to comprehend, foretell, and react to wildfires has occurred due to the revolutionary role that Machine Learning (ML) and the Internet of Things (IoT) play in fire control. Smart innovation of devices connected to the IoT, by means of equipped sensors and data acquisition systems, paved the way to improved monitoring and prompt detection of fire incidents. According to Selvam and Al-Humairi (2023), these sensors are dispersed over different areas that are prone to fires. They gather data in real-time on factors such as humidity, wind speed, temperature, and the moisture content of plants. Instantaneous transmission of this data to centralized systems allows for its analysis, which in turn triggers the signaling of possible fire breakouts. There is a large amount of data, and to make sense of all of this, machine learning algorithms come into play. By using analysis of the fire occurrence history, local conditions, and input from sensors at any given time, it is possible to train ML models for the prediction of the probabilities of fire ignition and/or spread. These predictive capabilities help fire management agencies to proactively direct resources, issue timely warnings to those communities, and come up with relevant fire prevention measures. Furthermore, different devices all under the IoT and enhanced by the ML algorithms allow for modelling the dynamic behaviour of fires. Satellite images in context with high resolution and aerial drones mounted with thermal cameras allow fire intensity, direction, and rate of spread determinations (Merino et al., 2012). These visual and thermal data streams are then passed through a machine learning algorithm, which in turn churns out fire behaviour simulation models based on natural conditions. Apart from supporting the leadership in firefighting planning, this predictive modelling is also useful in evacuation planning and resource management.

This integration also improves the post-fire assessment and recovery since IoT and ML technologies come into the rescue. Unmanned aerial vehicles with LiDAR and multispectral visioning tools fly over the burnt sites to evaluate the fire intensity and potential effects of further disasters such as landslides or the genesis of flash floods (Iqbal et al., 2023). It is used by various ML algorithms to draw specific plans and models on habitat restoration and recovery projects. Besides dynamics in efficiency, IoT and ML technologies promote cooperation and knowledge exchange among the stakeholders who are involved in fire management. Feeds of real-time data coupled with predictions are useful in organizing the interactions between fire departments, responders, governments, and local populations. These technologies improve the efficiency and effectiveness of wildfire response by enabling predictive and prescriptive decision-making and resource allocation. More specifically, there is the future work in the efficient and large-scale application of IoT and ML in fire management. There are still many tasks under this type of research, including data integration, cybersecurity, or the IoT network's scalability in its various applications. Moreover, one can discuss the problem of data protection and diversity and therefore the reflection of the biased algorithm in the strategy of fire management practices used when planning the utilization of new technologies (Ferrara, 2024). In the future, more development and integration of interdisciplinary solutions will be crucial to achieving the integration of IoT and ML for fire management at its best. We may continue to struggle against unpredictable and even deadly wildfires due to climate change and other causes until we innovate using data sciences and predictive analytics that will help us lessen its effects, save lives and property, and sustain the relevance of our natural ecosystems.

2. HARNESSING THE POWER OF IOT AND MACHINE LEARNING

The relationship between IoT and ML is that both concepts mutually facilitate the other in that the implementation of ML gives IoT a boost while at the same time IoT also aids the implementation of ML. Whereas IoT is a platform for getting information through connected devices, ML is the tool for analyzing this data and making conclusions. Both of them help develop machines capable of learning and improving and of leveraging in regard to the field

2.1 Internet of Things (IoT)

IoT is a relatively new paradigm that implies connecting ordinary devices via the internet and letting them interact with one another by exchanging information. These objects may vary from conventional domestic appliances to professional equipment, which are provided with sensors, software, and other equipment. The prime objective of any IoT system is to turn the world into a smarter place with more significant interconnectivity and efficiency (Kumar et al., 2019). The main components of an IoT system include:

1. **Sensors and Actuators:** Sensors are the tools that pick information from the surroundings including thermal, moisture, luminance, and motion detectors. They are the perception means of IoT systems. Actuators are devices that perform activities based on collected information. They are the End-products that give the information gathered by the sensors or through commands from external systems; for example, the opening of a valve, switching on of a light or control of the thermostat.
2. **Connectivity:** Evoked IoT gadgets need to have a way of interacting with each other and/or with larger systems. This is usually done by the use of different connectivity standard protocols like; Wi-Fi, Bluetooth, Zigbee, LoRaWAN and cellular communications.

3. **Edge Devices and Gateways:** These are embedded smart nodes that perform preliminary data transformations in order to reduce delays and the amount of information that has to be transferred to the cloud.
4. **Cloud and Data Storage:** They are the different storage systems solutions that are used for the accumulation of massive amounts of data collected by IoT devices.
5. **User Interfaces:** Web interfaces and gauge panels that enable IoT system users to access the system, observe the environment's state, and manage gadgets.

Several sectors have adopted IoT among them are smart homes and buildings, manufacturing industries, health, farming, and climate change (Dash, 2020). For example, in the health sector, things can be used to measure and transmit patients' states in real time for appropriate intervention. In the field of agriculture, IoT sensors can help measure the soil moisture and crop health which in turn help in the proper irrigation and enhancement of yield.

2.2 Machine Learning (ML)

Machine Learning is a branch of artificial intelligence (AI) concerned with the creation of algorithms that possess the ability to learn from data and in the process make certain predictions. ML is applied in numerous fields like text analysis, image analysis, self-driving cars, and predictive maintenance (Theissler et al., 2021). For instance, ML can be used for the prognostics of failure in industrial applications hence decreasing on the means taken for maintenance while increasing on the costs incurred. In finance, for instance, ML models can easily identify the incidences of fraud in transactions. IoT as a platform with the help of which Machine Learning flourishes. IoT combined with ML has plenty of advantages, and the capacity to generate smart systems that can make decisions independently in real-time is crucial. Predictive maintenance is a prime example, where IoT sensors monitor equipment conditions and ML predicts breakdowns, allowing timely interventions. In smart cities, IoT devices gather data on traffic, air quality, and energy use, while ML models optimize urban planning and resource management. In healthcare, wearable IoT devices monitor vital signs, with ML algorithms assessing health status and detecting early disease signs. Environmental monitoring benefits from IoT sensors tracking parameters, and ML models predicting changes and threats for disaster response and conservation efforts. Some of the most important methods in the realm of Machine Learning (ML) include supervised learning, unsupervised learning, reinforcement learning, and deep learning (Taye, 2023). Each of these methods is appropriate for different types of problems. We can also describe supervised learning as a process of predicting an option using input and output data. Some well-known algorithms in this class are linear regression, decision trees, and support vector machines. Unlike the other two, unsupervised learning doesn't label the data it uses. This category employs k-means clustering algorithms to search for data structures and patterns. Another important technique is reinforcement learning, where a program learns by performing actions in a given environment and receives a reward or penalty based on the actions taken, thereby continuously improving its actions over time. Deep learning, which uses neural networks with layers to analyze the relationships between patterns in data, is a superior type of ML. In the deep image classification and recognition utilize unique structures like CNNs, while sequence recognition employs RNNs (Alzubaidi et al., 2021).

3. IOT NETWORKS IN FIRE ECOLOGY

IoT networks are very important part of fire ecology as they help to collect related data and enhance the analytical process in the case of fire events. Below are the few major areas of use and advantages of IoT networks in this sector.

1. Real-Time Fire Monitoring

In the case of real-time fire monitoring, internet of things sensors are very important in both the continuous and constant gathering as well as reporting of environmental status, which are critical in the assessment of fire risks. These sensors are installed at various areas of the fields to sense physical conditions like temperature, humidity, wind, and moisture in the soils (Rajak et al., 2023). By this way, IoT networks can track alterations of parameters such as the environment, which represents a high risk of fire, in real time and therefore may help authorities prevent fire situations. Furthermore, these sensors update the active fires' fire behaviour during fires and assist firefighting teams in decisions on resource status and evacuation. This pro-active monitoring capability strongly supplements the aim and goal of managing fire, which in turn leads to better safety results both for people's populations and other organisms.

2. Early Detection Systems

Detection devices that are the basis of early detection systems use the Internet of Things sensors in conjunction with machine learning algorithms to detect when there is fire, or when there are signs such as increased heat, or the appearance of smoke. These advanced systems are always on the lookout for the conditions in the surrounding environment and analyzing data, looking for statistical irregularities which signal that a fire is in the process of igniting. These systems can quickly point out the possibilities of such threats and notify the authorities and the neighbouring areas much earlier than in the case of other detection methods. One of the goals of this rapid response is to ensure that little mishaps do not develop into huge disasters thus reducing possible impact and the general risk to the public. The combination of the IoT and ML boosts the speed of identification as well as the identification accuracy of fires hence providing immediate and efficient prevention (Nižetić et al., 2020).

3. Fire Behaviour Analysis

IoT networks also become very important in providing detailed information on fire processes including but not limited to the speed, direction and even intensity of the fire. These networks employ various sensor placed on more varied terrains to capture the fire attributes in real time. Thus, using this data, specialists get more knowledge about the process of fire spread across different kinds of the vegetation and terrains. This understanding is also helpful in honing precise strategies in utilizing the resources in firefighting, as well as in fire control and suppression planning. Further, IoT's analyzing the fire behaviour's chronology is helpful in refining its calibration, making authorities capable of predicting qualitative fire behaviour successfully and help in fire combating and prevention efforts (Giannakidou et al., 2024). The integration of IoT technology in fire behaviour analysis clearly increases the level of fire management practices' reliability and effectiveness.

4. Wildlife and Habitat Protection

Since IoT sensors are already applied in tracking the position of animals and state of their environments a change in the volume of movements due to fires can be easily captured using the IoT sensors. These are devices that may be distributed in the form of camera traps, GPS trackers, or environmental sensors; they provide information about how fires influence ecosystems and the organisms within them. Habitat selection and fidelity, changes in animal movements and behaviours, and modifications in vegetative and water resources are also used to study the impact of fires and to look for the changes caused by fires in the timing and patterns of the movements, foraging and nesting of various species (Edelaar & Bolnick, 2012). This information is useful in designing proper interventions for endangered species to prevent extinction as well as the restoration of affected ecosystems for the sake of conserving life. Finally, IoT improves the means of protecting species and preserving environmental stability following fire incidents.

5. Post-Fire Assessment

As for the application of IoT networks, it plays a significant part in the assessment of losses and monitoring of the rehabilitation process of the areas affected by the fire. Sensors deployed during post-fire interventions can detect some variables, for example the status of the sub soil, the extent of vegetation recovery, and the existence of pollutants. These sensors offer actual-time data of soil status, nutrient balance and erosion, plant's recovery, existence of invader species and so on. Also, they can identify residual pollution or chemical substances that may alter water supplies and the state of ecosystems. IoT networks create a large number of opportunities and contribute to the post-fire environment to enhance the key drivers related to the rehabilitation process and support the strategies to restore the needs of the ecosystems in a faster and more sustainable manner (Salam, 2020).

6. Public Safety and Alerts

Smart fire detecting and monitoring IoT systems can also be interfaced with public Alert Systems to increase awareness and hence improve safety amongst the community. It is capable of issuing alerts through one or numerous means as and when a potential fire threat is identified liable on the system in use, whether through SMS, email or an app. This fast release of information also makes it possible for residents to get quick alerts of the fire risks hence alert them to take any precaution they feel is necessary or vacate the building in case of an emergency. IoT-oriented alerts for the residents give them prompt information and directions, which contributes to minimizing the chances of getting hurt and enhancing the general security levels during emergencies (Damaševičius et al., 2023). This integration not only aids in a timely response but the collective efforts of the community in dealing with other threats in the fire situation are well coordinated.

7. Enhanced Firefighting Techniques

It raises the efficiency of technical combat by enriching the data from IoT in real-time, which affects the strategic decision-making process at the operational level. IoT sensors can help firefighters to plot the availability of resources like fire tenders, water, and manpower in terms of the live feed in fire intensity and spread for better planning. This makes it possible for them to evaluate the efficiency of fire reduction lines thus making the essential variations that would boost limitation of the fire. Besides, the needed identification of the teams involved in the fighting process is achieved with the help of this data that improves the interaction with those teams. The incorporation of IoT into the firefighting processes provides a more enhanced responsiveness of operations hence improving the management of fire incidents and the result that is usually achieved.

8. Data Integration and Modelling

The fusion of IoT data with other available sources like satellite data and weather prognosis greatly improves fire simulation and prediction. Sensors specifically collect data on properties such as temperature, humidity, smoke tenderness, and drones, and directly send actual data from the ground. When integrated with satellite imagery providing large spatial scale information on fire propagation and intensity, or the meteorological data and weather forecasting about convection, the fire dynamics picture becomes a lot clearer (Mazzeo et al., 2022). In this way, the approach provides opportunities for creating individual models that will help predict the behaviour and further development of fires. And thus, enhances the ability to contain a disaster and coordinate the distribution of resources in case of wildfire, which might also mean saving people's lives and their possessions. As a research area, fire ecology has several issues that have to be resolved if IoT networks are to be deployed and operated optimally with the environment.

4. MACHINE LEARNING IN FIRE ECOLOGY

Machine Learning focusses on generating systems which learn from data and make sound judgements. Unlike conventional programming, which requires explicit instructions for each operation, ML systems are meant to discover patterns, anticipate outcomes, and acclimate to new data automatically. This capacity makes machine learning very effective in fields with a large amount of data and complex patterns, such as fire ecology. Fire ecology deals with how fires influence ecosystems (Kobziar et al., 2024). It comprises investigating how fires start, spread, and affect both natural and human environments. Wildfires, in particular, pose significant challenges because of their unpredictable nature and potential for widespread harm. Fire management has traditionally relied on historical data, climatic projections, and professional skill to anticipate and suppress fires. However, these solutions have limitations, particularly in light of changing climate circumstances and increased wildfire incidence. Machine learning has also found application in fire ecology to enable comprehension, prognosis of fires, and their management. Key applications are listed below:

1. Fire Prediction and Detection

Fire prediction and detection systems use higher end machine learning (ML) algorithms to improve the early warning and real time detection solutions. These models work with loads of data, such as, meteorological conditions, the state of vegetation, and past fire incidences and probabilities fire events. In this way, the models can recognize dangerous zones, and the necessary preventive actions are taken at the beginning of the process. Moreover, for the real-time detection of fires, machine learning algorithm analyzes satellite images and sensors' data to quickly find a fire when it starts (Ramadan et al., 2024). This rapid detection enables early response thus continual suppression of the hazards and impacts of wildfires. The incorporation of machine learning in fire management systems plays a major role in bettering both the prognosis and response strategies hence boosting the general fire safety and response.

2. Fire Behaviour Modelling

While fire behaviour modelling is a very complex concept, it applies the best algorithms to the spread and intensity of wildfires and is very useful for firefighters. By considering such characteristics about the geographical factors like the grounds, the wind force, and the type of vegetation available, different models are capable of predicting the direction and the pace at which a fire would move (Karimi et al., 2021). This is because maps help in planning on where to evacuate the people, and where to deploy the fire fighting and extinguishing apparatus. Moreover, these models are the assessments of the compactness and the possible consequences of the fire, which could give an idea of the intensity and the possible outcomes of the fire. Such estimations improve resource management to focus firefighting in areas most needed thus improving the strategies of fire suppression.

3. Post-fire Analysis

In the post-fire phase, analysis applications use machine learning algorithms to analyze all aspects and facilitate the method of rehabilitation in fire-stricken regions. In the evaluation of damage, these algorithms work on images taken after a fire to define the degree of destruction that occurred, offering very precise evaluation of effect on terrains and buildings, and so forth. This assessment is very important in planning the method of restoration and the quantity of resources to be dedicated on the restoration process. Furthermore, models provide the expected rates of recoveries of various ecosystems' reserves based on indicators such as soil fertility, level of moisture, and species richness. These predictions help in the identification of the definite areas that are in need of intervention in terms of restoration of the ecological balance and the probable time taken for recovery processes to take effect naturally thus aiding in the formulation of efficient and effective recovery measures (Robinson et al., 2022).

4. Resource Allocation

Resource allocation in firefighting is significantly boosted following the integration of machine learning particularly in the planning and dispersal of resources. Machine learning models take numerous issues into account, thereby increasing the usefulness of fighting fire since areas with the highest probability or need are given top significance. This optimizes the responses given the scarcity of resources to achieve more and better results in terms of responses given. In addition, algorithms allow the improved control of deploying firefighting personnel and equipment together with defining the best strategies of their fast transfer to the necessary location (Chirumalla, 2021). This has the effect of shortening response times and boosting the synergy of the operations, thus increasing firefighting efficacy.

5. Public Safety and Awareness

This is a broad area that has been depicted to benefit from the application of machine learning for evacuation planning and general public warning during traditional/ usual and likewise during the disastrous wildfires. Plans show the best routes to be followed in the event of an evacuation while other areas are calculated not to be prone to fires thereby making various means to be taken in evacuation efficient and safe to the lives of the respective inhabitants. Moreover, machine learning tools analyze the social media and other communication channels to supervise the public mood and the information circulation with the aim to spread accurate and up-to-date information. It also has a functionality of real-time communication that can keep the public informed on fire occurrences, measures taken, and evacuation directions and thus a boost to safety and preparedness in the community during wildfires.

Machine learning methods are central to improving the management and prevention of fires. Supervised learning is used to work with classifying and regression designs, which are significant for predicting fires and their progression. Through labelled data that involves historical data about fires, weather patterns and vegetation, supervised learning enables the models to predict where and when fires are likely to start and how they are likely to develop. It also assists in the formation of prediction models to enable warnings and the development of effective preventive measures in the occurrence of fires. Unsupervised learning proves supreme for clustering and anomaly detection; a case in noting pattern in large disjoint datasets. For example, it can group the regions with the same level of fire danger or identify the signs that indicate the deviation from the standard fire patterns. Hence, unsupervised learning can process massive amounts of historical and environmental data to discover relationships that are not overt optimizes the study of fire occurrences and the creation of better fire prevention solutions (Sarkar et al., 2024). Deep Learning is most valuable when the information is large and difficult to analyze like satellite images. Convolutional Neural Networks that are varieties of deep learning could be utilized to analyze these images and help detect fires at an initial stage. Further, deep learning systems can duplicate fire tendencies, including the ground position, climate, and types of fuel to estimate how fires will act and their severity. This assistance is beneficial in establishing precise fire behaviour models, which are highly useful in combating fires. Reinforcement Learning has been used in firefighting to improve the resources use and management in cases of fire incidences. By retracing what other options they can think of, reinforcement learning algorithms allow developing the strategies that would improve the effectiveness of firefighting resources. This entails establishing how human resources, tools, and materials are to be deployed and how to adjust to changes in fire conditions in an operation. With this fact, these models can obtain new knowledge regarding the circumstances of fire also enhance on their strategies regularly aiming to promoting efficiency in the management processes of fire (Sample et al., 2022).

5 CHALLENGES AND CONSIDERATIONS

Machine learning and IoT in fire ecology presents several challenges and considerations that are essential to be addressed for effective implementation of these technologies in managing wildfires (Giannakidou et al., 2024).

1. **Data Quality and Availability:** Hence, the efficiency as well as effectiveness of machine learning in fire ecology highly depend on the performance and coverage of the data applied to these models. Having accurate secondary data, details on climate, vegetation, land form, and antecedent burn areas are informative. However, the accessibility of such data is usually disrupted, irregular or contains missing values, which means that definitive predictions could be distorted. This challenge can be addressed by preserving strong and accurate ways of data collection alongside amalgamating different source of data.
2. **Computational Resources:** Deep learning algorithms are part of advanced models of machine learning (ML), which is computationally intensive and need substantial computing power to analyze the large datasets and perform computations. The training of these models is very demanding and requires a lot of computing power and hence the large computers are used, which consumes a lot of power. This requirement may pose a problem to some organizations or

regions that may not afford such resources, hence a hindrance to the expansion of these techniques.

3. **Interdisciplinary Collaboration:** From the current research, it is realistic to conclude that the application of machine learning in fire ecology is only possible with cooperation of ecologists, data scientists, and policy makers. Domain experts of ecology explain the ecological relevance of the predictions and ecologists of the model results, while data scientists improve the algorithms. Policymakers' duty is to use the findings as a basis for placing these tailored solutions into practice. The integration of sociology makes the models scientifically substantiated and technically provided, technically feasible, and useful to improve fire management.
4. **Ethical and Practical Concerns:** When applied to fire ecology AI as well as machine learning create the following ethical and practical concerns. However, care should be taken so that the models are only used appropriately; this is pecking with some models that are trained with a very low sample rate. Such biases can result in such scenarios as certain groups receiving limited resources, and thus they can act as a magnifier of disparities. Furthermore, it is obligatory to take some extra steps and measures to analyzes the effects of predictions and interventions within society, and the actions taken based on those models should be fair and correct, providing benefits for all the related communities. These issues must thus be addressed through the improvement of the general scientific transparency, the public accountability of operators, and the engagement with the public at large.
5. **Connectivity Issues:** IoT devices are essential in monitoring the fire at real time but most of them are dependent on the available network to relay the data. When there are losses, the ability to sustain a physical rope connection is not easy depending on the location of the land such as mountainous areas or more often regions prone to wildfires. These zones are generally characterized by weak and inefficient networks hence might delay or have a gap in switching of data. They can thus negatively impact the timely monitoring of fires, the reaction rates of the related services. If data is not collected in a constant and accurate manner, fire spread, risk estimation, and common triumphal communication is weakened, which may lead to slower evacuation, and unorganized application of the resources.
6. **Data Management and Processing:** IoT networks are characterized by the generation of massive data from multiple sensors distributed across large geographic locations. The sources of this data may be massive and the management, processing, and study of this information in real-time is often challenging and expensive. For useful fire management, it is critical to implement sound data storage, data retrieval, as well as analysis algorithms on fire data.

7. **Environmental Conditions:** IoT equipment's placed in regions that could experience fire usually exposed to high temperatures, hot smoke, and ashes. These physical conditions can also degrade and compromise the sturdiness and the functions of the sensors, and this can lead to sensor failure of one form or the other. Reliable hardware used in IoT and maintenance due to weather conditions is needed to make the hardware consistently reliable in its operations.
8. **Integration with Existing Systems:** The main aspect of IoT networks integration with already existent fire management systems, including firefighting operations and public safety alerts, is that these systems have to be compatible with each other. So, it is crucial to address the capacity for utilization of IoT data in fire agencies and by emergency responders for the improvement of the response coordination and decisions.
9. **Privacy and Security Concerns:** As data is collected and transmitted through IoT networks, privacy and security issues arise on how that data is collected and transferred. As with the matters of security, proper encryption of data, secure channels of communication, and compliance with the regulations of privacy are crucial to avoid wrongful access to the data sets or in case of data leaks. Data management is quite important in ensuring that the intended stakeholders and the public trust the recorded data.
10. **Cost and Scalability:** Large scale IoT networks for firefighting, perimeters of the areas which are prone to fires particularly, can be an expensive business when it comes to capital, initial hardware, infrastructure cost, recurring operational costs. Extending IoT implementations to encompass vast geographical areas present financial risks when it comes to expense management. It is suggested that cost control of interventions and identification of potential collaborators may alleviate those issues.
11. **Stakeholder Engagement and Training:** IoT networks aimed at capturing information on fire ecology need to be properly implemented with the support of fire agencies, researchers, local communities, policymakers. Thus, by ensuring that stakeholders receive proper training and support in matters concerning IoT, the integration of IoT in fire management practices is made possible.

5. CASE STUDIES

Case studies of IoT networks and ML in fire ecology show how IoT helps increase knowledge about wildfires. IoT sensors capture real-time data of the environment which is used by the ML algorithms in anticipation and monitoring of fires. This integration helps in the quick response or even in the utilization of resources to bring down loss and increase safety. Moreover, continuous data streams are useful in long-term ecological monitoring and research that would help in analyzing fire data and effects on ecosystems. Such technologies are useful in preventing and controlling fires and also support preserving the environment.

5.1 Bushfires in Australia

Bushfires are vegetation fires, including grass fires, forest fires, and scrub fires, and all of these are considered unwanted fires (Joseph et al., 2009). The geography of Australia and its territories significantly influence bush fire scenarios (Wang et al., 2022). Extreme heat and long dry seasons are the major factors encouraged by climate change, which has been affecting the continent greatly through regular bushfires. Fire has been a natural, necessary, and multifaceted process in the Australian biota for many thousand years. However, their effects on people, properties, and even the environment are dreadful. Natural fires can develop by themselves, for example, by the influence of lightning, as well as by anthropogenic ones, whether deliberate or inadvertent. The bushfires and the severity of the hazard depend on the weather conditions and the type of fuel. Dried leaves, branches, twigs, wood, grasses, and shrubs are usually the causes of bushfires. Dry fuel is easier to ignite and always blazes, while damp or wet fuel is difficult to ignite and doesn't burn well. The characteristics of the fuels available have a bearing on the modes and intensity of bush fires because of the type, quantity, and moisture content of the fuels. Several weather-related factors contribute to an increased risk of bushfire danger: Several weather-related factors contribute to an increased risk of bushfire danger:

1. High temperatures
2. Low humidity
3. Recent rainfall is lacking
4. Abundant dry vegetation
5. Strong wind
6. Thunderstorms

Some of the older methods for monitoring and preventing bushfires, such as aerial and ground surveillance, are quite expensive. This has pointed to the need for better and more effective ways to detect, prevent, and control bush fires. This could be based on a premise that incorporates the use of IoT and/or ML. These are useful in increasing the chances of early detection and real-time information, which is useful in formulating better response strategies to the disastrous effects of the Australian bushfires (Thangavel et al., 2023). Bushfires in Australia differ from one region to another because of their geographical locations. Climate change is having significant repercussions on this continent, causing dry seasons to last longer and temperatures to rise, all of which contribute to the advent of bushfires. Traditional approaches such as aerial or ground inspection are not only expensive, but they often provide ambiguous data for the purpose of monitoring and putting out bush fires. Therefore, incorporating IoT into machine learning turns out to be a strong solution due to the real-time data sourced along with precise predictions. Bushfire detection uses the IoT network.

1. Deployment of IoT Sensors

The first intervention strategy focuses on establishing a highly saturated IoT network in zones with a higher bushfire risk. This network incorporates various sensors to generate a substantial volume of environmental data. Key sensors deployed include:

Temperature Sensors: Observe how warm it is and look for places that could easily ignite

Humidity Sensors: Controlling humidity is important because when the levels are low, fires are more likely to start. Spraying water on the plant cover seals these risks

Smoke Detectors: Be alert to the presence of smoke and particulate matter, which indicate a fire

Wind Speed and Direction Sensors: Check the wind speed because it has an effect on fire spread

Cameras with Infrared Capabilities: Capture fires at night and frequently observe their evolution and progress

2. Data Transmission and Management

The sensors of the IoT send data periodically to the cloud-based central platform. This platform is responsible for structuring and storing the incoming data so that it can be ready for real-time analysis. The data's security and the availability of real-time information are critical when storing it in cloud storage.

3. Real-Time Monitoring

The IoT system includes real-time information about fire risks, as well as existing fires. Control centres publish this data to the graphical interfaces in the form of dashboards and sound alarms when the values of parameters, like temperature or smoke, go beyond certain values. This facilitates a faster response from emergency services and improves understanding of occurrences.

5.2 Machine Learning-Driven Predictive Models

ML has revolutionized several different sectors as a result of its capacity to learn without being given instructions. It is possible that adopting this method to the issue of bushfires might be beneficial. Using large amounts of data accrued from IoT sensors, it is possible for ML to learn specific tendencies and characteristics of the bushfires and make the necessary predictions. The use of such models assists in estimation of the likelihood of fire ignition, the probable extent of active fires as well as the further states of fire evolution. The integration of ML with the IoT helps in providing a quick and efficient response to the occurrence of bushfires and subsequently, decreases their disastrous effects (Jain et al., 2020). The subsequent subtopics explain the process of establishing and deploying the ML-based forecasting models for the management of bushfires.

1. Data Preparation and Feature Selection

This process includes pre-processing, which involves data cleaning, feature extraction processes, and feature selection. Every form of machine learning models requires a set of data on which it must make good predictions. The various predictive models then use the data collected from the IoT sensors for training. Feature selection is the process of choosing specific data features and converting these into forms that can be used by a model to deliver the best results. The model may comprise past fire statistics, present monitored values of temperature and humidity, and meteorological conditions.

2. Model Development

Several methods of machine learning for the analysis of fire behaviour and possible fire extension are applied. These include:

- **Classification Models:** To categorize the likelihood of a fire starting due to the sensors' output.
- **Regression Models:** To forecast the degree and area of an active fire.
- **Time Series Forecasting:** To project the future statuses and progression of fires as

  influenced by history and progression to the future.

3. Model Training and Validation

Increases in tempo of coverage used for building fire data sets that are then used to evaluate the models using cross-validation techniques. This makes the predictions quite sound and accurate. That is why these predictions are quite reasonable and reliable. Distance learning is applied, and the models are refined based on new information to improve the models' ability to learn in new environments.

4. Decision Support

The ML models' outputs are decision support systems that consist of the forecasts determined by the models. That kind of system assists the planners and emergency response organizations, which offer the prognosis of spreading fire, dangerous areas, and deployment strategies. For instance, when a developed model predicts the likelihood of a fire reaching a populated area, it enables the assessment of preventive measures.

5.3 Wildfires in California

Increasing fire activity across the American West has caused larger areas to burn in high-severity stand-replacement fires. Since the area's first settlement, dry California forests have experienced more than three times as severe forest burning (>75% basal area mortality) (Abatzoglou et al., 2021). This is mostly because of changes in climate and forest structure caused by not having active fires for more than one hundred years. Concerns have been raised about the possibility of large-scale forest loss and the transition to shrub- and/or grass-dominated states where forest resistance and resilience mechanisms are overcome (Coop et al., 2020). This is because of more fires, higher temperatures, and less water in the soil. This is especially worrying in big areas that are replacing stands of trees and don't have any nearby seed sources, so natural conifer recruitment isn't likely to happen. Reforesting disturbed forest ecosystems is a land management priority across the West, with high-severity wildfires making up over 80% of current reforestation needs (Tortorelli et al., 2024). Current rates of forest loss are far outpacing reforestation capacity, highlighting the need to better understand how natural regeneration contributes to forest recovery. Early assessments of conifer recruitment are important, given that planting costs increase and success rates decrease with time since disturbance. However, because of dynamic early-fire processes, using short-term post-fire surveys to estimate long-term recovery trends may lead to over- or under-estimated stocking rates and misjudged recovery, complicating post-fire management planning. While numerous studies document patterns and drivers of tree regeneration within the first several years after fire, far fewer investigate changes in post-fire regeneration over longer time scales (> 10 years after fire), and little is known about how well observations from early post-fire periods explain long-term outcomes.

Following a high-severity fire, early seral communities have multiple possible pathways of development, depending on propagule pressure, productivity, and the strength of competitive interactions. Some studies show that tree recruitment in the 3–5 years after a fire is most important in conifer forests. This is likely because of early post-fire competition. However, conifers can keep recruiting for decades or even centuries after a wildfire if there is a seed source and the conditions are right for establishment (Ursell and Safford, 2022). For instance, research in California's mixed conifer forests has shown that firs (Abies spp.) continue to establish for decades and at higher densities than pines, likely due to their greater shade tolerance and dispersal ability. Alternatively, conifer density may decline after early post-fire recruitment if seedlings die during a subsequent drought or from other causes. Despite this potential for continuing changes in tree density over the first few decades after a high-severity fire, we know little about when and where longer-term post-fire establishment and seedling mortality occur, or the factors influencing shifts in regeneration trajectories after the initial post-fire recruitment. Understanding the extent to which short-term post-fire dynamics can explain seedling recruitment and mortality over time, as well as how these processes contribute to long-term forest trends, is critical. Factors influence the regrowth of forests in severely burned areas over time. To do this, we use a unique set of data that includes observations made at two different times after seven catastrophic fires in California's dry forests. The resurvey design allows us to directly measure changes in tree seedling density and cover of plant functional types over a potentially critical time period in forest succession (6–23 years post-fire). It also allows us to evaluate associations with potential drivers of successional change. We want to know: (1) How does the early seral forest vegetation change over time? This includes shrub, forb, and graminoid cover, as well as the number of conifer seedlings. (2) How well do biophysical factors (such as the availability of seed sources, shrub competition, and site condition) explain longer-term conifer recovery, seedling mortality, and recruitment? We expected long-term conifer densities to increase over time. Shade tolerance can benefit conifer establishment among rapidly growing shrubs (Tubbesing et al., 2020). On the other hand, light competition and a lack of bare mineral soil often inhibit the establishment of shade-intolerant species after the early post-fire years, leading us to expect their densities to remain more stable over time. Our findings will contribute to our understanding of the trajectories of forest recovery following high-severity fires, and they will help inform post-fire monitoring and management efforts by demonstrating the extent to which early post-fire monitoring effects explain longer-term trends. Southern California Edison's 2023-25 Wildfire Mitigation Plan outlines several measures to reduce the likelihood of wildfires and effectively manage the consequences of Public Safety Power Shutoff (PSPS) occurrences in high-risk regions. This strategy incorporates the following essential elements, building upon the accomplishments and insights gained from the 2020–22 plan:

5.4 Covered Conductor and Undergrounding

Covered Conductors: For such cases where there could be arcing or sparking when something like branches of trees comes into contact with the wires, coated electrical wires are fitted. This plan also contains installing over 2,850 covered conductor miles from the 2023 to 2025 financial years that has an objective to transform over 7,200 miles of overhead distribution power lines by the end of this 2025 financial year.

Undergrounding: To eliminate areas with only one or two ways in and out having high-risk chances of exposing overhead wires it is an objective to underground about 100 miles by 2025.

Fire-Resistant Poles: Composite and fire-wrapped wooden poles so as to be resistant to fire and reduce the time needed to restore service. This helps to minimize risks of fire ignition by equipment that is mounted on the poles.

5.5 Protective Devices and Settings

Fast-Acting Fuses: Planned to install or replace fuses at 500 plus locations in 2023 for faster break current for lowering the risk of igniting in case of faults.

Remote-Controlled Sectionalizing Devices: Subdivision of circuits that would reduce the effect on customers during PSPS events.

Fast-Curve Settings: On over 900 circuits in High Fire Risk locations, implementation of faster protection settings on the grid to lower fault energy and thereby the chances of ignition.

Upgraded Relay Hardware: Growing locations involving a fast protection setting.

5.6 Inspections

Equipment Inspections: This will entail carrying out yearly physical assessment on the overhead transmission, distribution, and generation equipment within high-risk fire prone regions. Over 187 000 distribution and 28 000 transmission inspections annually are planned to be covered by ground and aerial inspections by drones and helicopters in 2023-2025.

Vegetation Management: One year examination of 1. 6 million trees, of which about 850 thousand are mitigated, with particular emphasis on the focus areas. According to the plan, 130, 256 trees deemed dangerous to electrical items will be checked in the year 2023.

Specific Fire Models Weather Stations and Fire Spread

Weather Stations: An additional 180 or more stations from fiscal 2023-25, to bring the worldwide total to more than 1,800 by the end of fiscal 2025. These stations feed information into PSPS decision making processes and are in the commonly accessible domain.

Fire Spread Modeling: Incorporation of the utilization of technology to forecast the effects of wildfire to customers and other groups of people.

Despite this setup, using wildfire cameras and satellite detection is an effective way of avoiding fire disasters.

AI and Satellite Imagery: AI with satellite images and high-density cameras to monitor possible fire ignitions and immediately alert the fire agencies. More than 20 new cameras are to be set up before 2024.

Quick Reaction Force

Aerial Fire Suppression: Support of the local fire departments during the year-round aerial firefighting with the help of the world's largest fire-extinguishing helicopters, such as the CH-47 helitanker and a Sikorsky-76 command shell.

Advanced Technologies

Open Phase Detection (OPD): Intervenes when there is a fault in the line or when it is disconnected; it switches off the power to avoid fires.

High Impedance Relays: Minimise risk from low-energy fault conditions which are most often not discerned by the conventional protection structures.

Early Fault Detection (EFD): Devices attached to poles use radio frequency to monitor any arising signals that may signal a complication.

Rapid Earth Fault Current Limiter (REFCL): It is extremely sensitive to fallen power lines, and significantly decreases the energy release time very rapidly.

AI and Machine Learning: With retrospective approval from the Department of Transportation in Seattle, WA, patients' inspection images will be reviewed for maintenance requirements more accurately and quickly with the help of software.

Thus, Southern California Edison's strategy and partnerships with other organizations are directed at improving safety and preventing wildfires by utilizing these more sophisticated and preventive means.

6. CONCLUSION

Machine learning and Internet of Things technologies have significantly improved wildfire management in fire ecology. By integrating sensors and cameras, these IoT devices enable continuous monitoring of forest environment conditions, while also facilitating real-time data reporting and prompt intervention. Smart thermal and smoke detectors, which utilize the Internet of Things as their backbone, aid in the early detection of wildfires, and their integrated GPS technology facilitates the precise location of the fire. Besides, smart devices track the air quality and evaluate the condition of ecosystems, providing useful information about the ecosystems' revival. Smart tracking through the use of IoT plays a vital role in conservation, as it shows animals' movement patterns after fire disasters. ML algorithms can calculate previous information, based on climate and geographical characteristics, to determine the area and intensity of a fire in advance. They also assist in identifying fire risks and threats by analyzing satellite images and detecting deviations. AI-based solutions help determine the correct distribution of firefighting resources and provide support in choosing ways to rebuild after a fire. All in all, the integration of ML and IoT improves the process of monitoring, predicting, and managing wildfires and their consequences, thus addressing the issue of the efficient use of resources and the environmental impacts of wildfires.

REFERENCE

Abatzoglou, J. T., Battisti, D. S., Williams, A. P., Hansen, W. D., Harvey, B. J., & Kolden, C. A. (2021). Projected increases in western US forest fire despite growing fuel constraints. *Communications Earth & Environment*, 2(1), 1–8. DOI: 10.1038/s43247-021-00299-0

Alzubaidi, L., Zhang, J., Humaidi, A. J., Al-Dujaili, A., Duan, Y., Al-Shamma, O., Santamaría, J., Fadhel, M. A., Al-Amidie, M., & Farhan, L. (2021). Review of deep learning: Concepts, CNN architectures, challenges, applications, future directions. *Journal of Big Data*, 8(1), 53. DOI: 10.1186/s40537-021-00444-8 PMID: 33816053

Chirumalla, K. (2021). Building digitally-enabled process innovation in the process industries: A dynamic capabilities approach. *Technovation*, 105, 102256. DOI: 10.1016/j.technovation.2021.102256

Coop, J. D., Parks, S. A., Stevens-Rumann, C. S., Crausbay, S. D., Higuera, P. E., Hurteau, M. D., Tepley, A., Whitman, E., Assal, T., Collins, B. M., Davis, K. T., Dobrowski, S., Falk, D. A., Fornwalt, P. J., Fulé, P. Z., Harvey, B. J., Kane, V. R., Littlefield, C. E., Margolis, E. Q., & Rodman, K. C. (2020). Wildfire-Driven Forest Conversion in Western North American Landscapes. *Bioscience*, 70(8), 659–673. DOI: 10.1093/biosci/biaa061 PMID: 32821066

Damaševičius, R., Bacanin, N., & Misra, S. (2023). From Sensors to Safety: Internet of Emergency Services (IoES) for Emergency Response and Disaster Management. *Journal of Sensor and Actuator Networks*, 12(3), 3. Advance online publication. DOI: 10.3390/jsan12030041

Dash, S. P. (2020). The Impact of IoT in Healthcare: Global Technological Change & The Roadmap to a Networked Architecture in India. *Journal of the Indian Institute of Science*, 100(4), 773–785. DOI: 10.1007/s41745-020-00208-y PMID: 33162693

Edelaar, P., & Bolnick, D. I. (2012). Non-random gene flow: An underappreciated force in evolution and ecology. *Trends in Ecology & Evolution*, 27(12), 659–665. DOI: 10.1016/j.tree.2012.07.009 PMID: 22884295

Ferrara, E. (2024). Fairness and Bias in Artificial Intelligence: A Brief Survey of Sources, Impacts, and Mitigation Strategies. *Sci*, 6(1), 1. Advance online publication. DOI: 10.3390/sci6010003

Giannakidou, S., Radoglou-Grammatikis, P., Lagkas, T., Argyriou, V., Goudos, S., Markakis, E. K., & Sarigiannidis, P. (2024). Leveraging the power of internet of things and artificial intelligence in forest fire prevention, detection, and restoration: A comprehensive survey. *Internet of Things : Engineering Cyber Physical Human Systems*, 26, 101171. DOI: 10.1016/j.iot.2024.101171

Giannakidou, S., Radoglou-Grammatikis, P., Lagkas, T., Argyriou, V., Goudos, S., Markakis, E. K., & Sarigiannidis, P. (2024). Leveraging the power of internet of things and artificial intelligence in forest fire prevention, detection, and restoration: A comprehensive survey. *Internet of Things : Engineering Cyber Physical Human Systems*, 26, 101171. DOI: 10.1016/j.iot.2024.101171

Iqbal, U., Riaz, M. Z. B., Zhao, J., Barthelemy, J., & Perez, P. (2023). Drones for Flood Monitoring, Mapping and Detection: A Bibliometric Review. *Drones (Basel)*, 7(1), 1. Advance online publication. DOI: 10.3390/drones7010032

Jain, P., Coogan, S. C. P., Subramanian, S. G., Crowley, M., Taylor, S., & Flannigan, M. D. (2020). A review of machine learning applications in wildfire science and management. *Environmental Reviews*, 28(4), 478–505. DOI: 10.1139/er-2020-0019

Joseph, S. K. A., & Murthy, M. (2009). Forest fire in India: A review of the knowledge base. *Journal of Forest Research*, 14(3), 127–134. DOI: 10.1007/s10310-009-0116-x

Karimi, A., Eslamian, S., Nekooei, M., Ostad-Ali-Askari, K., Singh, V., Heidarian, A., Gholami, H., & Pazdar, S. (2021). Evaluating models and effective factors obtained from remote sensing (RS) and geographic information system(GIS) in the prediction of forest fire risk: A structured review. *Journal of Geography and Cartography*, 4, 40–48.

Kobziar, L. N., Hiers, J. K., Belcher, C. M., Bond, W. J., Enquist, C. A., Loudermilk, E. L., Miesel, J. R., O'Brien, J. J., Pausas, J. G., Hood, S., Keane, R., Morgan, P., Pingree, M. R. A., Riley, K., Safford, H., Seijo, F., Varner, J. M., Wall, T., & Watts, A. C. (2024). Principles of fire ecology. *Fire Ecology*, 20(1), 39. DOI: 10.1186/s42408-024-00272-0

Kumar, S., Tiwari, P., & Zymbler, M. (2019). Internet of Things is a revolutionary approach for future technology enhancement: A review. *Journal of Big Data*, 6(1), 111. DOI: 10.1186/s40537-019-0268-2

Mazzeo, G., De Santis, F., Falconieri, A., Filizzola, C., Lacava, T., Lanorte, A., Marchese, F., Nolè, G., Pergola, N., Pietrapertosa, C., & Satriano, V. (2022). Integrated Satellite System for Fire Detection and Prioritization. *Remote Sensing (Basel)*, 14(2), 2. Advance online publication. DOI: 10.3390/rs14020335

McLauchlan, K. K., Higuera, P. E., Miesel, J., Rogers, B. M., Schweitzer, J., Shuman, J. K., Tepley, A. J., Varner, J. M., Veblen, T. T., Adalsteinsson, S. A., Balch, J. K., Baker, P., Batllori, E., Bigio, E., Brando, P., Cattau, M., Chipman, M. L., Coen, J., Crandall, R., & Watts, A. C. (2020). Fire as a fundamental ecological process: Research advances and frontiers. *Journal of Ecology*, 108(5), 2047–2069. DOI: 10.1111/1365-2745.13403

Merino, L., Caballero, F., Martínez-de-Dios, J. R., Maza, I., & Ollero, A. (2012). An Unmanned Aircraft System for Automatic Forest Fire Monitoring and Measurement. *Journal of Intelligent & Robotic Systems*, 65(1–4), 533–548. DOI: 10.1007/s10846-011-9560-x

DOI: 10.1007/s10846-011-9560-x

Mishra, R. K., Agarwal, R., Mishra, R. K., & Agarwal, R. (2024). Sustainable Forest Land Management to Restore Degraded Lands. In *Sustainable Forest Management—Surpassing Climate Change and Land Degradation*.

DOI: 10.5772/intechopen.1004793

Nižetić, S., Šolić, P., López-de-Ipiña González-de-Artaza, D., & Patrono, L. (2020). Internet of Things (IoT): Opportunities, issues and challenges towards a smart and sustainable future. *Journal of Cleaner Production*, 274, 122877. DOI: 10.1016/j.jclepro.2020.122877 PMID: 32834567

Pandey, P., Huidobro, G., Lopes, L. F., Ganteaume, A., Ascoli, D., Colaco, C., Xanthopoulos, G., Giannaros, T. M., Gazzard, R., Boustras, G., Steelman, T., Charlton, V., Ferguson, E., Kirschner, J., Little, K., Stoof, C., Nikolakis, W., Fernández-Blanco, C. R., Ribotta, C., & Dossi, S. (2023). A global outlook on increasing wildfire risk: Current policy situation and future pathways. *Trees, Forests and People*, 14, 100431. DOI: 10.1016/j.tfp.2023.100431

Rajak, P., Ganguly, A., Adhikary, S., & Bhattacharya, S. (2023). Internet of Things and smart sensors in agriculture: Scopes and challenges. *Journal of Agriculture and Food Research*, 14, 100776. DOI: 10.1016/j.jafr.2023.100776

Ramadan, M. N. A., Basmaji, T., Gad, A., Hamdan, H., Akgün, B. T., Ali, M. A. H., Alkhedher, M., & Ghazal, M. (2024). Towards early forest fire detection and prevention using AI-powered drones and the IoT. *Internet of Things : Engineering Cyber Physical Human Systems*, 27, 101248. DOI: 10.1016/j.iot.2024.101248

Rethinking fire-adapted species in an altered fire regime—Tubbesing—2020—Ecosphere—Wiley Online Library. (n.d.). Retrieved August 4, 2024, from https://esajournals.onlinelibrary.wiley.com/doi/10.1002/ecs2.3091

Robinson, J. M., Aronson, J., Daniels, C. B., Goodwin, N., Liddicoat, C., Orlando, L., Phillips, D., Stanhope, J., Weinstein, P., Cross, A. T., & Breed, M. F. (2022). Ecosystem restoration is integral to humanity's recovery from COVID-19. *The Lancet. Planetary Health*, 6(9), e769–e773. DOI: 10.1016/S2542-5196(22)00171-1 PMID: 36087607

Salam, A. (2020). Internet of Things for Sustainable Forestry. In Salam, A. (Ed.), *Internet of Things for Sustainable Community Development: Wireless Communications, Sensing, and Systems* (pp. 147–181). Springer International Publishing., DOI: 10.1007/978-3-030-35291-2_5

Sample, M., Thode, A. E., Peterson, C., Gallagher, M. R., Flatley, W., Friggens, M., Evans, A., Loehman, R., Hedwall, S., Brandt, L., Janowiak, M., & Swanston, C. (2022). Adaptation Strategies and Approaches for Managing Fire in a Changing Climate. *Climate (Basel)*, 10(4), 4. Advance online publication. DOI: 10.3390/cli10040058

Sarkar, M. S., Majhi, B. K., Pathak, B., Biswas, T., Mahapatra, S., Kumar, D., Bhatt, I. D., Kuniyal, J. C., & Nautiyal, S. (2024). Ensembling machine learning models to identify forest fire-susceptible zones in Northeast India. *Ecological Informatics*, 81, 102598. DOI: 10.1016/j.ecoinf.2024.102598

Selvam, A. P., & Al-Humairi, S. N. S. (2023). The Impact of IoT and Sensor Integration on Real-Time Weather Monitoring Systems. *Systematic Reviews*.

DOI: 10.21203/rs.3.rs-3579172/v1

Shen, Y., Yu, Y., Lucas-Borja, M. E., Chen, F., Chen, Q., & Tang, Y. (2020). Change of soil K, N and P following forest restoration in rock outcrop rich karst area. *Catena*, 186, 104395. DOI: 10.1016/j.catena.2019.104395

Taye, M. M. (2023). Understanding of Machine Learning with Deep Learning: Architectures, Workflow, Applications and Future Directions. *Computers*, 12(5), 5. Advance online publication. DOI: 10.3390/computers12050091

Thangavel, K., Spiller, D., Sabatini, R., Amici, S., Sasidharan, S. T., Fayek, H., & Marzocca, P. (2023). Autonomous Satellite Wildfire Detection Using Hyperspectral Imagery and Neural Networks: A Case Study on Australian Wildfire. *Remote Sensing (Basel)*, 15(3), 3. Advance online publication. DOI: 10.3390/rs15030720

Theissler, A., Pérez-Velázquez, J., Kettelgerdes, M., & Elger, G. (2021). Predictive maintenance enabled by machine learning: Use cases and challenges in the automotive industry. *Reliability Engineering & System Safety*, 215, 107864. DOI: 10.1016/j.ress.2021.107864

DOI: 10.1016/j.ress.2021.107864

Tortorelli, C. M., Young, D. J. N., Reilly, M. J., Butz, R. J., Safford, H. D., Venuti, N. E., Welch, K. R., & Latimer, A. M. (2024). Post-fire resurveys reveal predictability of long-term conifer recruitment in severely burned California dry forests. *Forest Ecology and Management*, 566, 122100. DOI: 10.1016/j.foreco.2024.122100

Ursell, T., & Safford, H. D. (2022). Nucleation sites and forest recovery under high shrub competition. *Ecological Applications*, 32(8), e2711. DOI: 10.1002/eap.2711 PMID: 36161678

Vijayalakshmi, S. R., & Muruganand, S. (2017). A survey of Internet of Things in fire detection and fire industries. *2017 International Conference on I-SMAC (IoT in Social, Mobile, Analytics and Cloud) (I-SMAC)*, 703–707. https://doi.org/DOI: 10.1109/I-SMAC.2017.8058270

Wang, B., Spessa, A. C., Feng, P., Hou, X., Yue, C., Luo, J.-J., Ciais, P., Waters, C., Cowie, A., Nolan, R. H., Nikonovas, T., Jin, H., Walshaw, H., Wei, J., Guo, X., Liu, D. L., & Yu, Q. (2022). Extreme fire weather is the major driver of severe bushfires in southeast Australia. *Science Bulletin*, 67(6), 655–664. DOI: 10.1016/j.scib.2021.10.001 PMID: 36546127

Zeng, H., & He, B. (2023). Near Real-Time Wildfire Detection in Southwestern China Using Geo-Kompsat-2A Geostationary Meteorological Satellite Data. *IGARSS 2023 - 2023 IEEE International Geoscience and Remote Sensing Symposium*, 3391–3394. https://doi.org/DOI: 10.1109/IGARSS52108.2023.10283079

Chapter 10
Modernizing Wildfire Management Through Deep Learning and IoT in Fire Ecology

V. Valarmathi
Sri Krishna Arts and Science College, India

J. Ramkumar
https://orcid.org/0000-0001-9639-0899
Sri Krishna Arts and Science College, India

ABSTRACT

The increasing frequency and severity of wildfires present critical challenges to ecosystems, human safety, and property and underline the inefficiency of traditional methods of fire detection and management. This chapter will present how integration between DL and IoT could give way to a revolution in fire ecology by providing innovative tools for real-time fire prediction, detection, monitoring, and response. DL, in particular, through Convolutional and Recurrent Neural Networks, looks into terabytes of data ranging from historical fire data to weather patterns and topography to predict and assess wildfire risks. IoT aids this with real-time data that emanates from a network of sensors, drones, and cameras spread across susceptible areas. This synergy therefore offers DL and IoT more accurate, timely, and proactive fire management. Future technologies will focus on 5G, blockchain, and advanced robotics for more resilient fire management strategies.

DOI: 10.4018/979-8-3693-7565-5.ch010

1. INTRODUCTION TO MODERN WILDFIRE CHALLENGES

The outbreaks of wildfires are becoming urgent problems for ecosystems, human settlements, and even the global stability. Over the last two years, wildfires have increased in occurrence, intensity, and well, naturally resulting through the successful course in disastrous impacts both on the natural environment and on social man-made structures. Of the reasons, the most important factors are changes in unsettled climatic patterns, land usage, and shifts in vegetation management (Qayyum et al., 2024). The interaction of these elements creates conditions whereby wildfires easily occur, leading to more frequent, more destructive incidents.

The challenges that wildfires pose at present time require innovative solutions to traditional methods of management, which have been largely based on manual observation, elementary techniques of firefighting, and analysis of historical data. Real-time, correct information, and predictive capability are what contemporary wildfire management needs. This chapter introduces the current impacts of wildfires on ecosystems and societies, the shortcomings of existing fire detection and management systems, and technological advancements redefining this field (Zander et al., 2023).

1.1. Current Impact of Wildfires on Ecosystems and Societies

Wildfires have huge effects on landscapes and human societies. The ecological impacts of wildfires do not stop there with the simple loss of vegetation. Fires may degrade the soil, contaminate water, and cause loss of biodiversity, which unnaturally unbalances whole ecosystems. In most areas, fires actually play a very useful role by cleaning up accumulations of dead vegetation and giving way to the regeneration of new growth. However, with the upward trend in wildfire severity, these natural processes are now being pushed to an extent where it would overwhelm them and cause permanent ecological harm.

One of the worst fire effects is erosion. The heat from fire changes the structure of the soil, making it less capable of absorbing water and other nutrients needed to sustain vegetation. This degradation of topsoil could mean a loss of fertile land, which is usually hard to recover. Erosion may cause sedimentation in rivers and streams, which can harm aquatic ecosystems and impact the quality of water. Added to this is the loss of vegetation, which further aggravates the matter by having a lack of roots from plants make the soil so unstable that it can easily cause landslides or other forms of degradation (Rodriguez-Jimenez et al., 2023).

Apart from their ecological impacts, wildfires present a significant threat to human societies. Enormous economic losses result from the destruction of homes, infrastructure, and agricultural land, often amounting to billions of dollars. Of primary concern is the immediate threat to human life; annually, countless deaths and injuries are incurred due to wildfires (Groenenberg et al., 2023). The more profound psychological and social impacts of wildfires are communities that undergo trauma, displacement, loss of property, and disruption of livelihoods.

Wildfires also lead to air pollution, mainly from smoke and particulate matter that find their way into the atmosphere. Health problems directly related to such pollution are of a critical nature, more so to susceptible populations: children, the elderly, and those with pre-existing conditions related to respiratory health. On this consideration, long exposure to wildfire smoke would mean chronic health problems and therefore an added burden to health systems (Zhu et al., 2022).

The economic consequence does not only lie within the immediate cost of firefighting as well as the recovery drive. The loss of agricultural and forest land values might have perpetuation effects on the local economies, especially in regions that heavily depend on such industries. Added to this is transportation and energy infrastructure, as well as water supply systems, all of which have become seriously disrupted, further exacerbating the problems which these communities face economically.

Further, wildfires also feed global climate change by releasing huge volumes of carbon dioxide into the atmosphere. Conversely, loss of forests worsens this still further by acting as carbon sinks. This forms a feedback loop in which climate change raises the chances of wildfires and these, in turn, cause more climate change—a cyclic process. It is within this cycle that wildfires have important global implications, not just for localized regions in which fires actually occur, but also for environmental and climatic systems more broadly (Tunstill et al., 2022).

1.2. Shortcomings of Existing Fire Detection and Management Systems

Contemporary fire management techniques have, however, embraced so many changes but are still having substantial limitations to seriously undermine the need for effectiveness against modern wildfire challenges. Traditional methods of fire detection, like lookout towers and ground patrols, are labor-intensive and, most of the time, they also experience delayed response times since they depend on human observation, which has inherent errors and limitations, especially in a remote or heavily forested area.

At the same time, there has been some progress in the development of satellite-based detection systems that guarantee broader coverage and provide the possibility of detecting fires in vast areas. Satellite-based detection also has its inherent limitations. Cloud cover, smoke, and other atmospheric conditions affecting visibility to the ground are constraints for satellite imagery. The time intervals between passes also place delays on detecting new fires, which allows them to grow larger and more difficult to control before being identified.

This is further complicated by fire behavior being very unpredictable. Wind speed, humidity, and the amount of combustible material can all be responsible for spreading a fire quickly and can change direction on short notice. Traditional models of fire prediction—that is, those relying on historical data and static variables—oftentimes fail to account for these dynamic factors, leading to bad predictions and ineffective responses as a result (Platte & Torland, 2023).

Another major challenge to come to grips with is agency coordination. Wildfire management usually involves many organizations in general, which can range from local to states and federal levels, usually with different protocols and communication channels. The lack of standardized communication channels and data-sharing mechanisms can result in delays and inefficiencies, therefore hindering effectiveness in wildfire management.

Resource constraints also hobble the system of fire management. Resources, be it human, equipment-wise, or financial, are always overstretched, especially during high fire seasons. Allocation of these resources was made complex because prioritizing will be done in high-risk areas, a move which may leave other regions open (Villalba-Briones et al., 2023).

Another critical shortcoming of the present systems is that they predict fires and assessing risks based on historical data. In this regard, though past information on fires was able to give the trends, it may not represent the current and future risks more so considering the changing climatic conditions. Increasing variability of weather patterns and expansion of human settlement into fire-prone areas create new challenges ill-matched to historical data.

1.3. Technological Advancements: A Paradigm Shift

New paradigms characterizing the recent technological breakthroughs in the management of wildfires offer rather innovative solutions to challenges raised by traditional systems. Integration of advanced technologies, from Artificial Intelligence down to Machine Learning and Internet of Things, is able to revolutionize fire detection and prediction strategies and change the management paradigm.

Artificial intelligence and machine learning technologies can process huge volumes of data that are complex in nature. Applications in wildfire management involve the analysis of vast data amounts, which extends from weather patterns and vegetation types to topographic features to understand what might predict the likelihood and potential spread of wildfires. AI-driven models learn ceaselessly, and with more data added, keep improving on their accuracy (N. et al., 2024).

Real-time data provided by the Internet of Things improves the functionality for more precise predictions of natural disasters. The deployed sensors would include temperature and humidity-monitoring sensors, cameras, and drones to monitor the environment on a continuous basis in fire-prone areas. This should feed the system with real-time data, which can be used by the AI models to predict the fires as real-time and effective prediction inputs.

IoT technology has the prime advantage of monitoring large and remotely located areas that otherwise cannot be accessed through conventional means of surveillance. High-resolution data capture from regions inaccessible otherwise can be achieved by sensor- and camera-fitted drones, providing crucial information for fire detection in its early stages. Satellite images, together with IoT data, will give a view of fire dynamics at an overall level, enhancing situational awareness and hence strategic decision-making (García-Aguilar et al., 2023).

Using edge computing in collaboration with IoT technology improves the speed and efficiency of data processing. Within edge computing, information is handled closer to its source; this again reduces the duration it takes for information to be transmitted to some central server. In this respect, this approach will turn out to be particularly useful in wildfire management, wherein decision-making in real-time is extremely imperative for effective response efforts.

Advanced technologies have overcome many of the coordination challenges that have hampered the ability to manage wildfires in the past. IoT devices will help with seamless communication between various agencies and stakeholders by ensuring that all of them get to have the same real-time information. In this regard, improved coordination enhances the efficiency of firefighting efforts through reduced redundancies and allows for a more unified response to wildfire incidents.

Further evolution of these technologies may result in further steps forward made in wildfire management. Development of next-generation communications networks, including 5G, improves the power of IoT devices by facilitating faster and more reliable connections. This will enable the deployment of more sophisticated sensor networks that monitor a wide range of environmental variables in real time.

The integration of blockchain technology provides solutions to handle decentralized data securely while ensuring integrity and confidentiality of critical information. Advanced robotics, in particular, independent firefighting robots, is another promising avenue for innovation in wildland fire management. Such robots, equipped with sensors and AI algorithms, would be able to operate in hazardous conditions while keeping human firefighters out of harm's way and accomplishing fire suppression more precisely and effectively.

These advanced technologies bring into play a wonderful change in strategy with regard to the conduct of wildfire management. Making use of AI, IoT, and other innovative tools can mitigate to a great extent the inefficiencies within the traditional system and establish more proficient methodologies for the prediction, detection, and management of wildfires. This new paradigm can be used to handle present challenges and also offer an opportunity to project and take early mitigation measures against future risks, thus paving the way for a resilient and adaptive fire management approach.

2. DEEP LEARNING: UNPACKING ITS ROLE IN FIRE ECOLOGY

The complexities and unpredictability of wildfires often work as barriers to prevention and management. Deep Learning, being a subset of Machine Learning, is so advanced that it now holds the capability of taking up such challenges in the analysis of large datasets and generating accurate predictions. DL assists in the identification of complex patterns within environmental data and thus enhances the capability of coming up with more reliable predictions related to the occurrence, behavior, and spreading of wildfires pertaining to fire ecology (Ganapathy & Ajmera, 2024). The chapter is devoted to the basics of Machine Learning and Deep Learning and how those technologies are being used in fire ecology. It discusses which neural network architectures and algorithms have been shown to be effective for processing environmental data, principally regarding the application of this field in fire prediction and management.

2.1. Fundamentals of Machine Learning and Deep Learning

Machine Learning is the branch of Artificial Intelligence that allows computers to learn from data to make decisions or predict without any explicit programming on any particular tasks. In essence, it basically means developing algorithms that identify patterns in data and become more proficient in performing the pre-defined objective as they are more widely exposed to information. There are quite a number of ML types: supervised learning, unsupervised learning, and reinforcement learning. This is facilitated by a dataset with labeled inputs associated with the corresponding outputs that help the model learn the relation for input-to-output mapping. This type of learning is majorly employed in classification problems like recognizing images or detecting spam. Unsupervised learning deals with data that is not labeled, and one has to find the hidden patterns or structure in one's data. Commonly, one applies unsupervised learning using techniques such as clustering, which makes similar data points grouped together, or dimensionality reduction for the simplification of the complexity of data. Another type of ML is reinforcement learning, in which an agent learns to make decisions on its own initiative, based on interaction with an environment that signals back some form of reward or penalty (Shehab et al., 2024).

Deep learning is a niche segment of ML, using ANNs to model complex relationships in data. Unlike traditional models in ML, DL models can execute self-learning and representation of data, which no other techniques in machine learning can without the need for extraction of features. The term "deep" refers to the use of many layers of neurons in the neural network, with every consequent layer transforming the data in a more abstract way. For this reason, DL models are extremely good at picking out complex patterns and propelling effective predictions for such complex tasks as image classification and speech recognition, recently, environmental monitoring. A neural network in DL includes an input layer, one or more hidden layers, and an output layer. In the case of these hidden layers, they are composed of neurons that perform mathematical operations on the input. Multiplication of each input by a weight is done with the addition of bias to it before passing the result through an activation function. These functions introduce non-linearity into the model and allow it to learn and represent complex patterns. It is trained through backpropagation, an algorithm that changes the weights and biases in a way that lowers the difference between what the network predicts and what actually happens in the training data.

2.2. Critical Analysis of Neural Networks for Environmental Data

Neural networks, especially deep neural networks, can be very effective in the processing and analysis of environmental data, since they are able to model non-linear relationships and deal with large and complex datasets. Environmental data, as usual in fire ecology, ranges from meteorological and vegetation variables to satellite image and topographical features. It makes neural networks fit for wildfire prediction and management, given their ability to integrate diverse information and process it. Among these, in the context of wildfire management, Convolutional Neural Networks are good at analyzing spatial data. It has been shown that CNN is successful with visual data containing patterns, which makes this technique very applicable to satellite images, aerial photography, and other spatial data. In fire ecology, CNNs could identify potential fire hotspots, monitor changes in vegetation, and evaluate the potential spread with regard to terrain and other spatial factors. The major advantage here is that with as little manual preprocessing as possible, CNNs can automatically learn relevant features from raw data, especially in cases where relevant information needs to be extracted and delivered posthaste (Ganapathy & Ajmera, 2024).

RNNs, particularly LSTMs, are rather quite suitable for the analysis of temporal data, a main part in the development and spread of wildfires over time. RNNs process sequences and maintain a memory of previous inputs; thus, they are rather quite perfect at predicting time series. This ability is quite important in fire ecology, in which one needs a full understanding of how a fire will evolve from its historical data and current conditions in the development of effective response strategies. One of the reasons is that an LSTM network, being a specialized form of RNN, represents an approach to learning the long-term dependencies existing in the data and thus has a potential to be more accurate at predictions over more extended time periods. This makes LSTMs highly valuable in scenarios where predicting the future behavior of a wildfire, such as its direction and speed of spread, can save lives and property. However, while neural networks offer significant advantages in processing environmental data, they also come with challenges. One major limitation of the process is that huge amounts of labeled data are needed for training the proper models. Most environmental data are unstructured and unlabeled, elaborating preprocessing and annotation efforts so that they can be used to train neural networks. Besides, neural networks are computationally intensive; training requires a lot of processing power, and also time, especially when dealing with deep networks with many layers. Another challenge is the interpretability of neural networks. Such models are often referred to as "black boxes" because their inner workings are not that easy to grasp for humans. This raises problems regarding transparency, which can eventually become an important issue in environmental applications where the reason for a certain prediction of the model has to be known for decision-making(Ullah et al., 2023). Despite all odds, efforts in this regard are still on to make the models more interpretable and neural networks more efficient for environmental data analysis.

2.3. Specific Algorithms and Their Efficacy in Fire Prediction

Among the algorithms of deep learning, some have been proved effective on wildfire prediction and management; different capabilities can be applied to different parts of the task. The Convolutional Neural Network is the most used algorithm in fire prediction, more specifically when dealing with spatial data. These nets are designed to automatically detect patterns across images; therefore, they are quite suitable for satellite images and other visual data analyses in fire ecology. These networks can learn early signs of wildfire activities, like changing vegetation patterns or even smoke, and further assess possible fire spread according to lay. This is very useful in the domain of wildfire management, in which early detection and rapid response are critical. CNNs can process a large amount of spatial data quickly and with great accuracy (Vidal-Riveros et al., 2023).

Specifically, models of RNNs—Long Short-Term Memory (LSTM) networks have been found to predict rather well in simulating temporal dynamics for wildfires. As it was designed to remember long-term dependencies in sequential data, the LSTMs are really good at tasks concerned with time series prediction. For instance, in fire ecology, LSTM networks will run historical data, weather forecasts, and real-time sensor inputs and output information on how a wildfire will progress over time. This capability forms the root of early warning systems and informed firefighting strategies. The ability of the LSTM networks to model complex dependencies of factors, such as wind speed, temperature, and humidity, among others, involved in the fire behavior, results in the ability to generate more accurate predictions of fire spread and intensity (Shams et al., 2024).

The other deep learning algorithm toward environmental anomaly detection is represented by the autoencoder. Autoencoder is an unsupervised model in learning. The process follows the procedure where the input data is compressed toward the lowest possible dimensions; then, the compressor seeks to reconstruct the initial data out of this compressed form. Through the matching of reconstructed data with the original, the autoencoder detects anomalies, such as a fire outbreak that was not expected from normal patterns of incidence. This makes this technique useful for early detection of signs of wildfires in very large data sets, for which traditional techniques miss subtle indicators of fire activity. Specifically, autoencoders can be more useful in monitoring these vast and remote areas, for which manual observation is not feasible.

GANs have also been explored in the fire prediction and simulation domains. Basically, GANs involve two neural networks: a generator and a discriminator, both working adversarial with each other. A generator generates synthetic data, and a discriminator tries to classify the data as either real or synthetic. Such an adversarial process can be used to generate realism-based simulations of wildfire scenarios, further usable either in the training of predictive models or in the studies on the behavior of fires under different conditions. GANs are definitely a new frontier in fire prediction since this approach allows researchers now to create virtual environments that may be used to test and improve fire management strategies (Rathod et al., 2023).

Hybrid models for fire prediction are under development, where various neural networks are combined. These models use pros brought about by different algorithms to increase prediction accuracy and offer more comprehensive insights regarding wildfire behavior. For instance, a hybrid model could make use of a CNN to analyze spatial data from satellite images and an LSTM network to include temporal data like weather forecasts. Hence, such hybrid models will be able to provide a complete view of the factors affecting wildfire activity by combining these different types of networks and achieve more efficient prediction and management strategies.

As such, deep learning algorithms for fire prediction and management substantially rely on both the quality and quantity of the input data. For example, if a model has incomplete or inaccurate data, poor performance can be experienced. Hence, reliable sources of data and robust preprocessing techniques are very important. Since deep learning models are normally complex, the tuning of parameters like learning rates and layer configurations is paramount in arriving at optimum results. In spite of all odds, further development and improvement of deep learning algorithms are in the pipeline and hold immense potential for better prediction, early detection, and management of wildfires, providing more effective and timely responses to events that are happening ever more frequently and becoming ever more serious.

3. ADVANCED PREDICTIVE MODELS USING DEEP LEARNING

Advanced predictive models, using deep learning, are a big step change in the ability to predict, monitor, and manage wildfires. The models analyze hard data classes for the purpose of prediction and are capable of accurate, efficient, and real-time decision-making processes. Such models are suited for creating full outputs of wildfire behavior since they can integrate various neural networks such as Convolutional Neural Networks for spatial analysis and Recurrent Neural Networks for temporal prediction. This section discusses the design and training of CNNs for the spatial analysis of fires, the use of RNNs in tracking the development of fire over time, and the application of these models in high-risk areas in real life (Cheung et al., 2023).

3.1. Designing and Training CNNs for Spatial Analysis of Fires

CNNs are among the most critical components of deep learning models in relation to spatial analysis, especially on applications requiring the interpretation of visual data. They are applied to analyze satellite images and other spatial data sources with aerial photographs in an effort to recognize patterns and predict fire behavior. A CNN designed for this purpose would have a few major components: some convolutional layers for detecting local patterns in the data, some pooling layers for reducing the dimensionality of the data while keeping their essential features intact, and then the fully connected layers that at last integrate features extracted by the convolutional layer to make a final prediction (Bilgiç et al., 2023).

Training a CNN for fire prediction involves feeding the network large datasets of labeled images, where each image is associated with information about fire presence, location, and intensity. It learns to recognize patterns that are characteristic of fires by tuning its internal parameters so that its predictions are most similar to the ground-truth labels provided in the training data. This is supervised learning, and it is both time- and power-hungry—especially for large, high-resolution images.

To enhance the accuracy of the CNN, several techniques can be employed during the training process. Examples are data augmentation, where the diversity of a given training set is increased by creating modified versions of the training images through rotation or scaling; this improves generalization on new data. But probably the most vigorous of all is transfer learning: Fine-tuning a CNN already trained on a large dataset for the specific dataset of interest. This allows knowledge to transfer from other tasks to the model and can thus reduce both a dataset size used for training and computational resources required.

The effectiveness of the CNN also lies in how it is designed with its architecture. In fire prediction, deeper networks with more layers would be required to extract minute details pertaining to fire patterns across different types of terrain and vegetation. However, the deeper a network is, the more prone it is to overfitting—a case in which a model is excessively specialized to the training data and hence performs badly on new data. These risks can be reduced using regularization techniques such as dropout and weight decay to ensure the model will remain robust and perform well under real-world scenarios.

3.2. RNNs for Temporal Prediction: Tracking Fire Development Over Time

RNNs are highly specialized in solving the problems that involve sequence data. On the other hand, the nature of studying wildfires undoubtedly requires this functionality—the aspect of development over time. RNNs differ from other conventional neural networks due to their nature of inheriting memory capabilities. This makes them capable of retaining past information from inputs while processing the incoming data. In fire prediction, this is very important because how a fire spreads and changes in intensity, reacting to environmental factors, can be an important factor while predicting its behavior.

Perhaps the most used RNN network in wildfire prediction is the Long Short-Term Memory (LSTM) network. LSTM networks were designed properly to counter the limitations of the conventional RNNs due to their architecture. In particular, the use of the vanilla RNN brought the vanishing gradient problem into the learning process for long sequences. LSTMs can learn these dependencies in the training data which make them effective in modeling the complex and long-term processes involved in wildfire dynamics.

An LSTM network involves putting in a sequence of data points, for instance, temperature, wind speed, humidity, and past fire locations, to conduct a process on predicting fires. The network is learning to predict the next state of the fire from the current conditions but also from how those conditions have developed. This yields the capability to model temporal dependence, and hence, LSTMs make rather accurate predictions for the spread, direction, and intensity of a fire.

It is possible to further improve the accuracies of the RNNs in tasks related to fire prediction by including a variety of features that drive the behavior of fire. These may include meteorological data, vegetation indices, and topographical information—features supporting the movement of the fire. This addition of the different data sources enables an RNN to learn complex interactions between factors that drive wildfire dynamics.

The extent of layers and size of the memory cells are parameters that need to be fine-tuned in the design of this LSTM architecture itself for the model's complexity with respect to its performance. Too complex models will result in overfitting, and simple ones may not capture all the relevant patterns. Overall, techniques for regularization like recurrent dropout are pretty good at handling balance by preventing some features from or the overfitting of single data points.

3.3. Case Applications: Real-time Predictive Modeling in High-Risk Areas

In high-risk wildfire areas, the stakes are highest and the need for accurate, timely predictions is great; real-time predictive modeling based on deep learning is being applied. One of the most striking applications of these models is in regions like California and Australia where wildfires are frequent and often devastating. These deep learning models are integrated in these regions with a mesh of IoT devices, spanning from weather stations to drones to satellite sensors, that are continuously monitoring environmental conditions in order to obtain real-time wildfire occurrence predictions.

In California, for example, a deep learning-based predictive model is deployed on top of a sensor network that tracks key indicators of wildfire risks, which are temperature, humidity, and wind patterns. Also using CNNs in the satellite image classification domain, it detects indications of early fire activities such as smoke or heat signatures. Meanwhile, an LSTM network keeps track of changes in such indicators over time to establish how a detected fire can evolve at any particular moment according to current and historical data. This real-time modeling capability will therefore allow firefighting teams to be alerted and more effectively deploy resources to ensure that small fires do not develop into catastrophic ones (Zald et al., 2022).

Wildfire prediction with deep learning has also been embraced by Australia, which required methods applied in its vast and often remote landscapes. Deep learning models come into play, together with the drone-based surveillance systems, to guarantee continuous coverage in areas like the Outback, wherein the application of traditional monitoring methods might be challenging. High-resolution data are captured by infrared cameras mounted on drones, monitoring temperature and other environmental parameters, further fed into a CNN to identify possible fire hotspots. Consequently, these fires are then predicted by LSTM networks for spread and development, providing valuable insight that guides emergency response efforts.

These real-world applications underline deep learning's potential to transform wildfire management. Real-time, accurate prediction, with the ability to respond to fires before spiraling out of control, ultimately saves lives, property, and natural resources. Continued improvement and extension of these models, in conjunction with improved computing performance and data availability, will further increase the effectiveness of wildfire prediction and management in high-risk areas. The success of these applications underlines what deep learning can do to help meet one of humanity's most pressing environmental challenges.

4. IOT FRAMEWORKS FOR FIRE DETECTION AND RESPONSE

IoT has revolutionized environmental monitoring and disaster response, particularly in the aspect of fire detection. IoT frameworks build a strong structure for the real-time collection, processing, and decision-making of data in real instances. Therefore, responses to wildfires are improved and better done in a timely manner. IoT frameworks integrate massive sensor networks, communications, and data processing units to provide an uninterrupted assessment of the environmental conditions in regions that are prone to fires. This section will take into consideration IoT architectures for environmental sensing, how sensor networks will be placed strategically in regions vulnerable to wildfires, and mechanisms that ensure seamless integration of data for real-time response.

4.1. Architectures of IoT for Environmental Sensing

The architecture for an environmental-sensing IoT system is a multi-layered framework where several sensors and devices are connected to a central system that makes the continuous collection and analysis of environmental data possible. At the very center of the architecture is the sensor layer. It includes a wide array of devices making up diversified sensors in temperature, humidity, wind speed, and air quality measurement. All these should be installed at different locations across the landscape so as to get data at various points, thus offering an integrated view of environmental conditions in any specific location.

The other important component in the IoT architecture is the communication layer. On the other hand, the middle layer allows sensors to communicate data collected with the central processing units majorly using wireless communications technologies that are based on Zigbee, LoRaWAN, and cellular communications. The type of protocol to be used in communication, however, depends greatly on the distances between sensors and rates of data transfer and consumed power. Mesh networks or even satellite communication can form communication in a remote area or rugged terrains where one cannot establish a robust communication network (Ram Prasanna et al., 2023).

It is the layer that does the aggregation, processing, and analysis of the raw data from sensors. Such edge computing devices would have some initial processing done closer to the sensors in order to try to reduce latency and the amount of data that has to be sent to central servers. Edge computing comes in very handy in detecting fire anomalies, which timely identification—like a sudden surge in temperature or air composition—is critical to trigger an early warning. The processed data is then transmitted to servers in the cloud or to central control systems that have more advanced functionalities, such as predictive modeling and trend analysis.

The last IoT architecture constituent is the Application Layer. It is in this component that the processed data is exposed to the end-users through a number of diversified interfaces such as dashboards, mobile apps, or even an automated alert system. This layer may incorporate decision-support tools that enhance the authority's ability to interpret and hence make informed decisions on whether to deploy fire resources or issue public warnings. This application layer is user-friendly to enable access to critical information in real-time and make real-time decisions (García-Valls & Chirivella-Ciruelos, 2024).

A good architecture for IoT in environmental sensing will have these layers seamlessly integrate to provide an end-to-end flow of information from the sensor network to the decision-makers. This architecture has to support scalability in terms of additional sensors and growth of a network as time goes on. Moreover, it has to be resilient via redundant paths of communication and backup power sources in case bad events happen. Another important consideration for any IoT-based solution is that of security; an IoT system of this kind is going to protect sensitive environmental data against unauthorized access or tampering.

4.2. Strategic Deployment of Sensor Networks in Fire-Prone Zones

In essence, IoT frameworks in regard to fire detection and response are effective only based on how well sensor networks are strategically deployed in fire-prone zones. Accordingly, the emplacement of sensors has to be carefully planned for maximum coverage of high-risk areas, also considering that data being gathered from such areas is relevant and actionable. Topography, vegetation density, historical data on fires, and prevailing weather conditions are aspects which play vital roles in deciding where sensors will be located.

In the case of mixed landscapes, including forests, mountains, and urban-wildland interfaces, a suite of sensors will likely be required to effectively monitor the various environmental conditions. For example, on-ground temperature and humidity sensors in heavily forested areas are augmented by tower- or drone-mounted infrared cameras that detect heat signatures indicative of fires. Wind speed and direction sensors are important in mountainous areas, as these inform predictions of how fires might spread across ridges and valleys. On the other hand, in an urban-wildland interface where human activity is heightened, increasing the chances of a fire ignition, air quality sensors detecting smoke particles and toxic gases may alert quite early to outbreaks of fire.

It should also consider the element of accessibility and maintenance of the sensors. The application of autonomous sensor-carrying drones in challenging or inaccessible locations can offer a flexible and mobile solution. Such drones can easily cover large areas, acquiring data and flying back to respective base stations for recharging and transmission. On the other hand, fixed sensors need to be located at easily accessible places in order to do regular maintenance and battery replacement to enable them to continue working throughout the fire season.

In order to make a sensor network effective, the deployment strategy has to include redundancy. This is about increasing the number of sensors in order to capture data in case some of them fail, especially in sensitive locations. This approach is important in fire-prone zones, where equipment failure might lead to missed warnings. This is further reinforced by mesh networking, whereby each sensor node will be able to communicate with more than one other node, adding further reliability that reroutes the data in case of a sensor or communication link compromise.

This will also call for strategic placement of power sources within these sensors. Solar panels and wind turbines could offer renewable energy in keeping sensors running, particularly when they exist in remote areas where there is little or no access to the power grid. Low-power wide-area network technologies, like LoRaWAN, normally come into play in scenarios like these because they need very minimal power for sending data over a long distance. In this respect, battery life and power management become important for such sensors, so that they can at least make it through the fire season, thus reducing maintenance.

The data gathered by strategically deployed sensors provides the base for real-time monitoring and early warning systems. IoT frameworks can assure continuous, high-quality data coverage fundamental for effective fire detection and response through full area coverage prone to fires and including redundancy and reliability in the deployment strategy.

4.3. Data Integration and Real-Time Response Mechanisms

The critical components in tying a sensor data collection to actionable wildfire management would be data integration and mechanisms of real-time response. During data collection via a network of IoT sensors, integration becomes the challenge of putting together this data into some meaningful framework that will facilitate real-time analysis and decision-making. Basically, the integration process involves the collection of data from various sensors, standardization of data, and making available the same to different kinds of systems and personnel related to fire detection and response (Alghamdi et al., 2022).

Effective data integration starts with the aggregation of sensor data in a centralized or distributed data storage system. Cloud-based platforms are typical in this case, providing scalable storage and powerful processing capabilities able to cope with large volumes of data. Such platforms support integration of several types of data, numerical sensor readings, image, and video feeds, among others, in a way that allows for their comprehensive analysis. At this stage, standardization is very critical. It not only ensures comparability of data from the various sensors and sources but also presents them in consistent ways for further analysis. Standardization involves rendering the data into uniform format, applying calibration factors against them, filtering out noise, irrelevant information, etc.

All in all, real-time response mechanisms are based on the continuous monitoring of integrated data streams in order to identify early signs of fire activity. Advanced algorithms and machine learning models are applied to the data in order to detect patterns related to wildfire, which may include sudden temperature rises, changes in wind patterns, or the sighting of smoke particles. These algorithms usually run on edge computing devices or cloud servers and process incoming data for the discovery of anomalies in real time and with the lowest possible latency.

Upon detecting a potential fire, the IoT framework should trigger an instant response. This may include automated alerts to firefighting teams, local authorities, and probably affected communities. Most alerting systems are designed to prioritize the most critical information, such as localization of the fire, estimated size, and potential spread based on prevailing weather conditions. These warnings, in most cases, are coupled with a decision support system that guides the allocation of firefighting resources, whether it is ground teams, aerial firefighting units, or evacuation protocols.

It may also trigger automatic response mechanisms of the IoT framework in very high-risk scenarios. For example, in areas that have fire suppression systems, the event of a fire detection can trigger sprinklers or fire-retardant dispersal systems or other containment measures; drones with water or fire retardant can independently be sent to the scene of the fire and perform an immediate response in otherwise inaccessible locations by ground teams.

It allows for better situational awareness among decision-makers in terms of real-time data capture from multiple sources. The authority will be allowed to track a wildfire in real-time through real-time dashboards and visualization tools, with data updated live as new sensor readings come in. The tools provide a holistic view of the progression of a fire to the decision-makers for risk assessment and planning of interventions. Real-time integration and analyses for data coming from multiple sensors have become very critical in making informed decisions to mitigate the impact of wildfires and save lives and property.

In other words, IoT frameworks on fire detection and response are founded on strong architectures in environmental sensing, strategic deployment of the sensor networks, and effective data integration and response mechanisms. All these elements work together to offer an inclusive, real-time system capable of detecting wildfires and subsequently responding with unprecedented speed and accuracy. Especially considering the continuous development of these technologies, the general effectiveness of efforts for wildfire management will be raised, acting better in protecting natural environments and human communities (Kim et al., 2023).

5. IOT AND DEEP LEARNING SYNERGY FOR ENHANCED MONITORING

The integration of the Internet of Things and Deep Learning stands as a powerful synergy to monitor, detect, and manage wildfires. This sets up a collaboration between the real-time data acquisition capabilities of IoT and the advanced analysis powers of DL in making more accurate and timely predictions to improve response strategies. Going further, it fuses cross-platform data, flying drones, and automated systems for thorough data collection to maximize its flow and processing; the synergy ushers in a very strong framework toward handling the complexities of wildfire management. This section discusses how all these factors, when put together, provide a more efficient and responsive system for wildfire monitoring.

5.1. Cross-Platform Data Fusion Techniques

Cross-platform data fusion combines data from multiple IoT devices and platforms into an integrated framework suitable for advanced analytics and decision-making processes. In this regard, data fusion is equally followed in wildfire monitoring to help in integrating information from ground sensors, satellite images, weather stations, and drone surveillance. All these platforms deliver different types of data, and integration provides an all-inclusive understanding of the environmental conditions and fire behavior.

The data fusion process actually commences with raw data originating from heterogeneously networked IoT devices that solely monitor specific parameters relevant to detection and management of fire. Off-the-ground gadgets are like temperature, humidity, and air-quality monitors, while satellite systems offer images at the broad scale that capture vegetational and infrastructure characteristics. Drones carrying cameras and infrared sensors could offer high-resolution images in real time on some otherwise hard-to-reach areas.

Data, therefore, which is collected from these arrays, has to be standardized and aligned to enable analysis collectively. This will involve data format normalization, synchronization of time stamps, and spatial coordinate alignment. The heterogeneity at issue would be across data sources because devices may use different scales, frequencies, and resolutions. It is for this reason that sophisticated algorithms that can hold their own against such discrepancies are required for effective data fusion, ensuring that all relevant data has an accurate representation in this final dataset.

Subsequently, the fused data passes through Deep Learning models, which are specifically designed to detect any patterns in it and thus allow for making predictions on this integrated dataset. For instance, Convolutional Neural Networks may check fused satellite and drone imagery for fire hotspots, and Recurrent Neural Networks trace temporal patterns across a large number of data streams for fire behavior prediction. It is owing to the combination of these different data sources that, in this approach, what is achieved is multi-dimensional analysis that captures both spatial and temporal dimensions of wildfires to get more accurate and timely predictions.

The merits associated with cross-platform data fusion in monitoring wildfires are very relevant. Since data is sourced from multiple sources, it can detect early signs of wildfires that one single platform may miss out on. It delivers a finer and more comprehensive view of the environment, thus supporting improved decisions related to fire management and resource allocation. This further enhances the resilience of the system, as failure in one data source will be compensated for by the continuous operation of the others, ensuring that monitoring and detection capabilities remain robust under trying conditions.

5.2. Role of Drones and Automated Systems in Data Acquisition

In wildfire monitoring, drones and mechanized systems are quite indispensable at scale for interfacing with data and, therefore, allow unparalleled access to real-time information from hard-to-reach zones. The technologies augment conventional ground-based and satellite monitoring systems by providing high-resolution real-time data that can be used to analyze fire conditions and behavior instantly.

Equipped with high-definition cameras, thermal imaging, and LIDAR, drones are capable of flying over fire-prone areas and extracting detailed images and data. Since they can fly at low altitudes, they can acquire data that is close-up and precise; this is important for monitoring small-scale fire activities that satellites would be unable to detect. For instance, thermal cameras mounted on drones could detect heat signatures from fires even through smoke and provide a situational understanding of the location and size of a fire (Shiddiqi et al., 2023).

The flexibility and mobility of drones make them perfect for rapid, immediate deployment in emergency situations. Drones could quickly be launched to assess fire conditions in areas that ground crews have not reached or are unsafe to enter. This will allow situational awareness in real-time, helping guide decision-making in the very critical stages of a fire.

It extends the abilities with continuous, self-sustained operation. For example, drones fitted with AI-based navigation systems can be pre-programmed to automatically scan particular areas following regular surveys by sending data back to a central monitoring station. These systems work nonstop, removing much of the workload from emergency personnel and ensuring data collection is never stopped. Automated drones may also include self-charging and hence be able to return to base stations for recharging before turning out again to maintain persistent surveillance over long periods.

Integration of drones and automated systems with IoT networks enhances the efficiency of wildfire monitoring. Data that a drone collects can be streamed in real-time to IoT platforms, where it will be analyzed in conjunction with data from ground sensors and satellites. Such an integrated approach gives a fuller view of fire conditions, thus enabling more accurate predictions and more timely interventions.

Moreover, drones and automation systems find high relevance in remote or otherwise inaccessible areas where nothing associated with regular monitoring can be carried out as a function of human effort. It is in such environments that drones assume the role of a principal method of data collection, providing information of a critical nature that would otherwise remain unavailable. Their ability to function in hazardous conditions, such as dense smoke or extreme heat, further enhances their value in wildfire monitoring and management.

5.3. Optimizing Data Flow and Processing for Immediate Action

Optimizing data flow and its processing is necessary to ensure that information gathered by IoT devices and analyzed by Deep Learning models is followed up efficiently and effectively. In the context of wildfire management, speed becomes a critical factor in transmitting, processing, interpreting, and delaying valuable time that might be lost in containing and controlling fires.

This would begin with the effective transmission of data from the sensors and drones to their processing units. For that to take place, the state of communication networks will have to be strong enough to handle large volumes of data with minimal latency. Probably, technologies like 5G and edge computing will take a lead in this regard. 5G networks will be ultra-high in speed and low in latency, allowing real-time data transmission that quickens the relaying of information from the field levels up to the decision-making levels. Edge computing, in turn, allows for the processing of data closer to the source, thereby reducing time in analyzing data to come up with actionable insights (Kukkar et al., 2020).

In edge computing, data gets processed at the source of collection or close to it by local servers or devices and not transmitted to the cloud server. This is a very expensive local processing capability, especially when network connectivity may be limited in remote or high-risk areas. This increases the speed of analysis by reducing the quantity of data normally transmitted across the network by processing locally, thus reducing the risks of loss or degradation of data during transit.

Once the data is processed, it should enable its translation into real-time insights available across all hands responsible for wildfire management. This will necessitate the development of intuitive, real-time dashboards and alert systems that present critical information in a clear and actionable format. These dashboards can be integrated with decision support systems that will automatically prioritize the alerts based on the severity of the threat and recommend proper response actions—e.g., sending firefighting teams along, issuing public warnings, or activating the fire suppression system.

Additional data flow may be further improved by continuously updating the predictive models with new data so that it corrects its predictions in relation to changing conditions. This process keeps models accurate and relevant through a dynamic update, so the most recent information is given for decision-making. Machine learning algorithms can also be used to analyze trends in historical data combined with real-time data, showing patterns that might indicate the likelihood of future fire activity.

The final component of data flow and processing optimization is to make the information available to each and all the stakeholders in real-time. This may involve integration of the IoT and Deep Learning Systems with existing communication and coordination platforms that are presently used by teams responsible for emergency response, government agencies, and community organizations. Ensuring fluent information flows across these platforms, all those concerned with wildfire management become informed and are in a position to respond promptly to emerging threats.

The Deep Learning—IoT synergy enables one to increase monitoring and control on wildfires. Be it cross-platform data fusion, smart deployment of drones, automated systems, and optimized data flow and processing, this integrated approach offers an end-to-end solution in real-time for one of the most pressing environmental challenges of our times. These technologies, combined with their potential for continued growth and evolution, will be critical to saving ecosystems, communities, and lives from wildfires.

6. TECHNOLOGICAL EVOLUTION AND FUTURE PROSPECTS

The field of wildfire management is on the brink of dramatic technological change in a rapidly changing world. Now, emerging technologies like 5G, edge computing, blockchain, and advanced robotics hold the promise to redefine not just detection and monitoring, but the whole approach to wildfire management. Such new innovations promised better speed, accuracy, and security of wildfire response systems; major upshifts in efficiency and effectiveness are on the horizon. The development of these emerging technologies in wildfire management frameworks will most probably come in a more resilient and adaptive manner. This section reviews the impacts and prospects for future emerging technologies in wildfire management.

6.1. Emerging Technologies: Impact of 5G and Edge Computing

5G technology offers significant improvements in communication capabilities, particularly in wildfire management systems. It allows for fast data transfer, enabling real-time monitoring and seamless integration of IoT devices. Edge computing enhances 5G capabilities by processing data closer to the source, reducing latency and enabling faster decision-making. This decentralized approach speeds up response times and reduces the burden on centralized data centers, resulting in a more resilient and scalable system.

With 5G and edge computing, complex predictive models and AI algorithms could be field deployed for more sophisticated, localized analysis of fire risk. For example, a network of edge devices connected via 5G can monitor different environmental conditions, trace changes in weather, and evaluate with accuracy the possibility of outbreaks. This enables proactive resource allocation, such as prepositioning firefighting teams or pre-deployment of drones to higher-risk areas, enhancing the preparedness and responsiveness of the wildfire management system (Hoang et al., 2024).

With the expansion of 5G networks across the world, their role in wildfire management will only increase with the need for more integrated and intelligent systems developed to respond to the increasing challenges posed by climate change and related environmental factors. Therefore, future fire management will rely more on the smooth interplay between high-speed communication networks and edge computing in support of real-time decision-making and rapid responses through a robust infrastructure.

6.2. Blockchain for Secure and Decentralized Data Handling

Blockchain technology offers a decentralized, secure, and transparent method for handling data in wildfire-monitoring and response systems. It provides an immutable record of transactions, ensuring authenticity and integrity. This is particularly useful in real-time data tracking from IoT sensors, drones, and satellite imaging. Blockchain enhances transparency by removing the central authority, reducing the risk of manipulation or data loss. It is particularly relevant in large-scale disaster management scenarios where multiple organizations need to share information seamlessly. Smart contracts, self-executing agreements, can trigger specific actions based on real-time data inputs. Blockchain technology also has potential for organizing a decentralized marketplace for data and resources management, resulting in efficient resource allocation and coordination in large wildfire incidents.

6.3. Integration of Advanced Robotics for Automated Firefighting

Advanced robotics is revolutionizing wildfire management by providing new tools for challenging and dangerous environments. Firefighting robots, drones, and UGVs are equipped with various tools and sensors to support difficult conditions, reducing workload and risk to human firefighters. These robots can work independently, suppress flames, clear vegetation, and make firebreaks, while drones can dispense fire retardant or water to key locations, minimizing costs and environmental impacts. UGVs are increasingly involved in wildfire management, especially in hard-to-reach areas or hazardous situations. These vehicles can navigate through rough terrain, carry heavy loads, clear debris, and put out small fires before they escalate. Advanced robotics integration also offers benefits in data collection and situational awareness, enabling real-time data relay to command centers about fire progression and firefighting efforts. This integration of robotics in wildfire management will define the future of wilderness fire management, enabling a resilient and adaptive response system against rising wildfire challenges.

7. CONCLUSION

The integration of Deep Learning (DL) and the Internet of Things (IoT) offers a transformative approach to wildfire management, addressing the urgent need for more sophisticated and proactive strategies in fire detection, prediction, and response. As wildfires become increasingly frequent and intense, traditional methods are no longer sufficient. DL provides advanced analytical capabilities, enabling the processing of complex environmental data to predict wildfire risks with higher accuracy. IoT enhances these predictions by delivering real-time data from a network of interconnected sensors, drones, and other monitoring devices deployed in fire-prone areas. The synergy between DL and IoT facilitates the creation of a comprehensive, real-time monitoring system that significantly improves situational awareness and decision-making during wildfire events. Case studies from high-risk regions, such as California and Australia, have demonstrated the practical benefits of this approach, showcasing enhanced early detection, more effective resource allocation, and ultimately, better protection of both ecosystems and human communities. Moving forward, the incorporation of emerging technologies like 5G, blockchain, and robotics into this framework will likely further strengthen wildfire management efforts, leading to more resilient and adaptive systems capable of mitigating the increasingly complex challenges posed by wildfires in a changing climate.

REFERENCES

Alghamdi, W., Marchiandi, J., Szabo, D., Samandra, S., & Clarke, B. O. (2022). Release of per- and polyfluoroalkyl substances (PFAS) from a waste management facility fire to an urban creek. *Journal of Hazardous Materials Advances*, 8, 100167. https://doi.org/https://doi.org/10.1016/j.hazadv.2022.100167. DOI: 10.1016/j.hazadv.2022.100167

Bilgiç, E., Tuna Tuygun, G., & Gündüz, O. (2023). Development of an emission estimation method with satellite observations for significant forest fires and comparison with global fire emission inventories: Application to catastrophic fires of summer 2021 over the Eastern Mediterranean. *Atmospheric Environment*, 308, 119871. https://doi.org/https://doi.org/10.1016/j.atmosenv.2023.119871. DOI: 10.1016/j.atmosenv.2023.119871

Cheung, J. P. Y., Kuang, X., Zhang, T., Wang, K., & Yang, C. (2023). 5-Year progression prediction of endplate defects: Utilizing the EDPP-Flow convolutional neural network based on unbalanced data. *Journal of Orthopaedics*, 38, 7–13. DOI: 10.1016/j.jor.2023.03.001 PMID: 36910507

Ganapathy, S., & Ajmera, D. (2024). An Intelligent Video Surveillance System for Detecting the Vehicles on Road Using Refined YOLOV4. *Computers & Electrical Engineering*, 113, 109036. https://doi.org/https://doi.org/10.1016/j.compeleceng.2023.109036. DOI: 10.1016/j.compeleceng.2023.109036

García-Aguilar, I., García-González, J., Luque-Baena, R. M., & López-Rubio, E. (2023). Automated labeling of training data for improved object detection in traffic videos by fine-tuned deep convolutional neural networks. *Pattern Recognition Letters*, 167, 45–52. https://doi.org/https://doi.org/10.1016/j.patrec.2023.01.015. DOI: 10.1016/j.patrec.2023.01.015

García-Valls, M., & Chirivella-Ciruelos, A. M. (2024). VelogCPS: A safe blockchain network for cyber–physical systems leveraging block verifiers. *Journal of Systems Architecture*, 153, 103177. https://doi.org/https://doi.org/10.1016/j.sysarc.2024.103177. DOI: 10.1016/j.sysarc.2024.103177

Groenenberg, M., Crouthers, R., Yoganand, K., Banet-Eugene, S., Bun, S., Muth, S., Kim, M., Mang, T., Panha, M., Pheaktra, P., Pin, T., Sopheak, K., Sovanna, P., Vibolratanak, P., Wyatt, A. G., & Gray, T. N. E. (2023). Snaring devastates terrestrial ungulates whilst sparing arboreal primates in Cambodia's Eastern Plains Landscape. *Biological Conservation*, 284, 110195. https://doi.org/https://doi.org/10.1016/j.biocon.2023.110195. DOI: 10.1016/j.biocon.2023.110195

Hoang, D., Errahmouni, H., Chen, H., Rachuri, S., Mannan, N., ElKharboutly, R., Imani, M., Chen, R., & Imani, F. (2024). Hierarchical representation and interpretable learning for accelerated quality monitoring in machining process. *CIRP Journal of Manufacturing Science and Technology*, 50, 198–212. https://doi.org/https://doi.org/10.1016/j.cirpj.2024.02.010. DOI: 10.1016/j.cirpj.2024.02.010

Kim, S.-H., Dhakal, T., Yoon, T.-G., Cho, K. H., Kim, J.-Y., Kim, T.-S., Lee, D.-H., & Jang, G.-S. (2023). Occupancy model to unveil wildlife utilization at Yeongyang-gun wind farm management road, Korea. *Global Ecology and Conservation*, 48, e02692. https://doi.org/https://doi.org/10.1016/j.gecco.2023.e02692. DOI: 10.1016/j.gecco.2023.e02692

Kukkar, A., Mohana, R., & Kumar, Y. (2020). Does bug report summarization help in enhancing the accuracy of bug severity classification? *Procedia Computer Science*, 167, 1345–1353. https://doi.org/https://doi.org/10.1016/j.procs.2020.03.345. DOI: 10.1016/j.procs.2020.03.345

N., M., D.R., N., Reddy, B. E., Buyya, R., K.R., V., Iyengar, S. S., & Patnaik, L. M. (2024). Secure pharmaceutical supply chain using blockchain in IoT cloud systems. *Internet of Things, 26*, 101215. https://doi.org/https://doi.org/10.1016/j.iot.2024.101215

Platte, R. C., & Torland, R. E. (2023). Habitat Use of Wintering Mule Deer Across a Juniper-Treated and Burned Landscape. *Rangeland Ecology and Management*, 86, 18–25. https://doi.org/https://doi.org/10.1016/j.rama.2022.10.007. DOI: 10.1016/j.rama.2022.10.007

Qayyum, F., Samee, N. A., Alabdulhafith, M., Aziz, A., & Hijjawi, M. (2024). Shapley-based interpretation of deep learning models for wildfire spread rate prediction. *Fire Ecology*, 20(1), 8. Advance online publication. DOI: 10.1186/s42408-023-00242-y

Ram Prasanna, K., Mathana, J. M., Ramya, T. A., & Nirmala, R. (2023). LoRa network based high performance forest fire detection system. *Materials Today: Proceedings*, 80, 1951–1955. https://doi.org/https://doi.org/10.1016/j.matpr.2021.05.656. DOI: 10.1016/j.matpr.2021.05.656

Rathod, T., Patil, V., Harikrishnan, R., & Shahane, P. (2023). Multipurpose deep learning-powered UAV for forest fire prevention and emergency response. *HardwareX*, 16, e00479. https://doi.org/https://doi.org/10.1016/j.ohx.2023.e00479. DOI: 10.1016/j.ohx.2023.e00479 PMID: 37771320

Resop, L., Demarais, S., Strickland, B., Iglay, R. B., Nichols, R., & Lashley, M. (2023). Plant species-specific responses and community associations with fire season. *Forest Ecology and Management*, 529, 120724. https://doi.org/https://doi.org/10.1016/j.foreco.2022.120724. DOI: 10.1016/j.foreco.2022.120724

Rodriguez-Jimenez, F., Lorenzo, H., Acuña-Alonso, C., & Alvarez, X. (2023). PLS-PM analysis of forest fires using remote sensing tools. The case of Xurés in the Transboundary Biosphere Reserve. *Ecological Informatics*, 75, 102010. https://doi.org/https://doi.org/10.1016/j.ecoinf.2023.102010. DOI: 10.1016/j.ecoinf.2023.102010

Shams, M. Y., Tarek, Z., El-kenawy, E.-S. M., Eid, M. M., & Elshewey, A. M. (2024). Predicting Gross Domestic Product (GDP) using a PC-LSTM-RNN model in urban profiling areas. *Computational Urban Science*, 4(1), 3. Advance online publication. DOI: 10.1007/s43762-024-00116-2

Shehab, M. A., Khreich, W., Hamou-Lhadj, A., & Sedki, I. (2024). Commit-time defect prediction using one-class classification. *Journal of Systems and Software*, 208, 111914. https://doi.org/https://doi.org/10.1016/j.jss.2023.111914. DOI: 10.1016/j.jss.2023.111914

Shiddiqi, A. M., Yogatama, E. D., & Navastara, D. A. (2023). Resource-aware video streaming (RAViS) framework for object detection system using deep learning algorithm. *MethodsX*, 11, 102285. https://doi.org/https://doi.org/10.1016/j.mex.2023.102285. DOI: 10.1016/j.mex.2023.102285 PMID: 37533793

Tunstill, K., Grogan, L. F., Morrison, C., McCallum, H., & Lanctôt, C. (2022). Effects of two firefighting chemical formulations, Phos–Chek LC95W and BlazeTamer380, on striped marsh frog (Limodynastes peronii) tadpole survival, growth, development and behaviour. *Aquatic Toxicology (Amsterdam, Netherlands)*, 252, 106326. https://doi.org/https://doi.org/10.1016/j.aquatox.2022.106326. DOI: 10.1016/j.aquatox.2022.106326 PMID: 36270184

Ullah, F., Ullah, S., Srivastava, G., & Lin, J. C.-W. (2023). IDS-INT: Intrusion detection system using transformer-based transfer learning for imbalanced network traffic. *Digital Communications and Networks*. https://doi.org/https://doi.org/10.1016/j.dcan.2023.03.008

Vidal-Riveros, C., Souza-Alonso, P., Bravo, S., Laino, R., & Ngo Bieng, M. A. (2023). A review of wildfires effects across the Gran Chaco region. *Forest Ecology and Management*, 549, 121432. https://doi.org/https://doi.org/10.1016/j.foreco.2023.121432. DOI: 10.1016/j.foreco.2023.121432

Villalba-Briones, R., Piedrahita, P. M., Garces, D. O., & Monrós, J. S. (2023). Sensitisation, research and management for conservation within ESPOL University forests after the COVID-19 pandemic, Ecuador. *Neotropical Biology and Conservation, 18*(4), 283–303. https://doi.org/https://doi.org/10.3897/neotropical.18.e110615

Zald, H. S. J., Callahan, C. C., Hurteau, M. D., Goodwin, M. J., & North, M. P. (2022). Tree growth responses to extreme drought after mechanical thinning and prescribed fire in a Sierra Nevada mixed-conifer forest, USA. *Forest Ecology and Management*, 510, 120107. https://doi.org/https://doi.org/10.1016/j.foreco.2022.120107. DOI: 10.1016/j.foreco.2022.120107

Zander, K. K., Garnett, S. T., Ogie, R., Alazab, M., & Nguyen, D. (2023). Trends in bushfire related tweets during the Australian 'Black Summer' of 2019/20. *Forest Ecology and Management*, 545, 121274. https://doi.org/https://doi.org/10.1016/j.foreco.2023.121274. DOI: 10.1016/j.foreco.2023.121274

Zhu, Z., Tong, H., Wang, Y., & Li, Y. (2022). Enhancing bug localization with bug report decomposition and code hierarchical network. *Knowledge-Based Systems*, 248, 108741. https://doi.org/https://doi.org/10.1016/j.knosys.2022.108741. DOI: 10.1016/j.knosys.2022.108741

Chapter 11
Prospects and Challenges of Artificial Intelligence in Forest Protection

Laeeq Razzak Janjua
https://orcid.org/0000-0002-1948-5859
WSB University, Poland

Saquib Ahmed
https://orcid.org/0009-0008-1891-6910
Sharda University, India

Bhupinder Singh
https://orcid.org/0009-0006-4779-2553
Sharda University, India

ABSTRACT

The protection of Earth's ecology and balancing ecology rests heavily on forest preservation. Issues like trafficking of wildlife, illegal logging, and deforestation are still existing. Conventional methods of monitoring techniques and safety precautions have drawbacks and are inefficient to address these ecological problems. In the present era, Artificial Intelligence is advancing, and it has given a fresh hope in preserving the forest. The chapter examines the best possible use of Artificial Intelligence in forest preservation concentrating on certain situations like wildlife conservation, illegal logging surveillance and prediction of fire forests. Artificial intelligence (AI) technology makes available greater accuracy and efficiency than conventional techniques, allowing for quicker detection and reaction to forest damage activities. AI will become more significant in the future because of additional technical developments and the growth of application areas, opening new avenues

DOI: 10.4018/979-8-3693-7565-5.ch011

for the preservation and sustainable use of forests.

INTRODUCTION

In the current climate, forests are essential, particularly for reducing global warming. In addition to binding CO2, forests also improve water quality, help control urban temperatures, and most importantly are home to a variety of wildlife. There are many different tasks involved in forestry activities. Some of them are particularly focused on maintaining the health of forests and helping them reach their maximum potential, but this requires a lot of work. The majority of this work is now done by hand, which makes forest operations dangerous, expensive, and time-consuming (Watson et al., 2023).

The major categories of forestry include wildlife monitoring, forest inventory, forest management, search and rescue, and wildfire monitoring. Inventory and forest monitoring are quite closely related since the former refers to the field that collects and organizes data from forests. However, there are still several challenges that AI in conserving forests must overcome, including data collection, model training, and technology adoption (Kennedy et al., 2023). Tree species, height, diameter, number of trees of a particular kind, and so forth are a few examples. In addition to keeping track of the forest's existing condition, these activities are also intended to monitor its development and provide information for forest management initiatives (Singh & Kaunert, 2024).

The goal of forest management is to act based on monitoring data to guarantee the preservation and upkeep of wooded regions. Reforestation and the trimming of unhealthy plants are two methods used to achieve this (Dhali et al., 2023). The tasks involved in tracking the wildlife that lives in woods are logically described by wildlife monitoring. Search and rescue and wildfire monitoring are more closely related to crisis management since they are typically conducted after an accident or natural catastrophe occurs and efforts are made to either find individuals in need of help or minimize forest damage (Singh, 2023). Drones have improved in capability, affordability, and accessibility in recent years. This raises the question of how drones can support forestry initiatives such as the above-described practices. Drones designed to collect a range of image data on multiple spectra or other data collection methods, such as LiDAR, provide the foundation for integration into forestry initiatives, and when combined with continuously improving computer vision and machine learning algorithms, a number of benefits are evident (Hansen et al., 2023).

Forest is one of the most important parts of the ecosystem which support the wide variety of life forms and very crucial for soil conservation, regulation of climate and the water supply (Batisha, 2023). But as human activity grows, problems like illegal logging, poaching of wildlife, and forest devastation are getting worse. These problems endanger human society's long-term survival in addition to having a detrimental effect on ecosystems (Singh, 2024). The ineffectiveness of human patrols and data collection challenges are just two of the many drawbacks of traditional forest monitoring and preservation techniques. Therefore, it is more important than ever to set up a precise and efficient monitoring and protection plan (Kuenzel et al., 2016). The rapid advancement of artificial intelligence (AI) technology in recent years has enabled new methods to address this issue. AI can monitor forests comprehensively through deep learning, image recognition, and big data analysis, which will help detect and manage forest degradation operations more effectively (Qwaider et al., 2023). By offering new tools and ideas for forest protection, this increases hope for preserving Earth's valuable biological resources (Singh, 2024).

Causes and Effects of Forest Destruction

There are several reasons as why forest destruction is a major global concern for today: Firstly, illegal logging and deforestation are two of the biggest threats to forests; the uncontrolled tree harvesting by illegal loggers prevents large areas of woods from regenerating sustainably, but deforestation exacerbates the loss of forest resources by allowing illegal developers and wood traffickers to profit from illegally obtained timber(Singh, 2023). The second major cause of forest degradation is the expansion of agriculture, particularly large-scale planting, farming, and clearing operations that disrupt ecosystems and gradually reduce forest areas (Sachs, 2015).

Furthermore, overgrazing, plant damage, soil erosion, and land degradation are all consequences of livestock farming and grazing methods that increase the rate of forest loss. Furthermore, the considerable land clearing and deforestation caused by infrastructure development projects like roads, dams, and electricity lines may result in significant ecological changes and degradation (Sobha, 2023). Ultimately, because the removal of rare species can result in ecological imbalance and species extinction, wildlife pillage directly threatens biodiversity. The repercussions of forest extinction are profound and far-reaching (Muhmad Kamarulzaman et al., 2023). First, the loss of forests causes a major environmental imbalance by destroying and disrupting ecosystem services that affect soil quality, the availability of water resources, climate regulation, and other important ecological processes (Bogers, 2023).

Both the natural environment and human society are straightly and significantly impacted by this. The second important effect is the deterioration of biodiversity (Singh, 2023). The extinction of plant and animal species due to forest deprivation poses an unadorned danger to biodiversity and has the possibility to cause ecosystem collapse and worsen environmental issues (Nahar, 2024).

Moreover, the loss of trees upsurges carbon dioxide discharges and decreases carbon storage ecosystem, which worsens the negative effects of climate change and creates new environmental issues. Extinguishing forests can lead to land degradation, loss of livelihoods, recurrent natural disasters, and other problems that affect the stability of neighboring communities and society as a whole, both directly and indirectly (Desai et al., 2023).

Limitations and Challenges

Problems with forest annihilation and protection, outdated forest protection techniques have several drawbacks and difficulties, such as the following:

a. Human examinations are the backbone of traditional forest security, despite their various drawbacks. First, due to limitations in terms of work force, time, and location, human watches are ineffective and provide only a limited amount of monitoring coverage. This frequently leads to inadequate monitoring outcomes. Second, because human patrolling requires significant financial, material, and human resources, it is expensive. In addition to increasing the cost of forest upkeep, this restricts the frequency and potential of monitoring and protection efforts (Sharma & Singh, 2022).
b. Traditional forest protection relies on real-time data support, but data collection is challenging. First of all, the difficult terrain and harsh environment of forest regions often make it difficult to obtain accurate monitoring data in real time (Ashukem & Sama, 2023). This challenge increases the cost and risk of data gathering while limiting the frequency and extent of monitoring actions. Second, it is more difficult to respond swiftly to emergencies involving forest loss due to the long cycle of data updates and delayed information transfer that characterize traditional forest conservation (Cornford et al., 2023).
c. Threats and challenges: There are numerous risks and difficulties associated with traditional forest conservation techniques, such as limited resources, human interference, and a lack of specialized knowledge. First of all, it is challenging to effectively combat contemporary, sophisticated criminal approaches due to the comparatively low technological level of conventional procedures (Singh, 2022).
d. Traditional surveillance and protection systems are now comparatively antiquated and unable to precisely identify and address the needs of forest destruction operations in the light of rapid technological innovation. Second, it is difficult for protection personnel to do their duties efficiently because of the frequent disruptions and interference by criminals. Human intervention affects the sustainability and efficacy of protection operations, endangering the safety of

protection personnel while also complicating protection activities (Koh et al., 2023).

Artificial Intelligence

The diverse discipline of artificial intelligence includes natural language processing, computer vision, machine learning, and expert systems, to name a few. Giving computer systems human-like abilities such as perception, comprehension, learning, and reasoning—will enable them to do a wide range of tasks and tackle difficult problems (Sheng & Lian, 2023). Machine learning, one of the core technologies of artificial intelligence, aims to improve computer system performance through data learning and algorithm optimization. Computer vision aids computers in understanding and interpreting image or video data, whereas natural language processing allows computers to understand and generate natural language (Singh, 2022). In order to handle difficult problems, expert systems also mimic the expertise and mental processes of human specialists. Giving computers more intelligence so they can imitate or even surpass human thought and behaviour is the overall aim of AI development. This will enable computers to solve a wide range of real-world problems (Ashukem, 2024).

In the early 1950s, the field of artificial intelligence began to take shape. Because of computers' limited capabilities at the time, research focused mostly on reasoning and symbolic processing. Between the 1960s and the 1980s, artificial intelligence (AI) experienced a golden age of development that yielded many significant scientific findings and technological advancements, including rule-based reasoning systems and expert systems. By the late 1980s and early 1990s, however, the “AI Winter” had hit the area of artificial intelligence. Due to skepticism and disenchantment with AI caused by technology and computer performance limitations, funding and research resources were reduced (Singh & Kaunert, 2024).

During this period, notable applications and technology like support vector machines, neural networks, speech recognition, and picture recognition came into being. Since the start of the twenty-first century, artificial intelligence (AI) has made considerable strides, particularly in the fields of machine learning and deep learning (Prasanna et al., 2023). The advent and widespread application of deep learning technology, which has led to significant improvements in computer systems in areas like speech recognition, image recognition, and natural language processing, has ignited a fresh wave of interest in artificial intelligence (Terry et al. 2024).

Artificial Intelligence and Forest Protection

Monitoring Forest Fire

Monitoring and early warning of forest fires is one of the most significant uses of AI in forest conservation. Real-time monitoring is made possible by AI's ability to recognize forest fire indicators like smoke and flames in satellite photos and drone footage using image recognition capabilities. Additionally, AI can improve fire incidence prediction models built on historical data and environmental circumstances, enabling early fire warnings and suitable preventive actions. This technique has the potential to significantly improve the effectiveness of forest fire monitoring and response by minimizing harm to ecosystems and human communities (Ashukem & Sama, 2023).

Monitoring Illegal Logging

One notable use of AI in forest preservation is monitoring, providing close attention, and preventing illicit logging. Artificial intelligence (AI) can swiftly detect illicit logging activities by using image analysis technologies to look at high-resolution satellite images or drone footage to detect and identify changes in forest cover. Additionally, AI is able to control whether illicit logging is occurring by analyzing the behavioural characteristics of individuals or vehicles captured by security cameras (Ashukem & Sama, 2023). By assisting in the prompt identification and eradication of illicit logging operations, this knowledge can enhance the effectiveness of monitoring and safeguarding forest resources, preserving the stability and integrity of ecosystems AI (Ozkavaf Senalp, 2023).

Wildlife Protection and Monitoring

Monitoring and conserving wildlife is one of the most important uses of artificial intelligence. AI can recognize and classify wild animals using image recognition and classification technologies, allowing it to monitor population size, activity ranges, and dispersion. AI, acoustic sensors, and sound recognition technology can all be used to recognize the sounds produced by the wild animals. By tracking wildlife movements and patterns and offering practical tools and technological support for survival and protection, this application helps ensure endangered species (Cumming et al., 2023).

This means that AI use in forest monitoring and conservation has a vast amount of value and relevance. It drastically improves efficiency with regards to early warning through the possibility of timely monitoring and providing warnings, such as illegal logging, animal poaching, and forest fires. For instance, sound recognition technology monitors wildlife activity, picture recognition technology monitors indications of fires or illicit logging, and data analysis and pattern recognition help evaluate the health of ecosystems and predict possible issues. AI can also provide emergency preparations and intelligent decision support in response to any need, accelerating the response to unexpected events and overall effectiveness of the response (Cork et al. 2023).

Forest Fire Prediction

One of the very advanced prediction technologies related to the area of forest fires: Google Earth Engine & apos; in conjunction with the use of machine learning algorithms, particularly Convolutional Neural Networks (CNN).' This method builds a model of the probability of a fire using satellite images and trains this model using past data. This model in specific can be of significant assistance for references in firefighting and prevention by recording real-time vegetation quality, weather amongst other factors, which shall help predict where the probable fires might occur (Gadsden et al., 2023).

Illegal Logging Monitoring

The Global Forest Watch platform is a very powerful example of forest monitoring with its combination of AI image-processing technology and high-resolution satellite imagery. The software automatically detects the indications of logging operations, road building, and changes in global forest cover through continuous monitoring of change. It reports quickly to the concerned authorities any possible unlawful logging activities for immediate action, especially in the area of detection. This helps to provide necessary technical support in the preservation of forest resources (Lamalle & Stoett, 2023).

Wildlife Protection

Examples include monitoring and protection of animals on land or water using AI picture recognition technology and mobile applications, for instance, Instant Wild. Users take pictures of animals from their mobile devices and upload them onto the website, while the AI algorithms automatically classify the species seen in the pictures. It will enable the conservationists to identify endangered species quickly so that whatever measure is required can be taken. They can monitor the status of population and understand the location and numbers of wildlife very fast.

The Instant Wild platform provides a practical method and instrument for animal conservation (Dunlap, 2023).

Conclusion Such applications of AI in theoretical preservation of forests, tracking illegitimate logging, predicting forest fires, and protecting animals have also been demonstrated. These examples, involving a variety of data sources and creative algorithms, do provide substantial support and help for projects in forest protection (Newman et al. 2024).

Advantages of Artificial Intelligence in the Forest

One of the most obvious advantages of AI technology against traditional forest preservation is that it is effective. Using AI automated monitoring, the efficacy of monitoring and response times increase significantly. As compared to traditional protection methods for forests, which involve human patrols and, therefore, are less effective, due to covering little area in terms of monitoring, because of the amount of time and human resources required. Moreover, with big data analysis and pattern recognition, AI can be more precisely used for the betterment of forecasting and alerting them of such possibilities as occurrence of forest fires, illegal logging, and so on. AI is an extremely valued tool in forest protection because it is effective and gives enormous support in the timely identification and handling of incidents related to damage in the forests (Ghosh et al., 2023).

Another critical advantage of AI technology in forest protection is precision. AI performs incredibly well in the accurate identification of species of animals, documentation of events, and the like, especially in the dimensions of auditory and visual recognition that reduced false alarm and misses remarkably (Shinde et al., 2023). Such accuracy helps advance in time the recognition and response to various forms of activities degrading the forests, improving the reliability of monitoring results. Beyond that, the application of modern technologies such as deep learning would enhance the legitimacy and universality of monitoring outcomes and make the selection of models of monitoring adaptable for different environments and conditions (Cardwell, 2023).

One important advantage of AI technology for the conservation of forests is sustainability. For one, AI lends itself to continuous learning and optimization, hence continually improves performance through continuous reliable support in long-term forest conservation projects. Through routine analysis of new data and situations, AI systems may adjust their models and algorithms to fit changes in the environment and needs. This involves conserving human and material resources (Aryal et al., 2023). AI can replace the traditional human patrolling and monitoring techniques since it can do the task of monitoring and analytical jobs without human intervention. This enhances the sustainability and economic effectiveness of activities in preserving forests through a reduction in the use of human labour inputs and the judicious utilization of available resources using innovative techniques (Ngadze, 2023).

Disadvantages of Artificial Intelligence in Forest

Gathering large volumes of high-quality data is a significant issue for the efficacy and accuracy of AI technology (Sofi et al., 2023). The harshest climate conditions coupled with complex topography in forest areas make it relatively difficult to obtain high-quality data concerning forest conservation, hence making it inconvenient to collect data (Shukla et al. 2023). Some other locations may not have adequate update methods for real-time data, leading to inadequate timeliness in the data. In some instances, lack of adequate or missing data can result in not so well-trained AI models; this will affect how well they work in practice (Singh & Kaunert, 2025).

Another challenge in AI concerning forest conservation relates to the maturity of the technology. While the idea of picture recognition and data analysis has improved significantly, there are still certain technical obstacles and difficulties (Singh et al., 2025). Moreover, some AI models call for significant time and resource commitment for training and debugging, which increases technical costs (Singh et al., 2025). This is why protection of forests using AI technology requires continuous optimization and development of models and algorithms for further stabilization and dependability in handling various environmental and scenario obstacles (Singh & Kaunert, 2024). Further strengthening of the safety and dependability of the technology as well as healthy development of AI technology in forest protection will require a strengthening of regulations and management.

CONCLUSION

Forestry is one of the most biodiverse ecosystems on earth, playing a great role in preserving ecological balance and allowing human life. However, its hazards, such as illegal logging, hunting animals, and wildfires, severely strain both habitat and stability in the ecosystem of the forests. Protection of the forests by using the AI technology is the only way to improve this situation, since traditional protection techniques often fail to properly take care of these complex issues.

This paper explores the prospects of applying AI in forest protection. This paper begins by bringing to the attention some of the challenges associated with forest protection and inadequacies of traditional methods, thus underlining the urgency and necessity to introduce AI technology. Secondly, it defines the research questions and objectives of this paper and will try to search for the possible applications and trends of development for AI in forest. Protection is one of the solutions that involves more impactful ways to conserve the forest.

In analysis of the main content, the paper defines the reasons and effects of forest degradation, on one hand and limitations of the traditional approaches toward forest protection, on the other. It also initiates the basic concepts, key technologies and practical application cases of AI exploring the potential role of AI technology in forest protection. From then on, comparing the advantages and disadvantages of AI technology versus traditional methods, the paper emphasizes the importance of merging both approaches to achieve more full and effective protection of the forest.

The study concludes by predicting the future development trends of AI in the field of forest conservation and making suggestions like fostering social education and ethical standards, increasing technological research and innovation, and fortifying international cooperation. These recommendations are meant to encourage the extensive use and ongoing advancement of AI technology in forest preservation, offering Important tips and information for additional study and application of AI technology in this field.

REFERENCES

Aryal, K., Maraseni, T., & Apan, A. (2023). Examining policy– institution– program (PIP) responses against the drivers of ecosystem dynamics. A chronological review (1960–2020) from Nepal. *Land Use Policy*, 132, 106789. DOI: 10.1016/j.landusepol.2023.106789

Ashukem, J. C. N. (2024). Land grabbing in the Anthropocene and the "questionable" pursuit of sustainability in Africa–highlighting ecological disproportionality. In *Human Rights and the Environment in Africa* (pp. 276–294). Routledge.

Ashukem, J. C. N., & Sama, S. M. (Eds.). (2023). *Domestic and Regional Environmental Laws and Policies in Africa: A Research Companion*. Taylor & Francis. DOI: 10.4324/9781003382256

Ashukem, J. C. N., & Sama, S. M. (Eds.). (2023). *Human Rights and the Environment in Africa: A Research Companion*. Taylor & Francis. DOI: 10.4324/9781003382249

Batisha, A. (2023). A lighthouse to enhance the quality of life in the Nile River basin. *Environmental Economics and Policy Studies*, 1–29. DOI: 10.1007/s10018-023-00380-2

Bogers, M. (2023). *Integrating global sustainability governance: How the Sustainable Development Goals impact institutional and policy integration at the global level* (Doctoral dissertation, Utrecht University).

Cardwell, M. (2023). Results-based agri-environmental scheme design: Legal implications. *Environmental Law Review*, 25(4), 260–288. DOI: 10.1177/14614529231185678

Cork, S., Alexandra, C., Alvarez-Romero, J. G., Bennett, E. M., Berbés-Blázquez, M., Bohensky, E., & Wyborn, C. (2023). Exploring Alternative Futures in the Anthropocene. *Annual Review of Environment and Resources*, 48(1), 25–54. DOI: 10.1146/annurev-environ-112321-095011

Cornford, R., Spooner, F., McRae, L., Purvis, A., & Freeman, R. (2023). Ongoing over-exploitation and delayed responses to environmental change highlight the urgency for action to promote vertebrate recoveries by 2030. *Proceedings of the Royal Society B, 290*(1997), 20230464. DOI: 10.1098/rspb.2023.0464

Cumming, G. S., Adamska, M., Barnes, M. L., Barnett, J., Bellwood, D. R., Cinner, J. E., Cohen, P. J., Donelson, J. M., Fabricius, K., Grafton, R. Q., Grech, A., Gurney, G. G., Hoegh-Guldberg, O., Hoey, A. S., Hoogenboom, M. O., Lau, J., Lovelock, C. E., Lowe, R., Miller, D. J., & Wilson, S. K. (2023). Research priorities for the sustainability of coral-rich western Pacific seascapes. *Regional Environmental Change*, 23(2), 1–15. DOI: 10.1007/s10113-023-02051-0 PMID: 37125023

Desai, B. H. (2023). Land-Soil Sustainability: India's International Obligations. In *Soil Law and Governance in India* (pp. 3–24). Springer International Publishing. DOI: 10.1007/978-3-031-32360-7_1

Dhali, M., Hassan, S., & Subramaniam, U. (2023). Comparative analysis of oil and gas legal frameworks in Bangladesh and Nigeria: A pathway towards achieving sustainable energy through policy. *Sustainability (Basel)*, 15(21), 15228. DOI: 10.3390/su152115228

Dunlap, A. (2023). The Structures of Conquest: Debating Extractivism (s), Infrastructures and Environmental Justice for Advancing Post-Development Pathways. *International Development Policy| Revue internationale de politique de développement.*

Fopa, V. K., Bayir, N., & Özdal, D. (2023). Assessing the status and spatial-temporal dynamics of the Bamenda Mountains (BM), North West region of Cameroon. *Environmental Monitoring and Assessment*, 195(9), 1053. DOI: 10.1007/s10661-023-11630-x PMID: 37589789

Gadsden, G. I., Golden, N., & Harris, N. C. (2023). Place-Based Bias in Environmental Scholarship Derived from Social–Ecological Landscapes of Fear. *Bioscience*, 73(1), 23–35. DOI: 10.1093/biosci/biac095 PMID: 36643594

Ghosh, S., Dinda, S., Chatterjee, N. D., & Bera, D. (2023). Linking ecological vulnerability and ecosystem service value in a fast-growing metropolitan area of eastern India: A scenario-based sustainability approach. *Environment, Development and Sustainability*, •••, 1–31. DOI: 10.1007/s10668-023-03966-8

Hansen, H. H., Bergman, E., Cowx, I. G., Lind, L., Pauna, V. H., & Willis, K. A. (2023). Resilient rivers and connected marine systems: a review of mutual sustainability opportunities. *Global Sustainability*, 1-71.

Kennedy, C. M., Fariss, B., Oakleaf, J. R., Garnett, S. T., Fernández-Llamazares, Á., Fa, J. E., Baruch-Mordo, S., & Kiesecker, J. (2023). Indigenous Peoples' lands are threatened by industrial development; conversion risk assessment reveals need to support Indigenous stewardship. *One Earth*, 6(8), 1032–1049. DOI: 10.1016/j.oneear.2023.07.006

Koh, T., Lye, L. H., & Lum, S. (Eds.). (2023). *Peace With Nature: 50 Inspiring Essays On Nature And The Environment.* World Scientific. DOI: 10.1142/13561

Lamalle, S., & Stoett, P. (Eds.). (2023). *Representations and Rights of the Environment.* Cambridge University Press. DOI: 10.1017/9781108769327

Muhmad Kamarulzaman, A. M., Wan Mohd Jaafar, W. S., Mohd Said, M. N., Saad, S. N. M., & Mohan, M. (2023). UAV Implementations in Urban Planning and Related Sectors of Rapidly Developing Nations: A Review and Future Perspectives for Malaysia. *Remote Sensing (Basel)*, 15(11), 2845. DOI: 10.3390/rs15112845

Nahar, S. (2024). Modeling the effects of artificial intelligence (AI)-based innovation on sustainable development goals (SDGs): Applying a system dynamics perspective in a cross-country setting. *Technological Forecasting and Social Change*, 201, 123203. DOI: 10.1016/j.techfore.2023.123203

Newman, R. J., Enns, C., Capitani, C., Thorn, J. P., Courtney-Mustaphi, C. J., Buckton, S. J., Om, E. S., Fazey, I., Haji, T. A., Nchimbi, A. Y., Kariuki, R. W., & Marchant, R. A. (2024). 'Kesho' Scenario Development for Supporting Water-Energy Food Security under Future Conditions in Zanzibar. *Land (Basel)*, 13(2), 195. DOI: 10.3390/land13020195

Ngadze, F. (2023). Applying the Safe and Just Operating Space (SJOS) framework to Sustainable Development in Zimbabwe.

Özkavaf Şenalp, S. I. L. A. (2023). Exploring resilience of socio-ecological productive landscape through understanding change, impact and response among farmers: The case of Northwestern ankara.

Prasanna, S., Ameen, A., Saurav, K., Singh, D. B., Tripathy, A., Sharma, P., ... Sanjaya, K. (2023). *Empowerment and equality navigating human rights law in a complex world.* Institute of Legal Education.

Qwaider, S., Al-Ramadan, B., Shafiullah, M., Islam, A., & Worku, M. Y. (2023). GIS-Based Progress Monitoring of SDGs towards Achieving Saudi Vision 2030. *Remote Sensing (Basel)*, 15(24), 5770. DOI: 10.3390/rs15245770

Sharma, A., & Singh, B. (2022). Measuring Impact of E-commerce on Small Scale Business: A Systematic Review. *Journal of Corporate Governance and International Business Law*, 5(1).

Sheng, S., & Lian, H. (2023). The spatial pattern evolution of rural settlements and multi-scenario simulations since the initiation of the reform and opening up policy in China. *Land (Basel)*, 12(9), 1763. DOI: 10.3390/land12091763

Shinde, V. R., Mishra, R. R., Bhonde, U., & Vaidya, H. (Eds.). (2023). *Managing Urban Rivers: From Planning to Practice*. Elsevier.

Shukla, G., Bhat, J. A., Chakravarty, S., Almutairi, A., Li, M., & Iyer-Raniga, U. (2023). *Floristic Diversity: Biology and Conservation*. BoD–Books on Demand. DOI: 10.5772/intechopen.100674

Singh, B. (2022). Relevance of Agriculture-Nutrition Linkage for Human Healthcare: A Conceptual Legal Framework of Implication and Pathways. *Justice and Law Bulletin*, 1(1), 44–49.

Singh, B. (2022). COVID-19 Pandemic and Public Healthcare: Endless Downward Spiral or Solution via Rapid Legal and Health Services Implementation with Patient Monitoring Program. *Justice and Law Bulletin*, 1(1), 1–7.

Singh, B. (2023). Blockchain Technology in Renovating Healthcare: Legal and Future Perspectives. In *Revolutionizing Healthcare Through Artificial Intelligence and Internet of Things Applications* (pp. 177-186). IGI Global.

Singh, B. (2023). Federated Learning for Envision Future Trajectory Smart Transport System for Climate Preservation and Smart Green Planet: Insights into Global Governance and SDG-9 (Industry, Innovation and Infrastructure). *National Journal of Environmental Law*, 6(2), 6–17.

Singh, B. (2024). Lensing Legal Dynamics for Examining Responsibility and Deliberation of Generative AI-Tethered Technological Privacy Concerns: Infringements and Use of Personal Data by Nefarious Actors. In Ara, A., & Ara, A. (Eds.), *Exploring the Ethical Implications of Generative AI* (pp. 146–167). IGI Global., DOI: 10.4018/979-8-3693-1565-1.ch009

Singh, B. (2024). Social Cognition of Incarcerated Women and Children: Addressing Exposure to Infectious Diseases and Legal Outcomes. In Reddy, K. (Ed.), *Principles and Clinical Interventions in Social Cognition* (pp. 236–251). IGI Global., DOI: 10.4018/979-8-3693-1265-0.ch014

Singh, B. (2024). Evolutionary Global Neuroscience for Cognition and Brain Health: Strengthening Innovation in Brain Science. In Prabhakar, P. (Ed.), *Biomedical Research Developments for Improved Healthcare* (pp. 246–272). IGI Global., DOI: 10.4018/979-8-3693-1922-2.ch012

Singh, B., & Kaunert, C. (2024). Computational Thinking for Innovative Solutions and Problem-Solving Techniques: Transforming Conventional Education to Futuristic Interdisciplinary Higher Education. In Fonkam, M., & Vajjhala, N. (Eds.), *Revolutionizing Curricula Through Computational Thinking, Logic, and Problem Solving* (pp. 60–82). IGI Global., DOI: 10.4018/979-8-3693-1974-1.ch004

Singh, B., & Kaunert, C. (2024). Revealing Green Finance Mobilization: Harnessing FinTech and Blockchain Innovations to Surmount Barriers and Foster New Investment Avenues. In Jafar, S., Rodriguez, R., Kannan, H., Akhtar, S., & Plugmann, P. (Eds.), *Harnessing Blockchain-Digital Twin Fusion for Sustainable Investments* (pp. 265–286). IGI Global., DOI: 10.4018/979-8-3693-1878-2.ch011

Singh, B., & Kaunert, C. (2024). Adventure in High Altitude of Mountainous Topographies and Health Impacts: Lensing Tourism Sustainability via Reducing Ecological and Sociocultural Footprint and Health Emergency and Medical Assistance Management. In Meraj, G., Hashimoto, S., & Kumar, P. (Eds.), *Navigating Natural Hazards in Mountainous Topographies. Disaster Risk Reduction*. Springer., DOI: 10.1007/978-3-031-65862-4_15

Singh, B., & Kaunert, C. (2025). Leveraging IoT for Patient Monitoring and Smart Healthcare: Connected Healthcare System. In B. S & S. Kadry (Eds.), *Revolutionizing Healthcare Systems Through Cloud Computing and IoT* (pp. 27-46). IGI Global. https://doi.org/DOI: 10.4018/979-8-3693-7225-8.ch002

Singh, B., & Kaunert, C. (2025). Cloud Computing and IoMT in Disease Screening and Diagnosis: AI Approaches in Transmuting Healthier Homes. In B. S & S. Kadry (Eds.), *Revolutionizing Healthcare Systems Through Cloud Computing and IoT* (pp. 99-120). IGI Global. DOI: 10.4018/979-8-3693-7225-8.ch005

Singh, B., Kaunert, C., & Singh, G. (2025). Scaling Legal Framework for Plastic Pollution and Advancing Cutting Edge Water Governance: Reducing and Eliminating Marine Pollution in Alignment With SDG 14 (Life Below Water). In Gaur, N., Sharma, E., Nguyen, T., Bilal, M., & Melkania, N. (Eds.), *Societal and Environmental Ramifications of Plastic Pollution* (pp. 197–222). IGI Global., DOI: 10.4018/979-8-3693-9163-1.ch010

Singh, B., Lal, S., Arora, M. K., & Kaunert, C. (2025). Mounting Legal-Driven Solutions for Plastic Pollution Focusing on Environment and Coastal Management: Eradicating Marine Pollution in Alignment With SDG 14 (Life Below Water). In Gaur, N., Sharma, E., Nguyen, T., Bilal, M., & Melkania, N. (Eds.), *Societal and Environmental Ramifications of Plastic Pollution* (pp. 223–252). IGI Global., DOI: 10.4018/979-8-3693-9163-1.ch011

Sobha, P. (2023). *Future Energy Landscapes in Northern Sweden: Sustainable Transition Scenarios for Municipalities* (Doctoral dissertation, Luleå University of Technology).

Sofi, I. I., Shah, M. A., & Ganie, A. H. (2023). Integrating human footprint with ensemble modelling identifies priority habitats for conservation: A case study in the distributional range of Arnebia euchroma, a vulnerable species. *Environmental Monitoring and Assessment*, 195(8), 914. DOI: 10.1007/s10661-023-11528-8 PMID: 37395941

Terry, N., Castro, A., Chibwe, B., Karuri-Sebina, G., Savu, C., & Pereira, L. (2024). Inviting a decolonial praxis for future imaginaries of nature: Introducing the Entangled Time Tree. *Environmental Science & Policy*, 151, 103615. DOI: 10.1016/j.envsci.2023.103615

Varughese, C., Henry, L., Morris, A., Bickerton, S., Rattenbury, N., Mankelow, C., Gorman, A., Katavich-Barton, S., & Dhopade, P. (2023). The intersection of space and sustainability: The need for a transdisciplinary and bi-cultural approach. *Acta Astronautica*, 211, 684–701. DOI: 10.1016/j.actaastro.2023.07.009

Watson, J. E., Ellis, E. C., Pillay, R., Williams, B. A., & Venter, O. (2023). Mapping Industrial Influences on Earth's Ecology. *Annual Review of Environment and Resources*, 48(1), 289–317. DOI: 10.1146/annurev-environ-112420-013640

Chapter 12
Role of Machine Learning and Deep Learning in Forest Management Through Data Analytics

Ravish
Meerut College, India

Anurag Singh
https://orcid.org/0009-0005-1322-7380
Meerut College, India

Anjali Raghav
https://orcid.org/0009-0004-0248-7956
Sharda University, India

Bhupinder Singh
https://orcid.org/0009-0006-4779-2553
Sharda University, India

Christian Kaunert
https://orcid.org/0000-0002-4493-2235
Dublin City University, Ireland

Tarun Kumar Kaushik
https://orcid.org/0000-0002-3279-8307
Sharda University, India

ABSTRACT

Everyone must create a good framework to handle environmental, financial, and social issues if we are to move towards sustainability. Many researchers have used artificial intelligence (AI) and machine learning to forward sustainable development goals by building a highly efficient system that supports a circular economy and aligns the needs of the current generation while safeguarding the capacities of future generations. Artificial intelligence has made significant progress in recent years, leading to significant changes in various industries including healthcare, transportation, agriculture, energy, and media. This paper offers a comprehensive

DOI: 10.4018/979-8-3693-7565-5.ch012

analysis of the junction between artificial intelligence (AI) and sustainability together with several possible directions for the next investigation.

INTRODUCTION

The scientific community is anticipated to endorse the efforts undertaken by policymakers. The Carbon Management Plan: PAS 2060 Qualifying Explanatory Statement provides a detailed explanation of the commitment to carbon neutrality and the strategies employed by industries or countries to attain and validate it. This study aims to showcase the global endeavors undertaken by firms to negotiate a contract that enhances energy efficiency and reduces greenhouse gas (GHG) emissions. Sectors such as transportation, building, and mining rely on non-renewable resources, which have adverse effects on the environment and present substantial issues for governments, such as economic crises and societal concerns. Global forums on sustainable development employ diverse indicators, competitive indices, and rankings to assess and monitor sustainable development objectives on a large scale. Nevertheless, the responsibility of stakeholders and lawmakers is limited due to the lack of effective investigative frameworks (Wicki et al., 2023).

Environmental sustainability encompasses a wide range of factors and involves making difficult choices, which can sometimes lead to the dismissal of difficulties through unduly simplified, satisfactory, and self-centered solutions. Trade can have positive as well as negative consequences on the surroundings. Growing pollution, resource depletion, and the environment's vulnerability to economic development are correlated. Solving problems of environmental sustainability depends on artificial intelligence in great part. Trade in several spheres of our surroundings can be profoundly affected by artificial intelligence (AI) (Potner et al., 2023). Using blockchain technologies, for example, can improve the token-based trade of digital assets or intangible goods. Ensuring regulatory compliance, quickening transaction speed, and enabling the usage of digital tokens for quick transactions help one to reach this. At last, these developments affect the scene of world trade. Likewise, there is an obvious trade-off between the early investments and the long-term advantages to the environment as well as between the general cost of production and the impact on the environment. Furthermore, traded off are the total cost and CO2 emissions (Watson et al., 2023).

Companies are working together to strike a balance between many factors vital for environmental sustainability and development. The multi-objective optimization approach presented by the authors helps one to understand the trade-offs resulting from considering multiple constraints of projects and their environmental feasibility, such cost, time, and quality. The use of artificial intelligence (AI) has reportedly the ability to improve the environmental sustainability of goods and services (Kennedy et al., 2023).

A nation's economic development depends much on artificial intelligence (AI), which also has the potential to support a worldwide economy of up to $15.7 trillion by 2030, according to 2018 PricewaterhouseCoopers (PwC) research. Analytical and symbolic approaches as well as many technologies and techniques constitute artificial intelligence. Artificial intelligence aims to replicate human cognitive capacities or show many features of human intelligence by means of task performance and decision-making procedures. These qualities mostly consist of analytical, intuitive, and sympathetic intelligence. Artificial intelligence (AI) continues to encounter numerous obstacles across different domains (Singh & Kaunert, 2024).

For instance, the relationship between continuous environmental states and optimal control decisions, techniques for predicting crashes, timely detection of hazardous traffic condition formation short-term crash risk results that can reveal crash hotspots and inform city planning large-scale entrepreneurship development projects, malware detection accurate wind power forecasting and prediction and other similar topics. Nevertheless, AI is limited and focused on domains of the environment and processes. For instance, cryptocurrencies consume an equivalent amount of energy to sustain the entire country of Finland. Therefore, it is imperative to examine the influence of AI on sustainability, since it plays a pivotal role in tackling the socioeconomic behavioral difficulties that are linked to economies (Dhali et al., 2023).

I. PROGRESSION OF ARTIFICIAL INTELLIGENCE

AI has developed over the last few years into a potent tool enabling machines to replicate human thought and action. AI has intellect, learning, and reasoning capacity if it is to serve a goal. In artificial intelligence, intelligence is the capacity to participate in thinking, problem solving, learning, and the integration of several human capacities like perception, attention, memory, language, and planning. Reasoning in artificial intelligence helps robots to analyze current data, facts, and knowledge to produce accurate predictions and sound conclusions by using deductive, inductive, abductive, and monotonic approaches. Learning is another aspect of artificial intelligence since it is the means of accumulating knowledge. Four forms of this process are distinguished: supervised learning, semi-supervised learning, unsupervised learning, and reinforcement learning. Although artificial intelligence depends on learning, it does not guarantee that the statistical rules learnt by a machine will be able to adapt to fast-changing surroundings (Singh, 2023).

Therefore, it is crucial to implement an artificial intelligence (AI) approach that can dynamically adapt, learn, and modify itself according to the requirements of sustainable development. Supervised learning is the process in data mining when a function or class is inferred from labelled training data. Semi-supervised learning involves utilizing a relatively modest quantity of labelled data alongside a substantial quantity of unlabeled data. Unsupervised learning involves the discovery of concealed structures and patterns within data that lacks labelling. Reinforcement learning models are increasingly becoming popular as robots acquire knowledge through trial-and-error approaches, mirroring the way people perform tasks and make decisions based on positive or negative rewards (Hansen et al., 2023).

Adaptation is a significant foundational aspect of AI, in which machines acquire knowledge by engaging in physical actions, gathering information from their senses, manipulating their surroundings, and cultivating an understanding of their surroundings. Artificial intelligence (AI) emulates the learning and adaptive abilities of biological creatures by employing mathematical functions to imitate true intelligence. This is evident in the behaviour of living organisms and animals, which utilize specific algorithms or a combination thereof. The literature suggests that AI has made great progress by adapting established approaches. Traditional methods such as Artificial Neural Networks (ANN) and decision trees are not efficient in adjusting to diverse environments. Emerging models such as reinforcement learning, Convolutional Neural Networks (CNN), Federated Model, TinyML, and Deep Neural Networks (DNN) have the capability to address complex challenges when combined with the use of AI in sustainability (Singh et al., 2022). This study focused on summarizing the methodologies of artificial intelligence (AI) that have been employed to investigate difficulties connected to sustainability (Batisha, 2023).

II. ECOLOGICAL SUSTAINABILITY

Four basic domains might help one generally classify the junction of artificial intelligence and sustainability: Human, social, economic, environmental. Using social, environmental, and financial data, artificial intelligence (AI) can perform quantitative research on the effects of various policies in many sectors and produce forecasts in related fields. Based on a lot of studies written in the scientific magazine Nature, artificial intelligence (AI) might help 79% of the Sustainable Development Goals (SDGs) be reached as reported by Vinuesa et al. in 2020. Many businesses have implemented artificial intelligence (AI) into their product development processes to improve their sustainable development practices following thorough investigation on the impact of AI on sustainability (Singh, 2024).

For builders and specialist seismic contractors, efficient tools and equipment management in building projects including accelerometers is a great obligation. This is so because it significantly affects many economic aspects including project cost and business profitability as well as environmental variables including CO2 emissions and NOX levels and health and safety issues. Government agencies are actively developing several strategies to address the obstacles related to surrounding sustainable development issues (Qwaider et al., 2023).

AI-enabled medical drone applications in one country's healthcare sector can have a significant impact on other countries' healthcare sectors, therefore helping to meet sustainable development goals, especially those pertaining to climate (SDGs 3, 8, and 13). Provided are strong healthcare, clean water and sanitation facilities, easily available and ecologically friendly energy sources, sustainable economic plans, and sensible consumption patterns. By following simple Sustainable Development Goals (SDGs), one can solve problems about biological life, innovation and infrastructure, and climate change. Among these methods are reducing carbon emissions and using noise-free drones for transportation of goods and services (Singh, 2024).

Driving innovation in several fields has been shown to be much benefited by artificial intelligence (AI) capable of efficiently processing and analyzing vast amounts. The AI-powered software estimates future consumer buying trends and can lower inventory management costs, offer reasonably priced storage and rental options, and increase tenants' income. Countries that practice supply chain management, for instance, use both physical and digital chain networks, manage high volumes of work, distribute resources effectively, meet deadlines by using AI technologies for optimization, and coordinate networks among many service partners (Singh, 2023).

III. SIGNIFICANT FACTORS FOR SUSTAINABLE DEVELOPMENT WHILE UTILIZING ARTIFICIAL INTELLIGENCE INCLUDE ISSUES, POSSIBILITIES, AND BARRIERS

To facilitate the sustainable development of AI-based technologies and address the increasing implications of AI, it is crucial to have the requisite regulatory knowledge and support. Failure to do so may lead to potential consequences such as compromised transparency in information dissemination, safety concerns, and ethical dilemmas (Sobha, 2023).

Challenges and concerns persist in the future integration of AI with sustainable development, despite our progress towards this goal. For instance, the growth of AI has had significant socioeconomic effects. However, the practicality of these services is still in the virtual stage and is encountering numerous implementation hurdles. While there is agreement on the significant socioeconomic effects of AI's current and future advancements, there is still much debate on the specific form of these advancements. After conducting a content analysis of the 287 articles, we have categorized barriers into the following segments (Muhmad Kamarulzaman et al., 2023):

- Environmental
- Social
- Technical factors are listed in Table 1.

Table 1. An overview of the obstacles and their descriptions from the chosen articles

Groups	Barriers	Description
1. Environment	• Factors such as meteorological circumstances, unfamiliar background conditions, labour competence and demand, and site logistics (such as traffic and pollution rules) can all influence the situation. • Inadequate management, security, stack, network, cost, software, hardware. • The topics of interest include technology, security, information sharing, user compliance, cost saving, ethical considerations for artificial intelligence technologies, and the lack of legislation/ regulation. • Topics include construction, household behaviour, consumption patterns, solar and wind energy, congestion, pollution, global warming, and noise. • Efficiency measure refers to the evaluation of how effectively resources are used to achieve a desired outcome. Environmental pressure refers to the impact or stress exerted on the environment due to human activities. Energy-intensive manufacturing refers to industries or processes that use a significant amount of energy to produce goods or services. Waste management involves the proper handling, disposal, and treatment of waste materials to minimise their negative impact on the environment. Dumping refers to the illegal or improper disposal of waste materials, often in unauthorised locations.	• Real-time transmission of accurate data for production management is crucial for industries that face risks such as rework, machine and tool failures, resource shortages, and unforeseen disasters, which can result in delayed and over-budget product delivery • Policymakers must integrate environmental indicators throughout the design phase. • Emerging agricultural technologies aimed at enhancing efficiency may give rise to several obstacles, including issues related to security, stacking, inter-operability, networking, cost management, and software and hardware maintenance • Blockchains are commonly employed to address security concerns in the process of exchanging information. Public stakeholders, such as the authorities, may express concerns regarding environmental issues such as traffic congestion, pollution, climate change, and noise. These concerns should be addressed without compromising urban development planning and the provision of employment opportunities • Ensuring awareness of factors such as stream connection and the presence of unmapped dams is crucial for understanding emerging human activities such as culverts. The environmental impact of the rebound-based levy is determined by the emission factor, which represents the amount of emissions per unit of expenditure in a specific consumption category. Understanding this relationship is crucial for addressing environmental concerns. The energy consumption pattern not only lowers gasoline costs but also mitigates adverse environmental effects.

continued on following page

Table 1. Continued

Groups	Barriers	Description
Impact on society	• Advanced technology is required for the seamless integration of digital content with the physical environment. This technology should be capable of accurately measuring resource efficiency, conserving energy, tracking real-time locations, protecting against cyber-attacks, and supporting strategic decision-making. Additionally, it should enable the use of Internet of Things (IoT) enabled products and handle unstructured data. • Knowledge representation, logical deduction, strategic decision-making, pattern recognition, handling uncertain information, and interactive communication	• Supply chain management professionals can utilize digital twins to analyze the impact of society, environment, and technology on supply chain management. AI in education encompasses the application of AI techniques to address several challenges in the field, such as knowledge management, reasoning, planning, prediction, training, uncertainty in knowledge, and natural language processing.
Effects on the economy	• Enhanced data transparency, accessibility, and consumption, as well as the execution of policies, enable accurate forecasting of energy consumption patterns. • Challenges related to the availability, organisation, accessibility, and trustworthiness of data, as well as challenges related to management. • Designing and managing operations, measuring the advantages	• Service firms possess valuable data resources from utility distributors that aid city planners in assessing the sustainability and carbon reduction strategies within cities. • Along with the application of measures to improve the accuracy of energy demand forecasting and so extend our knowledge of urban energy dynamics, the use of data analysis tools to investigate trends of energy consumption and the consequences of policies. • Stakeholders must properly address the growing acceptance and maintenance-related issues. • Another element that calls for consideration is performance problems. Including a feedback mechanism within the system where GPS may offer comments or suggestions for development would help to increase acceptance, use, and maintenance capability. Future planning cycles should routinely go over these remarks and recommendations and apply them. Furthermore, one should improve and thoroughly monitor the use.

Potential in this field: Artificial intelligence can help society and businesses in many different respects. Artificial intelligence (AI) can be used by companies to maximize the use of current assets, distribute resources more effectively, and enhance data and information management and sharing across many fields (Bogers, 2023). Using intelligent chatbots for customer service in the banking sector, artificial intelligence (AI) offers individualized services to individuals and customized client care, therefore providing by accurately forecasting weather patterns and fast spotting power infrastructure weaknesses, artificial intelligence might be used to forecast future power outages. This evaluation helps to minimize environmental disasters by allowing users to proactively stop outages and react fast to any such events (Singh, 2023).

IV. THE ENVIRONMENTAL IMPLICATIONS OF ARTIFICIAL INTELLIGENCE

Artificial Intelligence in the Field of Water Management

In the fields of treatment, transportation, and water recycling especially, sustainable solutions are much sought. Given rising pollution levels and declining fresh, uncontaminated water availability, we must create technology to guarantee the long-term sustainability of this resource. One of the main focus of the 17 Sustainable Development Goals (SDGs) globally is water management. Different toxicity levels of the particulate matter present in water call for appropriate actions to minimize their impact on the surroundings and society. Machine learning and artificial intelligence (AI) have shown to be effective approaches for spotting and fixing dangerous particles in water (Nahar, 2024).

The selection of algorithms and methodologies employed in wastewater treatment can have profound consequences for the well-being of humanity. A computational framework powered by artificial intelligence is required to identify many options for wastewater management. Artificial intelligence can continuously adjust and modify itself in response to new sources of knowledge. By integrating the latest AI service algorithms, it is possible to develop a novel strategic framework to enhance wastewater management and increase its efficiency. The understanding of artificial intelligence harnesses the capabilities of water experts (Desai et al., 2023).

Nevertheless, there is a problem with the artificial intelligence (AI) system used in water management regarding the type of data it receives and the amount of comprehension that the management has of that data. Human cognitive abilities are limited when it comes to comprehending machine language and effectively using that understanding to solve issues. Currently, the comprehension of AI still relies on professionals, but as time progresses, it is possible that human involvement may become unnecessary. The objective of AI is not to be exemplary, but rather to facilitate the efficient management of resources (Sharma & Singh, 2022).

Artificial Intelligence and its Use in the Field of Agriculture

Given the current global birth rate, the global population is projected to reach 9.1 billion by 2050, and the demand for food would increase to 56% from the current 35%. Therefore, it is imperative to establish a system that not only enhances output but also monitors diseases and makes forecasts. In the past, soil quality and crop health were assessed by manual methods and human observation. Nevertheless, recent algorithms are more suitable and precise. We employ unmanned aerial vehicles (UAVs), also known as drones, to take high-altitude photos (Ashukem & Sama, 2023). Subsequently, we train our model to accurately analyze the acquired data. Visual sensing AI can analyze and interpret data to assess crop health, predict yields with precision, monitor agricultural diseases, and do climate analysis for crop growth.

The initial step in building the AI database involves determining the 30 most crucial diseases that can impact crops and are considered significant for the AI training model. Given good quality photos, the model can effectively recognize around 99% of them. Still, this efficiency declines to 32% with poor image quality. While image analysis can be helpful for small-scale forecasts, a more comprehensive technique is needed for large-scale predictions that transcend only depending on images. Still, the most recent artificial intelligence-based model uses low-quality data to perform analysis, allowing small-scale farmers to improve their agricultural output by means of this use. Consequently, low-income farmers with limited resources can use artificial intelligence (AI) to improve the quality of their produced agricultural products. Artificial intelligence could help to close the gap between developed countries and developing ones (Cornford et al., 2023).

Artificial Intelligence (AI) and its Impact on Transportation

Driving the technical developments of autonomous vehicles (AVs), which are regarded as a major achievement in the world of transportation, artificial intelligence (AI) is crucial. To run effectively, an autonomous car depends on an exact awareness of its surroundings. In the automobile sector, the Internet of Things (IoT) is a decentralized network that, via data collecting, analysis, and data execution, supports group learning. One may see the growing volume of data and the evolution of artificial intelligence as advances in knowledge-sharing among smart automobiles. Though a lot of data is readily available for study, knowledge sharing runs against many major challenges (Singh, 2022). Regarding data transfer, the current transportation system has major dependability problems and lacks sufficient security mechanisms. By sending erroneous signals, incorrect information, or interfering with knowledge reception, destructive vehicles can greatly impede the process of knowledge sharing. Privacy is seriously threatened by compiling all data at the central server to handle the worldwide training model on the Internet of Vehicles. This is so because the data contains personal information of consumers, such driving behaviours and location choices, which indirectly link to users' safety and driving circumstances (Koh et al., 2023).

Thus, certain challenges must be resolved before commercial promotion of the use of autonomous cars for wide acceptance. The several consequences and technological difficulties related to DL-based autonomous cars will be discussed in this part. Apart from technological challenges, there are other issues that need to be taken care of include user acceptance, cost effectiveness, ethical issues for artificial intelligence, and the absence of laws and rules for autonomous cars. The ADLCM model created after combines different decision thresholds to examine how driving conditions affect driving decisions (Sheng & Lian, 2023). The study produced a new variable called the "congestion index" (CI), which takes flow and speed together at every detector point. Traffic encryption issues provide further difficulties for traffic management. Different machine learning techniques are applied to address the challenges related to autonomous cars, such statistical flow characterizing Studies have shown that, independent of traffic volume and speed, the danger of collision is most evident after midnight. At signalized intersections, there is more chance of mishaps. Previous research has not, however, considered the effect of non-signalized junctions, weather circumstances, or road surface conditions on the occurrence of accidents at each intersection (Singh, 2022).

Artificial Intelligence and its Applications in the Healthcare Industry

Artificial intelligence (AI) and machine learning (ML) have great promise in the medical field. AI has the remarkable capacity to foretell diseases with perfect accuracy without human involvement. Technology helps people to combine better practices and embrace a healthy way of life. By means of improved healthcare services and better feedback, guidance, and assistance to patients in preserving their well-being, artificial intelligence (AI) improves the performance of healthcare providers (Ashukem, 2024).

Using prediction methods, the AI prediction model can foresee dependable conditions and effectively evaluate the environmental air quality (AQ). Urban partners involved in important and necessary activities depend on a precise and customized air quality index since the possible hazards to human life call for it. The DL-based model might be utilized to precisely monitor and evaluate wireless signals in indoor environments, therefore enabling the identification of several human activities like absence, working, and sleeping. In the realm of artificial intelligence, electrical resistivity measurement (ERM) is prone to several major uncertainty resulting from many different sources (Singh & Kaunert, 2024). Like other algorithms, the XGBoost one provides a consistent and predictable way to combine ERM results to forecast and assess the health of several buildings including water/cement ratio, chloride permeability, and settlement time. (Prasanna et al., 2023). Artificial intelligence is extensively used in healthcare largely for the establishment of a comprehensive understanding and classification of clinical documentation by means of Natural Language Processing (NLP). Artificial intelligence (AI) helps the medical sector to better grasp daily needs and classification of people who call for healthcare services. This helps to provide better direction, comments, and required support so enabling individuals to keep their well-being (Terry et al. 2024).

AI in healthcare might cover a broad spectrum of tasks in the not too future. In the field of healthcare, daily responsibilities include answering phone calls, analyzing patient records, projecting population health trends, evaluating the effects of drugs, creating healthcare tools, interpreting radiography images, rendering clinical diagnosis, and rendering treatments.

Artificial Intelligence and its Application in the Industrial Industry

The manufacturing sector is a crucial pillar of every nation and possesses the ability to exert an enormous effect on the country's economy, environment, and society. Sustainable manufacturing in the economy leads to significant changes in the company model, fosters higher economic growth, enhances the reputation of trade services across the entire production process, and supports the development of diverse economic modes of trade. Sustainable manufacturing reduces the consumption and waste of raw materials, enhances the efficient utilization of resources, and slows down the emission of pollution from an environmental standpoint (Ashukem & Sama, 2023).

Augmented reality (AR), a component of artificial intelligence (AI), has proven to be an effective tool for reducing cognitive load by bridging the gap between the task at hand and the necessary knowledge. It achieves this by providing users with the essential information they require. Augmented reality is an advantageous algorithm utilized in the manufacturing sector to efficiently carry out a diverse array of tasks such as product assembly, product and process proliferation, product distribution, and maintenance of different manufacturing sections, all while minimizing costs. AR can optimize manufacturing processes by minimizing costs and enhancing production efficiency throughout all stages of the product life cycle (Fopa et al., 2023).

The manufacturing process involves a wide array of entities, including urban logistics, production enterprises, suppliers, wholesalers, retailers, logistics corporations, stakeholders, and governments. Various stakeholders are involved in manufacturing sectors, such as users or consumers, investors or shareholders, suppliers, government agencies, and the broader community. Typically, governments have a role in creating policies, organizing activities, and developing strategies for the efficient distribution of commodities in metropolitan areas to promote sustainable manufacturing practices. Public stakeholders actively engage in environmental matters related to pollution, global warming, and noise, so ensuring minimal harm to the environment (Ozkavaf Senalp, 2023).

Human resource industry. Private stakeholders, such as logistic service providers, manufacturing enterprises, and retailers, closely monitor the economic efficiency and benefits of the manufacturing sector. They believe that the economic well-being of nations, along with fulfilling social responsibilities and protecting the environment, contributes to the sustainability of urban logistics (Cumming et al., 2023).

(Cork et al. 2023) emphasized that digital twins enable owners to detect operational failures, improve product quality, and minimize disruptions prior to incurring any economic losses. Additionally, the Internet of Things (IoT), simulation algorithms, machine learning (ML), artificial intelligence (AI), cloud computing, and other similar technologies play a crucial role in enabling a digital twin. By extending the digital twin algorithm's scope to include small businesses as well as human supply chain participants, owners and stakeholders will be able to make quick decisions in the supply chain (Gadsden et al., 2023). Through the concept of the circular economy, the writers hope to underline that, by means of reclaiming long-lasting goods already at the end of their life cycle, environmental sustainability is achievable. Metal additive manufacturing technology has attracted increasing attention under terms like remanufacturing, repair, and restoration, which have spurred different manufacturing developments. Optimizing the basic design is essential to improve manufacturing efficiency in the manufacturing sector (Lamalle & Stoett, 2023). The industrial sector needs a well-defined decision support system to properly control the supply chain and guarantee sustainability. This structure should help companies manage the social, environmental, and financial sides of the supply chain in a good way. This structure helps a multi-party cooperative approach across several networks to evolve to improve sustainability in the delivery of completed products. In industrial sectors, the energy consumption pattern is a fundamental component of sustainability and hence a major focus of research to grasp the several consequences on the environment and global climate change. Models like sensitivity analysis and parametric analysis help one to expose the link between energy use and other properties (Dunlap, 2023).

V. POTENTIAL AREAS FOR FUTURE INVESTIGATION

The paper's findings suggest that most of the study focuses on Transportation, Communication, Electric, Gas, and Sanitary services. However, there is a scarcity of research in sectors such as Mining, Wholesale Trade, Public Administration, agriculture, forestry, and fishery. These domains have the potential to be future research focuses in the field of AI and sustainability. Researchers can investigate several algorithms, such as Adaptive Neuro-Fuzzy Inference Systems (ANFIS), Binary Tree, Gradient Boosting, Random Forest, Clustering, Ant Colony, Heuristic, and Sensitivity Analysis, to analyze the influence of AI on sustainable development methods. Additionally, in reference to Figure. Researchers can investigate these techniques by applying them to use cases such as Sequence Mining, Anomaly Detection, Natural Language Processing, Ensemble, Association, and Deep Learning. A suggested future direction of the impact of AI on sustainability has been determined by a comparative examination of chosen literature (Newman et al. 2024).

Our analysis indicates that the NLP algorithm is mostly utilized in the manufacturing sector. On the other hand, the LSTM algorithm finds application in the manufacturing, non-classifiable (IT, Environment, water, agriculture), and transportation sectors. In these sectors, LSTM is employed for tasks such as reinforcement learning, classification and regression, and association. In several industries such as transportation and non-classifiable sectors, the DNN (Deep Neural Network) and RNN (Recurrent Neural Network) algorithms are employed for classification and regression purposes. Still, the usage of Deep Neural Networks (DNN) has not been looked at in the service sector and Recurrent Neural Networks (RNN) have not been looked at in the construction business. From manufacturing to services to other areas that cannot be readily categorized, SEM, or structural equation modelling, is used in many different fields (Ghosh et al., 2023).

It is used in concert with association methods, classification, and regression. Common applications of the Taguchi approach are in fields including manufacturing for classification and regression needs. Common applications of the Support Vector Regression (SVR) technique include in areas including construction and others that cannot be readily categorized. In Natural Language Processing (NLP), regression analysis, and ensemble learning especially so it is highly valuable. Based on the count. Investigating new algorithms in numerous fields and using them in many situations helps researchers to draw conclusions for next research. Within the discipline of natural language processing (NLP), the manufacturing sector is among the unexplored ground still. Likewise, companies in mining, services, administration, and others are still in early phases of investigating LSTM algorithms. This understanding can be rather helpful for next studies and help to advance real sustainable development.

In the future, research methodologies such as Federated learning and Tiny ML will be utilized. There are numerous opportunities in the field of sustainability. Federated learning operates effectively on distributed servers that store substantial local datasets without the need for data exchange between them. It minimizes network burden and guarantees anonymity for both devices and organizations. Typically, a Machine learning algorithm consume a significant amount of energy and necessitate substantial computer capacity. For instance, the execution of a single deep learning model can produce carbon dioxide emissions equivalent to the total emissions generated by five cars during their entire lifespan (Li et al., 2020). This could potentially give rise to significant environmental concerns. Federated learning can be a viable method for implementing sustainable development strategies in such circumstances (Shinde et al., 2023)

Tiny ML has enabled the deployment of machine learning capabilities at the forefront of progress. The primary motive behind our shift in attention towards TinyML is sustainability. This is a convergence of low-power embedded systems and machine learning, where the computational processing takes place at the edges. It is referred to as the next generation of artificial intelligence (Cardwell, 2023). Due to the enormous increase in carbon emissions resulting from advancements in AI, it is necessary to design more energy-efficient computational algorithms to promote sustainable development practices. To address these problems, TinyML excels in achieving maximum benefits through efficient computing. It provides a viable approach to decrease our carbon use and utilization in a way that can be maintained over time. Until now, it has garnered less attention due to certain limitations linked to microcontroller units. Nevertheless, this has the potential to enhance the efficiency of existing devices in the long term (Aryal et al., 2023).

VI. ARTIFICIAL INTELLIGENCE (AI) AND ITS USE IN THE FIELD OF CONSTRUCTION

The building sector uses artificial intelligence (AI) to provide instantaneous communication between many entities including workers, machines, raw materials, and managers. It guarantees additionally real-time application of safety precautions, construction fault identification, and production level monitoring. Ensuring sustainability in construction depends on exact prediction of energy consumption. Broadly analyzing the energy consumption pattern helps improve general energy efficiency (Ngadze, 2023). Thus, energy-efficient buildings are essential in advancing sustainability and improving the energy consumption patterns in society. Building sectors' energy consumption models call for different energy consumption degrees. For example, the use of distributed energy sectors or combined energy systems helps to identify factors influencing the features related with building a structure, which is essential for the design and running of such a system. The physical model created with sophisticated software such as Energy Plus identifies the effects that building occupants can have on electric lighting, water heaters, and water coolers. Additionally, meteorological data has a significant impact on heat energy consumption (Sofi et al., 2023).

The diverse materials utilized in construction, such as concrete, provide a high level of durability. Fireproof, cost-effective, and durable. Therefore, this building site is garnering significant interest from academics to implement sustainable development objectives. Due to urbanization and industrialization, there has been a consequent growth in the need for housing (Shukla et., 2023). Therefore, a significant number of raw materials is required for global concrete production, resulting in the production of industrial wastes, agricultural wastes, and other forms of solid wastes that might potentially cause environmental problems. To mitigate the adverse effects of concrete debris produced because AI can be utilized to effectively manage the consumption of smart materials, such as complementary cementitious material, which can contribute to the environmental sustainability of industries and national urban initiatives like smart cities (Malekpour et al., 2023).

VII. CONCLUSION

In this study, we have endeavored to encompass all the sectors of Session Initiation Protocol (SIP)-based communication (Singh & Kaunert, 2025). This study examines the impact of artificial intelligence (AI) on industries and its influence on their sustainable management practices. The industries are categorized, and the matching utilization of AI in these sectors is recognized (Singh & Kaunert, 2025). The quantitative data indicates that the industries that can be classified, as well as those that cannot be classified, such as manufacturing, construction, and services, have extensively embraced AI-based solutions to enhance their SDG practices (Singh et al., 2025). Among all the articles focused on AI and sustainable practices, our findings indicate that 74% of sectors employ both supervised and unsupervised algorithms. Nevertheless, the suggested framework is still in its early stages as sustainable management approaches necessitate time, and implementing these methods using AI necessitates significant computational efficiency (Singh & Kaunert, 2025). We have compiled a comprehensive list of all the technologies that have been utilized thus far for the purpose of sustainable management practices (Singh et al., 2025). Researchers and practitioners are seeking to advance research in this field. Based on the information provided, researchers can identify deficiencies in existing publications on the implementation of artificial intelligence in sustainable development goals (SDGs) and can conduct future studies.

REFERENCES

Anekwe, I. M. S., Zhou, H., Mkhize, M. M., & Akpasi, S. O. (2024). Addressing Climate Change Challenges in South Africa: A Study in KwaZulu Natal Province. *Climate Crisis: Adaptive Approaches and Sustainability*, 475-496.

Aryal, K., Maraseni, T., & Apan, A. (2023). Examining policy– institution– program (PIP) responses against the drivers of ecosystem dynamics. A chronological review (1960–2020) from Nepal. *Land Use Policy*, 132, 106789. DOI: 10.1016/j.landusepol.2023.106789

Ashukem, J. C. N. (2024). Land grabbing in the Anthropocene and the "questionable" pursuit of sustainability in Africa–highlighting ecological disproportionality. In *Human Rights and the Environment in Africa* (pp. 276–294). Routledge.

Ashukem, J. C. N., & Sama, S. M. (Eds.). (2023). *Domestic and Regional Environmental Laws and Policies in Africa: A Research Companion*. Taylor & Francis. DOI: 10.4324/9781003382256

Banerjee, O., Cicowiez, M., Malek, Ž., Verburg, P. H., Vargas, R., Goodwin, S., Bagstad, K. J., & Murillo, J. Á. (2023). Banking on strong rural livelihoods and the sustainable use of natural capital in post-conflict Colombia. *Environment, Development and Sustainability*, 26(10), 1–22. DOI: 10.1007/s10668-023-03740-w

Basu, M. (2024). *The Routledge Handbook of International Environmental Policy*. Taylor & Francis. DOI: 10.4324/9781003440574

Batisha, A. (2023). A lighthouse to enhance the quality of life in the Nile River basin. *Environmental Economics and Policy Studies*, 1–29. DOI: 10.1007/s10018-023-00380-2

Bogers, M. (2023). *Integrating global sustainability governance: How the Sustainable Development Goals impact institutional and policy integration at the global level* (Doctoral dissertation, Utrecht University).

Bustamante, M., Roy, J., Ospina, D., Achakulwisut, P., Aggarwal, A., Bastos, A., ... & Zscheischler, J. (2023). Ten new insights in climate science 2023/2024. *Global Sustainability*, 1-58.

Cardwell, M. (2023). Results-based agri-environmental scheme design: Legal implications. *Environmental Law Review*, 25(4), 260–288. DOI: 10.1177/14614529231185678

Cork, S., Alexandra, C., Alvarez-Romero, J. G., Bennett, E. M., Berbés-Blázquez, M., Bohensky, E., & Wyborn, C. (2023). Exploring Alternative Futures in the Anthropocene. *Annual Review of Environment and Resources*, 48(1), 25–54. DOI: 10.1146/annurev-environ-112321-095011

Cornford, R., Spooner, F., McRae, L., Purvis, A., & Freeman, R. (2023). Ongoing over-exploitation and delayed responses to environmental change highlight the urgency for action to promote vertebrate recoveries by 2030. *Proceedings of the Royal Society B, 290*(1997), 20230464. DOI: 10.1098/rspb.2023.0464

Cumming, G. S., Adamska, M., Barnes, M. L., Barnett, J., Bellwood, D. R., Cinner, J. E., Cohen, P. J., Donelson, J. M., Fabricius, K., Grafton, R. Q., Grech, A., Gurney, G. G., Hoegh-Guldberg, O., Hoey, A. S., Hoogenboom, M. O., Lau, J., Lovelock, C. E., Lowe, R., Miller, D. J., & Wilson, S. K. (2023). Research priorities for the sustainability of coral-rich western Pacific seascapes. *Regional Environmental Change*, 23(2), 1–15. DOI: 10.1007/s10113-023-02051-0 PMID: 37125023

De Vries, B. J. (2023). *Sustainability science.* Cambridge University Press. DOI: 10.1017/9781009300230

Desai, B. H. (2023). Land-Soil Sustainability: India's International Obligations. In *Soil Law and Governance in India* (pp. 3–24). Springer International Publishing. DOI: 10.1007/978-3-031-32360-7_1

Dhali, M., Hassan, S., & Subramaniam, U. (2023). Comparative analysis of oil and gas legal frameworks in Bangladesh and Nigeria: A pathway towards achieving sustainable energy through policy. *Sustainability (Basel)*, 15(21), 15228. DOI: 10.3390/su152115228

Dunlap, A. (2023). The Structures of Conquest: Debating Extractivism(s), Infrastructures and Environmental Justice for Advancing Post-Development Pathways. *International Development Policy| Revue internationale de politique de développement.*

ESCAP. (2023). The changing landscape of regional cooperation in Asia and the Pacific.

Fopa, V. K., Bayir, N., & Özdal, D. (2023). Assessing the status and spatial-temporal dynamics of the Bamenda Mountains (BM), North West region of Cameroon. *Environmental Monitoring and Assessment*, 195(9), 1053. DOI: 10.1007/s10661-023-11630-x PMID: 37589789

Gadsden, G. I., Golden, N., & Harris, N. C. (2023). Place-Based Bias in Environmental Scholarship Derived from Social–Ecological Landscapes of Fear. *Bioscience*, 73(1), 23–35. DOI: 10.1093/biosci/biac095 PMID: 36643594

Ghosh, S., Dinda, S., Chatterjee, N. D., & Bera, D. (2023). Linking ecological vulnerability and ecosystem service value in a fast-growing metropolitan area of eastern India: A scenario-based sustainability approach. *Environment, Development and Sustainability*, •••, 1–31. DOI: 10.1007/s10668-023-03966-8

Gill, D. A., Blythe, J., Bennett, N., Evans, L., Brown, K., Turner, R. A., Baggio, J. A., Baker, D., Ban, N. C., Brun, V., Claudet, J., Darling, E., Di Franco, A., Epstein, G., Estradivari, , Gray, N. J., Gurney, G. G., Horan, R. P., Jupiter, S. D., & Muthiga, N. A. (2023). Triple exposure: Reducing negative impacts of climate change, blue growth, and conservation on coastal communities. *One Earth*, 6(2), 118–130. DOI: 10.1016/j.oneear.2023.01.010

Hansen, H. H., Bergman, E., Cowx, I. G., Lind, L., Pauna, V. H., & Willis, K. A. (2023). Resilient rivers and connected marine systems: a review of mutual sustainability opportunities. *Global Sustainability*, 1-71.

Jarvelainen, M. (2023). Evaluating the Effectiveness of Forest Conservation as a CO2 Offset Measure for Urban Development: A Case Study. *Forest Ecology and Management.*

Kennedy, C. M., Fariss, B., Oakleaf, J. R., Garnett, S. T., Fernández-Llamazares, Á., Fa, J. E., Baruch-Mordo, S., & Kiesecker, J. (2023). Indigenous Peoples' lands are threatened by industrial development; conversion risk assessment reveals need to support Indigenous stewardship. *One Earth*, 6(8), 1032–1049. DOI: 10.1016/j.oneear.2023.07.006

Khan, I. A. (Ed.). (2024). *Security and Global Governance*. Interdisciplinary Institute of Human Security & Governance.

Koh, T., Lye, L. H., & Lum, S. (Eds.). (2023). *Peace With Nature: 50 Inspiring Essays On Nature And The Environment*. World Scientific. DOI: 10.1142/13561

Lamalle, S., & Stoett, P. (Eds.). (2023). *Representations and Rights of the Environment*. Cambridge University Press. DOI: 10.1017/9781108769327

Malekpour, S., Allen, C., Sagar, A., Scholz, I., Persson, Å., Miranda, J. J., Bennich, T., Dube, O. P., Kanie, N., Madise, N., Shackell, N., Montoya, J. C., Pan, J., Hathie, I., Bobylev, S. N., Agard, J., & Al-Ghanim, K. (2023). What scientists need to do to accelerate progress on the SDGs. *Nature*, 621(7978), 250–254. DOI: 10.1038/d41586-023-02808-x PMID: 37704766

Muhmad Kamarulzaman, A. M., Wan Mohd Jaafar, W. S., Mohd Said, M. N., Saad, S. N. M., & Mohan, M. (2023). UAV Implementations in Urban Planning and Related Sectors of Rapidly Developing Nations: A Review and Future Perspectives for Malaysia. *Remote Sensing (Basel)*, 15(11), 2845. DOI: 10.3390/rs15112845

Nahar, S. (2024). Modeling the effects of artificial intelligence (AI)-based innovation on sustainable development goals (SDGs): Applying a system dynamics perspective in a cross-country setting. *Technological Forecasting and Social Change*, 201, 123203. DOI: 10.1016/j.techfore.2023.123203

Neimark, B. (2023). *Hottest of the Hotspots: The Rise of Eco-precarious Conservation Labor in Madagascar*. University of Arizona Press. DOI: 10.2307/jj.7941382

Newman, R. J., Enns, C., Capitani, C., Thorn, J. P., Courtney-Mustaphi, C. J., Buckton, S. J., Om, E. S., Fazey, I., Haji, T. A., Nchimbi, A. Y., Kariuki, R. W., & Marchant, R. A. (2024). 'Kesho' Scenario Development for Supporting Water-Energy Food Security under Future Conditions in Zanzibar. *Land (Basel)*, 13(2), 195. DOI: 10.3390/land13020195

Ngadze, F. (2023). Applying the Safe and Just Operating Space (SJOS) framework to Sustainable Development in Zimbabwe.

Özkavaf Şenalp, S. I. L. A. (2023). Exploring resilience of socio-ecological productive landscape through understanding change, impact and response among farmers: The case of northwestern Ankara.

Papas, M. (2023). *Capacity-Building and the Water-Energy-Food Nexus: Rethinking Integration in the Asia-Pacific*. Taylor & Francis. DOI: 10.4324/9780429342905

Platjouw, F. M., & Pozdnakova, A. (Eds.). (2023). *The Environmental Rule of Law for Oceans: Designing Legal Solutions*. Cambridge University Press. DOI: 10.1017/9781009253741

Pörtner, H. O., Scholes, R. J., Arneth, A., Barnes, D. K. A., Burrows, M. T., Diamond, S. E., Duarte, C. M., Kiessling, W., Leadley, P., Managi, S., McElwee, P., Midgley, G., Ngo, H. T., Obura, D., Pascual, U., Sankaran, M., Shin, Y. J., & Val, A. L. (2023). Overcoming the coupled climate and biodiversity crises and their societal impacts. *Science*, 380(6642), eabl4881. DOI: 10.1126/science.abl4881 PMID: 37079687

Prasanna, S., Ameen, A., Saurav, K., Singh, D. B., Tripathy, A., Sharma, P., ... Sanjaya, K. (2023). *Empowerment and equality navigating human rights law in a complex world*. Institute of Legal Education.

Quamar, M. M., Al-Ramadan, B., Khan, K., Shafiullah, M., & El Ferik, S. (2023). Advancements and Applications of Drone-Integrated Geographic Information System Technology—A Review. *Remote Sensing (Basel)*, 15(20), 5039. DOI: 10.3390/rs15205039

Qwaider, S., Al-Ramadan, B., Shafiullah, M., Islam, A., & Worku, M. Y. (2023). GIS-Based Progress Monitoring of SDGs towards Achieving Saudi Vision 2030. *Remote Sensing (Basel)*, 15(24), 5770. DOI: 10.3390/rs15245770

Santarsiero, V. (2023). Spatial analyzes and remote sensing for land cover change dynamics: assessing in a spatial planning.

Sarker, M. N. I., Hossain, B., Shi, G., & Firdaus, R. R. (2023). Promoting net-zero economy through climate-smart agriculture: Transition towards sustainability. *Sustainability Science*, 18(5), 2107–2119. DOI: 10.1007/s11625-023-01379-0

Sharma, A., & Singh, B. (2022). Measuring Impact of E-commerce on Small Scale Business: A Systematic Review. *Journal of Corporate Governance and International Business Law*, 5(1).

Sheng, S., & Lian, H. (2023). The spatial pattern evolution of rural settlements and multi-scenario simulations since the initiation of the reform and opening up policy in China. *Land (Basel)*, 12(9), 1763. DOI: 10.3390/land12091763

Shinde, V. R., Mishra, R. R., Bhonde, U., & Vaidya, H. (Eds.). (2023). *Managing Urban Rivers: From Planning to Practice*. Elsevier.

Shukla, G., Bhat, J. A., Chakravarty, S., Almutairi, A., Li, M., & Iyer-Raniga, U. (2023). *Floristic Diversity: Biology and Conservation*. BoD–Books on Demand. DOI: 10.5772/intechopen.100674

Singh, B. (2022). Relevance of Agriculture-Nutrition Linkage for Human Healthcare: A Conceptual Legal Framework of Implication and Pathways. *Justice and Law Bulletin*, 1(1), 44–49.

Singh, B. (2022). COVID-19 Pandemic and Public Healthcare: Endless Downward Spiral or Solution via Rapid Legal and Health Services Implementation with Patient Monitoring Program. *Justice and Law Bulletin*, 1(1), 1–7.

Singh, B. (2023). Blockchain Technology in Renovating Healthcare: Legal and Future Perspectives. In *Revolutionizing Healthcare Through Artificial Intelligence and Internet of Things Applications* (pp. 177-186). IGI Global.

Singh, B. (2023). Federated Learning for Envision Future Trajectory Smart Transport System for Climate Preservation and Smart Green Planet: Insights into Global Governance and SDG-9 (Industry, Innovation and Infrastructure). *National Journal of Environmental Law*, 6(2), 6–17.

Singh, B. (2024). Lensing Legal Dynamics for Examining Responsibility and Deliberation of Generative AI-Tethered Technological Privacy Concerns: Infringements and Use of Personal Data by Nefarious Actors. In Ara, A., & Ara, A. (Eds.), *Exploring the Ethical Implications of Generative AI* (pp. 146–167). IGI Global., DOI: 10.4018/979-8-3693-1565-1.ch009

Singh, B. (2024). Social Cognition of Incarcerated Women and Children: Addressing Exposure to Infectious Diseases and Legal Outcomes. In Reddy, K. (Ed.), *Principles and Clinical Interventions in Social Cognition* (pp. 236–251). IGI Global., DOI: 10.4018/979-8-3693-1265-0.ch014

Singh, B. (2024). Evolutionary Global Neuroscience for Cognition and Brain Health: Strengthening Innovation in Brain Science. In Prabhakar, P. (Ed.), *Biomedical Research Developments for Improved Healthcare* (pp. 246–272). IGI Global., DOI: 10.4018/979-8-3693-1922-2.ch012

Singh, B., & Kaunert, C. (2024). Computational Thinking for Innovative Solutions and Problem-Solving Techniques: Transforming Conventional Education to Futuristic Interdisciplinary Higher Education. In Fonkam, M., & Vajjhala, N. (Eds.), *Revolutionizing Curricula Through Computational Thinking, Logic, and Problem Solving* (pp. 60–82). IGI Global., DOI: 10.4018/979-8-3693-1974-1.ch004

Singh, B., & Kaunert, C. (2024). Revealing Green Finance Mobilization: Harnessing FinTech and Blockchain Innovations to Surmount Barriers and Foster New Investment Avenues. In Jafar, S., Rodriguez, R., Kannan, H., Akhtar, S., & Plugmann, P. (Eds.), *Harnessing Blockchain-Digital Twin Fusion for Sustainable Investments* (pp. 265–286). IGI Global., DOI: 10.4018/979-8-3693-1878-2.ch011

Singh, B., & Kaunert, C. (2025). Featuring Healthcare, Environment, and Human Rights: Applications of Artificial Intelligence in the Health Domain. In Chakraborty, S., & Satapathy, S. (Eds.), *Gender, Environment, and Human Rights: An Intersectional Exploration* (pp. 45–74). IGI Global., DOI: 10.4018/979-8-3693-6069-9.ch004

Singh, B., & Kaunert, C. (2025). Intelligent Machine Learning Solutions for Cybersecurity: Legal and Ethical Considerations in a Global Context. In Thangam, D. (Ed.), *Advancements in Intelligent Process Automation* (pp. 359–386). IGI Global., DOI: 10.4018/979-8-3693-5380-6.ch014

Singh, B., & Kaunert, C. (2025). Innovative Business Models in Digital and Virtual Reality for the Gig Economy: Space-Age Sustainable Development Projecting Smart Business Practices. In Taherdoost, H., Drazenovic, G., Madanchian, M., Khan, I., & Arshi, O. (Eds.), *Business Transformation in the Era of Digital Disruption* (pp. 127–156). IGI Global., DOI: 10.4018/979-8-3693-7056-8.ch005

Singh, B., Kaunert, C., & Ravesangar, K. (2025). Artificial Intelligence and Employee Well-Being: Projecting Psychological and Physical Healthcare. In Yadav, P. (Ed.), *Human Resource Strategies in the Era of Artificial Intelligence* (pp. 109–134). IGI Global., DOI: 10.4018/979-8-3693-6412-3.ch005

Singh, B., Singh, A., Kaunert, C., Arora, M. K., Lal, S., & Ravesangar, K. (2025). Unleashing the Ethical and Legal Implications of E-Business Revolution: Consumer Privacy and Security Concerns in Phase of Digital Disruption. In Taherdoost, H., Drazenovic, G., Madanchian, M., Khan, I., & Arshi, O. (Eds.), *Business Transformation in the Era of Digital Disruption* (pp. 157–180). IGI Global., DOI: 10.4018/979-8-3693-7056-8.ch006

Sliuzas, R., Mwesigye, P., Kisembo, T., Wamboga, J., & Ajambo, E. (2023). Land Governance for Climate Resilience.

Sobha, P. (2023). *Future Energy Landscapes in Northern Sweden: Sustainable Transition Scenarios for Municipalities* (Doctoral dissertation, Luleå University of Technology).

Sofi, I. I., Shah, M. A., & Ganie, A. H. (2023). Integrating human footprint with ensemble modelling identifies priority habitats for conservation: A case study in the distributional range of Arnebia euchroma, a vulnerable species. *Environmental Monitoring and Assessment*, 195(8), 914. DOI: 10.1007/s10661-023-11528-8 PMID: 37395941

Terry, N., Castro, A., Chibwe, B., Karuri-Sebina, G., Savu, C., & Pereira, L. (2024). Inviting a decolonial praxis for future imaginaries of nature: Introducing the Entangled Time Tree. *Environmental Science & Policy*, 151, 103615. DOI: 10.1016/j.envsci.2023.103615

Varughese, C., Henry, L., Morris, A., Bickerton, S., Rattenbury, N., Mankelow, C., Gorman, A., Katavich-Barton, S., & Dhopade, P. (2023). The intersection of space and sustainability: The need for a transdisciplinary and bi-cultural approach. *Acta Astronautica*, 211, 684–701. DOI: 10.1016/j.actaastro.2023.07.009

Watson, J. E., Ellis, E. C., Pillay, R., Williams, B. A., & Venter, O. (2023). Mapping Industrial Influences on Earth's Ecology. *Annual Review of Environment and Resources*, 48(1), 289–317. DOI: 10.1146/annurev-environ-112420-013640

Wellmann, T. (2023). Remote Sensing for sustainable and resilient. *Cities (London, England)*.

Wicki, S., Black, B., Kurmann, M., & Grêt-Regamey, A. (2023). Archetypes of social-ecological-technological systems for managing ecological infrastructure. *Environmental Research Letters*, 19(1), 014038. DOI: 10.1088/1748-9326/ad1080

Yousuf, A. H. (2023). Integrating ecosystem–based management and marine spatial planning for sustainable ocean governance in the Bay of Bengal.

Zhang, Z. (2023). *Mitigating Investment Risks in Nature-Based Solutions: A Strategic Approach Towards Sustainable Project Implementation* (Doctoral dissertation, Massachusetts Institute of Technology).

Chapter 13
Role of UAV-IoT Networks in Future Wildfire Detection

Ujjwal Agrawal
https://orcid.org/0000-0002-2604-7953
UM-DAE Centre for Excellence in Basic Sciences, Mumbai, India

ABSTRACT

Wildfires have posed a significant threat to ecosystems and human settlements throughout history, and their frequency and intensity have only increased in modern times due to climate change and other anthropogenic factors. This chapter explores the evolution of wildfire management, from early historical approaches to modern-day challenges. It emphasizes the growing need for prevention over reactive firefighting efforts, as prevention strategies offer more sustainable and effective long-term solutions. The chapter introduces the role of Unmanned Aerial Vehicles (UAVs) integrated with Internet of Things (IoT) networks as a transformative approach to wildfire detection and monitoring. It examines how real-time data collection, sensor networks, and machine learning algorithms enable early detection and prediction of wildfires, allowing for rapid response and prevention. The potential and limitations of these technologies are discussed, along with the challenges of implementation, such as data processing, scalability, and resource allocation.

INTRODUCTION

Wildfires are still one of the most dangerous threats to any ecosystem, way of life, or economy, with frequency and intensity soaring dramatically in recent times. Wildfires were, at one time, seen as signs of divine punishment or forces beyond

DOI: 10.4018/979-8-3693-7565-5.ch013

control. However, they are now studying science and technology. Factors such as climate change, urbanization, and alterations in land use practices aggravated wildfire activity around the world.

Amidst this backdrop, technology-driven solutions, particularly UAVs and IoT, have become indispensable for detecting, monitoring, and combating wildfires. UAVs have aerial images, detect heat anomalies, and cover unapproachable areas; IoT sensors give real-time status information about environmental conditions to help ensure timely actionable responses (Baena et al., 2017; Cui et al., 2021). ML increases their efficacy by allowing possible predictive modeling in terms of fire behavior, fire risk zone prediction, and resource allocation optimization.

The rise of UAV-IoT networks represents a paradigm shift in wildfire management from combating fires to preventing them. The chapters in the present study shall substantiate case studies accompanying each technological leap and ponder myriad challenges and opportunities in its relative global context.

2. HISTORICAL CONTEXT OF WILDFIRE MANAGEMENT

Over the centuries, wildfire management practices have changed dramatically with cultural beliefs and technological advancements. From the lost civilizations in the past to classical civilization, where the chances of their departure at times appealed mostly to the action of gods, to the modern one where they are scientifically interpreted and controlled, development, particularly of control over wildfire, has been highlighted through human adaptability and technological progression. This section will describe the evolution of wildfire management practices through time, from ancient through medieval, into the Industrial Revolution, and the modern era with important practices, innovations, and changes in understanding signaled.

Ancient Fire Practices

For thousands of years, wildfires were thought to represent uncontrollable forces beyond human mastery. Early civilizations developed primitive fire management methods to safeguard agrarian communities in Mesopotamia and Egypt. While some of these included firebreaks in Mesopotamia by leaving barren patches of land to check the intensity of the fire, controlled burning was popular in the Nile Delta in Egypt for removing the dry brush and preventing the accumulation of potential fires.

In Greece and Rome, wildfires were mythologically represented as acts for which the fire gods, Hephaestus and Vulcan, were responsible. While these societies viewed fire as a predominantly supernatural force, localized fire control was in practice with firebreaks and controlled burning. Such initial practices formed a background for later fire management practices, showing that fire can be destructive as well as beneficial if it is controlled.

Medieval to Early Modern Practices

The medieval period was the period of most restrictions on the development of urban fire management in England when a common occurrence proved most devastating. Thus, it was the age of the Great Fire of London of 1666, which destroyed a fairly large area of the city, leaving over 13,000 homes and several public buildings in ruins. It was clear that during those days, firefighting was restricted to primitive means of displaying great consummation on fires; basically, buckets, axes, and fire hooks did suggest extinguishing such unbridled flames. The fire couldn't be contained; thus, because of that, they were inclined not to spare fire engines and organized brigades to detain fires on end, made worse by closely knit wooden structures that served to fuel their happy flames.

After the Great Fire, awareness started to infuse the realization that firefighting must be organized, and fire management techniques improved. The fire brigades themselves began to appear in this era; they usually consisted of community volunteers rushing to the scene with whatever means were available. Unfortunately, such brigades did not normally provide their members with adequate equipment, coordination between brigades often left much to be desired, and sometimes water supply became insufficient.

The 18th and 19th centuries were marked by the Industrial Revolution, which gave a new momentum to almost every aspect of wildfire management, especially in the case of urban areas. The steam engine was a radical innovation in firefighting during the early 19th century. The steam engines carried water and firemen with much speed, thereby greatly reducing their reaction time. With the use of steam engines, more water could be drawn almost instantaneously, and the quantity of water used against blazes would add a new power of efficiency to firefighters instead of earlier methods. In addition, the advent of professional fire brigades, such as the London Fire Engine Establishment (founded in 1833), was the harbinger of more organized and systematic methods of fire management in the cities.

While urban fire management was aided by tech advancements, wildfires in cities and forests still remained a problem besetting forest managers. In this phase of human history, wildfires were still considered a primarily destructive force that must be suppressed at all costs. The acceptance of the positive and important role fire could play in the maintenance of the ecosystem was not enjoyed, and such notions prevailed during the years when it was believed that every fire should be snuffed out as soon as it threatened to outrun human control.

Fire Ecology in the 19th-20th Century

From the late 19th to the early 20th century, with fire ecology rising as a new scientific discipline, the understanding of wildfires underwent a remarkable transformation. Richard Fernow and Aldo Leopold's that most wildfires are, in fact, pernicious or should be put out challenged the view that all wildfires are harmful and should, therefore, be suppressed. Rather, they contend that fire is normal and pivotal to many ecosystems in sustaining biodiversity, controlling vegetation growth, and promoting ecological equilibrium.

Bernhard Fernow, among the first professional foresters in the United States, championed prescribed burns-controlled fires that are intentionally set to parallel the natural fire cycle. Fernow's work contributed significantly to a change in the attitude toward fire management, particularly in forested areas where fire suppression practices had led to the accumulation of massive fuel loads of dead vegetation and such that may give rise to more destructive fires. His ideas laid a firm basis for modern practices in fire management involving ecological health and fire suppression.

Aldo Leopold, the ecological statesman and environmentalist, elaborated on fire use as a management tool for conservation purposes. The crux of his work changed to fire, which plays a significant role in the health of ecosystems in the western United States. Wildfires shape plant communities and act against the overgrowth of exotic species. Leopold argued that fire suppression tolerance must have strict and different logic attached when dealing with fire-related management; a shift away from fire suppression toward an integrated landscape approach should be achieved with the understanding of the ecological positive attributes of fire in certain settings.

Transitioning to the 2nd half of the 20th century, fire use as a management tool was embraced, more so in forest management. The emergence of fire ecology lent much credence to this latter move by switching from an older school of thought of total fire suppression to an integrated approach that recognizes the advantages of fire in certain ecologies. This realization then overshadowed the evolvement of modern wildfire management policies emphasizing the balance between fire suppression and normal rachels of fires.

The Modern Era

The modern age of wildfire management was characterized by the onset of more sophisticated technologies and policies that can perceive the multifarious role of fire, both as a destroyer and sustainer of ecosystems. One of the biggest developments during the 20th century was aerial firefighting, which used airplanes to drop water and retardants on raging fires. With that, controlling large-scale wildfires, especially in remote regions, was made incredibly easier since they were not easily reachable by ground crews. Aerial firefighting became one of the key fire management tactics capable of contradicting the others by bringing about immediate and effective responses to massive fires.

In the same breath, emerging policies like the "Let It Burn" policy reflected a growing awareness of fire's myriad benefits. Under this policy, fires within certain areas, notably wilderness and forested portions, would be allowed to burn on their own as long as they posed no imminent danger to human life or property. The "Let It Burn" method was inspired by the understanding that fire usually plays a key role in the maintenance of healthy ecosystems by burning fuel loads and renewing the regeneration of fire-maintained plant species. Natural burning of certain fires could offer some fire managers the opportunity to lessen the buildup in fuel lights and cause bigger fires in the future.

However, Global warming and increasing urbanization intensified the frequency and severity of wildfires, particularly at the urban-wildland interface (UWI)—the zone where human development coincides with the natural landscape. The increasing presence of homes and infrastructure in these fire-prone areas has complicated wildfire management since they now burn residential communities that traditionally burned through large uninhabited forests. This has resulted in the development of more advanced wildfire detection and management strategies involving real-time data collection, predictive analytics, and efficient resource allocation.

In the 21st century, the integration of unmanned aerial vehicles (UAVs), IoT sensors, and machine learning has radically transformed wildfire management. They allow for real-time monitoring of zones whose actual threats become identified and modeled with increased accuracy and precision now hidden from human view. With the incorporation of this range of modern firefighting aids, fire managers are thus empowered further in controlling, mitigating, and managing wildfires in urban and rural environments.

3. UAV-IOT NETWORKS: A TECHNOLOGICAL LEAP

The rapid advancement of Unmanned Aerial Vehicles (UAVs) and the Internet of Things (IoT) is going to revolutionize wildfire detection and management. Each, on its own or together, provides real-time, accurate monitoring, allowing emergency teams to detect fires earlier, predict fire behavior, and better manage firefighting resources. In this, we will look at UAVs' basic components, their role in wildfire detection, integration of UAVs and IoT systems, and examples in real life involving this mighty duo.

Introduction to UAVs

UAVs, also known as drones, are aircraft without a human pilot on board, remotely controlled by a human operator on the ground or autonomously by the onboard systems. The rapid establishment of UAVs became indispensable to a wide variety of fields, such as wildfire management, due to the UAV's ability to extend to the areas that would be difficult to reach, obtain real-time data, and ferry multiple sensors, which provide critical information about the surrounding environment.

Core Components of UAVs

1. Thermal Imaging Cameras
    - A thermal camera is one of the most beneficial components of UAVs for wildfire detection. These cameras detect infrared radiation emitted as heat. Through the characteristic signature of the areas that are prone to fire, thermal imaging provides UAVs access to the detection of hot spots, which can be effective even under thick smoke and poor visibility before flames become visible to the naked eye. Early detection is an invaluable capability and is critical to ensuring a quick response from fire management teams to emerging fires (Baena et al., 2017).
2. LiDAR Sensors
    - Light Detection and Ranging (LiDAR) sensors create high-resolution 3D maps of the environment through laser beams. LiDAR is, thus, highly useful in wildfire management, measuring vegetation density, fuel loads, and roughness, parameters crucial for explosive wildfire modeling. Providing a detailed topography view, LiDAR sensors would assist fire management teams in predicting how a fire will behave regarding the particulars of the terrain and vegetation (Akhloufi et al., 2020).
3. High-Resolution Cameras
    - UAVs can capture detailed snapshots of the affected areas with high-resolution cameras. Such cameras can provide real-time pictures and videos that help substantiate damage assessments, enabling authorities to assess fire severity and allocate resources efficiently. High-resolution imagery is also used to assess the area where the fire damage has occurred, analyze the extent of environmental degradation, and plan for recovery efforts.
4. Multispectral and Hyperspectral Sensors
    - These sensors detect across a wide range of wavelengths beyond visible light, thereby providing insight into the health and moisture content of vegetation. Using multispectral or hyperspectral sensors for analy-

sis across various spectral bands, UAVs can ascertain vegetation stress, which is a key indicator of the risk of fire. This data is useful in determining areas more prone to ignition due to drought or other environmental constraints (Rodriguez et al., 2018).

In addition, UAVs are fitted with GPS systems and altimeters as well as all kinds of navigation equipment, enabling them to operate over vast areas autonomously. This makes UAVs particularly useful for the monitoring of large, remote landscapes susceptibility to wildfires.

Role of UAVs in Wildfire Detection

UAVs have emerged as a game changer in wildfire detection due to their ability to monitor live events almost instantaneously, covering large locations in a brief time with utmost efficiency. Traditional methods of wildfire detection-such as lookout towers, ground patrols, and satellite imagery-were too slow to provide real-time, high-resolution data on Wildfire-risk areas. With the advent of UAVs, the wildfire management stratagem is being changed to a rapid and thorough data-gathering ordeal in almost every part of the world.

Real-Time Monitoring

- One of the biggest advantages of UAVs is that they provide real-time surveillance of fire-prone areas. These unmanned aerial platforms, powered by thermal cameras and high-quality sensors, continuously monitor large areas and detect small heat anomalies that may signal the presence of a wildfire. This capability for early detection is vital in preventing small fires from developing into large and uncontrolled wildfires (Cui et al., 2021). UAVs are being deployed during the firefighting seasons in California to maintain surveillance on the movement of fires, thus enabling the emergency response teams to modify their strategy according to the situation at hand, as indicated by real-time data.

Case Study: California Wildfires

• In the 2020 California wildfires, UAVs equipped with thermal imaging and real-time data transmission capabilities were deployed to monitor fire behavior, assess the damage, and provide continuous updates to firefighting teams on the ground. With UAVs, authorities could identify hotspots that had not been previously detectable and redirect resources to critical areas before the fires could spread any further. This intervention was timely enough to save lives and property (Akhloufi et al., 2020).

Case Study: Australian Bushfires

In one of the most disastrous fire seasons on record, namely the 2019-2020 Australian bushfires, UAVs proved invaluable in detecting and monitoring fires in remote and heavily forested areas. The UAVs were deployed as eyes in the sky to monitor fire fronts and provide live thermal images and video, allowing firefighters to more effectively develop their suppression strategies. Also, these UAVs helped to undertake post-fire damage assessments, with authorities utilizing them to estimate the ecological impact of the fires and to identify damaged areas requiring immediate recovery efforts (Filkov et al., 2020).

In both case studies, UAVs exhibited their usefulness for early detection, real-time monitoring, and strategic resource allocation, and thus proved that the technologies have become typical tools for modern wildfire management.

Integration with IoT

Coupling UAVs with IoT technologies would further enhance the capability of wildfire detection and management systems. IoT refers to an interconnected group of devices exchanging relevant data in real-time while collecting and analyzing data to manage the systems that would be far more intelligent and efficient. In terms of wildfire management, IoT devices like temperature sensors, humidity sensors, smoke detection systems, and soil moisture sensors help in monitoring essential environmental conditions to evaluate predicted fire risks.

IoT Sensors in Wildfire Management

1. **Temperature Sensors**:
    - Temperature sensors measure the ambient temperatures in fire-prone areas and provide the first warning when temperatures exceed a certain threshold. Coupled with UAV data, temperature sensors estimate the probability of a fire outbreak in a certain region.
2. **Humidity Sensors**:
    - Humidity sensors play a key role in assessing the flammability of vegetation. Low humidity dries vegetation, increasing its fire susceptibility. Humidity sensors, once placed outdoors, can offer real-time data for fire-risk assessment (Rodriguez et al., 2018).
3. **Smoke Detectors**:

- Among the first signs of a fire is smoke. IoT-enabled smoke detectors can find very small amounts of smoke, giving fire management teams a chance to stop a wildfire before it spirals out of control. Smoke detectors can be placed in isolated areas or in the heart of dense forests to complement UAVs, enabling them to monitor huge expanses of land where humans may find it difficult or unfeasible (Cui et al., 2021).

4. **Soil Moisture Sensors**:
   - Soil moisture content is directly linked to vegetation's health. Dry soil most often times indicates an increase in wildfire risk as the dry vegetation serves as fuel for a fire. Soil moisture sensors allow the continuous monitoring of soil moisture levels, and the data obtained can be employed to identify fire-prone areas and prevent wildfires.

How UAVs Work with IoT

The UAV and IoT can build a strong interactive data-gathering system for the efficiency of fire management teams by monitoring real-time conditions across large landscapes, whether smooth or remote. The UAVs, equipped with IoT sensors, can plainly enter fire-prone areas and capture data about local environmental conditions such as temperature, humidity, and vegetation health. The data collected will combine with that of the ground-based IoT devices that monitor the moisture of the soil, wind directions, and atmospheric conditions. Data from UAVs and ground sensors are communicated to command centers and analyzed in order to assess the state of fire risk, track current fires, and optimize the allocation of resources (Akhloufi et al., 2020).

In a wildfire situation, the UAVs flying with thermal cameras will spot heat anomalies, and the ground IoT sensors will provide the wind speed, humidity, and soil moisture measurements. All this data is processed in real-time by machine learning algorithms that predict the fire's potential expansion and give fire managers actionable insights. This interaction of UAVs with IoT allows for automatic data gathering and real-time assessments, allowing decisions to be made quickly, something critically needed to prevent wildfires from turning into larger disasters.

Real-World Applications

The practical implications in wildfire management in terms of resource allocation, fire modeling, and decision-making from UAV-IoT systems are immense. These technologies allow fire management teams to use accurate, up-to-the-minute data so as to allocate their resources efficiently, minimize risks, and improve the effectiveness of fire suppression efforts.

Resource Allocation

- UAV-IoT experiences phenomenal benefits in terms of personalization to firefighting resource allocation. By sending information in real-time about behavior in fire, fuel loads, and environmental conditions, fire managers are allowed to prioritize high-risk areas and allocate resources where needed. Hence, efficiency is attained because resources are targeted to avoid costly firefighting operations and to increase the chances of event containment before they escalate (Rodriguez et al., 2018).

Accurate Fire Modeling

- UAV-IoT systems also play a crucial role in fire modeling, which involves predicting the movement of fire based on factors such as wind speed, vegetation type, and the topography of the area. The UAVs and IoT sensor data are channeled into machine learning models, where the information is analyzed and accurate predictions about the fire's movement are built.The models enable fire management teams to respond to fire spread dynamics and help them position firefighting teams and equipment more efficiently (Jain, Wang, & Flannigan, 2020).

Informed Decision-Making

- The real-time data that UAV-IoT systems provide enhance decision-making processes at every stage of wildfire management, from early detection through fire suppression and postfire recovery. Emergency response teams use this information to make informed decisions about where to position personnel, how to best allocate resources, and when to order evacuations of at-risk communities. The combination of UAV and IOT not only ensures decision-makers have the best and most real-time information available, thereby minimizing the chances of human error, but significantly increases the firefighting outcomes (Cui et al., 2021).

4. MACHINE LEARNING INTEGRATION IN WILDFIRE MANAGEMENT

As wildfires continue to grow larger and more intense due to climate change, expanding urban areas, and land-use changes, wildfires urgently require more accurate predictive methods for their control and governance purposes. The data reveals that machine learning (ML) has emerged to be one of the most formidable technologies for assisting wildfire management systems' forecasting capability. Machine learning algorithms are capable of detecting patterns in enormous datasets collected from UAVs, IoT sensors, and historical wildfire data, predicting fire behavior, and optimizing firefighting strategies through implication analysis. Herein, the role of ML will be discussed in the detection of wildfires, focusing on particular algorithms and real-time analyses, along with the challenges continued and possible opportunities in the area.

Introduction to Machine Learning (ML)

Machine learning is one of the aspects of artificial intelligence that allows computers and systems to use data and, over time, improve performance without being explicitly programmed. The use of machine learning in wildfire management allows for the analysis of huge amounts of data coming from different sources, such as UAVs, IoT sensors, satellite imagery, and meteorological data, to understand fire risks, predict fire spread, and advise decision-making.

The ability of machine learning models to apply lessons learned from historical data and continuously learn and update their predictions in real time has completely altered how wildfires are detected and managed. In the past, the prediction of fires relied heavily on human expertise and limited datasets. However, ML models have the capacity to analyze much bigger datasets, including environmental conditions such as wind speed, temperature, vegetation type, and humidity, and make more accurate and timely predictions. This methodology, which is based on data, significantly enhances the precision and efficacy of wildfire management, allowing emergency teams to take pre-emptive measures that can save human lives, property, and natural ecosystems (Rodriguez et al., 2018).

Machine learning algorithms can analyze complex relationships among variables and recognize patterns that are too difficult for human analysts to see. Such capability is especially important in wildfire management, where interacting in complicated ways are many factors affecting fire behavior, such as topography, vegetation, fuel load, and weather conditions. An added advantage is that fire managers can apply these advanced modeling techniques to predict very accurately the location and time of wildfires and the rate of spread, along with the type and extent of resources needed for effective firefighting intervention efforts.

ML Algorithms in Fire Prediction

Many approaches to wildland fire management, mainly via machine-learning techniques such as detection, prediction, and resource allocation, have been adopted. They are aimed at spotting the patterns in data and making predictions based on past trends and present conditions. The most widely used algorithms in wildfire management include decision trees, random forests, and deep learning models.

1. Decision Trees:
    - Decision trees use a set of decision rules to classify data into various categories. They may be utilized to classify areas with respect to wildfire risk in wildfire management. For example, an algorithm could evaluate environmental features like temperature, humidity, wind speed, and vegetation type to classify an area as high or low risk for a wildfire. These predictions enable fire management teams to prioritize areas for closer monitoring or resource allocation for firefighting efforts when fires threaten to occur (Cui et al., 2021).
2. Random Forests:
    - Random forests are an extension of decision trees and are therefore very useful for wildfire predictions because they improve their accuracy through the aggregation of results provided by many decision trees. In relation to wildfire management, random forest models can process high-dimensional data from various sources, including UAVs, IoT sensors, and historical records of fire incidents, to predict the probability of fire starting in a determined area. Random forests also support predicting potential spread rates for wildfires, assisting emergency response teams in deciding how to allocate resources to concerted control efforts to avert uncontrollable fire spread (Jain, Wang, & Flannigan, 2020).
3. Deep Learning Models:
    - Deep learning is an important aspect of machine learning where the employed model often uses an artificial neural network in order to describe intricate relations between inputs and outputs. In the domain of wildfire management, especially the convolutional neural networks (CNN), deep learning models may be leveraged in analyzing satellite images, UAV footage, and sensor data for the extant detection of wildfires. These CNNs have proven especially effective in identifying patterns in images, such as pinpointing areas of vegetation with the highest potential for ignition or identifying heat signatures, which indicate the presence of an active fire (Rodriguez et al., 2018).

These models could predict fire behavior through the analysis of past wildfire data and present environmental conditions. For instance, a deep learning model would analyze historical data on wind patterns, fuel loads, and rates of spread to predict how a fire would behave under similar conditions. This prediction ability is invaluable for emergency response teams since it allows response teams to cascade resources and firefighting strategies more meaningfully.

Real-Time Analysis

One of the most vital contributions of machine learning to wildfire management is the provision of real-time analysis. By continuously processing data from UAVs, IoT sensors, and satellite imagery, machine learning models can offer near-real-time insights into the state of the environment and wildfire movement. These models work on both structured data (temperature readings) and unstructured data (images and videos) to predict possible fire risks before they occur.

Real-time wildfire detection benefits greatly from using CNNs for image analysis. They are simplified sets of algorithms that further extract visual data by recognizing simplistic and advanced patterns and features in an image. In managing wildfires, CNNs analyze satellite or UAV imagery, detecting the earliest signs of a fire, such as smoke or heat anomalies, often before it becomes visible to human observers. These first notifications allow the fire management team to react swiftly and effectively to contain the fire early before any uncontrollable spreading (Baena et al., 2017).

For instance, CNNs can analyze images from UAVs equipped with thermal cameras to detect heat signatures that indicate the presence of a fire. These models can differentiate between heat from the sun and heat from an active fire, allowing for more accurate detection. Once a fire is detected, CNN can provide real-time updates on the fire's location, size, and movement, enabling fire managers to make informed decisions about how to respond.

Recurrent Neural Networks (RNNs) and **Long Short-Term Memory (LSTM) Networks** can be used extensively for fire behavior prediction over time in wildfire management. RNNs are specially tailored for handling time-series data, which is particularly useful for tracking fire spread and predicting further progression of the fire. Long short-term memory networks, a subtype of RNNs, learn long-term dependencies in data and thus can predict future fire behavior based on past trends (Cui et al., 2021).

For example, LSTM networks can analyze time series data for wind speed, temperature, and humidity to predict how a fire will behave over the next few hours or days. This allows fire management teams to anticipate changes in fire behavior and make adjustments to tactics as necessary. LSTM models help with long-term predictions associated with the fire and can carry out management decisions by diverting the fire from populated areas or critical infrastructure.

Challenges and Future Improvements

Although machine learning has several advantages in wildfire management, a host of challenges remain. One of these is the challenge of adapting machine learning models to such uneven terrains, climates, and ecosystems. Wildfires behave differently in different environments, depending on the vegetation type, topography, and weather conditions. Hence, machine learning models that perform well in one area may not do so in another.

Another challenge is the availability and quality of data. Machine learning models usually require huge quantities of reliable and high-quality data to produce accurate estimates. In some regions, however, there is neither appropriate historical fire data nor live environmental conditions data, which means it is difficult to train effective models. Besides, the records from these UAVs and IoT sensors may vary in quality according to the individual sensor types, environmental conditions, and the data transmission infrastructure used. To this end, researchers have developed various strategies to improve the adaptability and accuracy of ML models in wildfire management. An interesting area of transfer learning would be applying a model trained in one environment to another that bears similar conditions. Since transfer learning allows models to benefit from knowledge of previous fire events to enhance predictions in other areas, for instance, a model trained on California fire data can be adapted to address the Australian case using climatic and vegetation patterns even when the regions are different (Jain et al., 2020).

Moreover, along with the improvements in different data collection technologies, which include utilizing higher-order Unmanned Aerial Vehicles (UAVs), IoT sensors, and satellite systems, providing high-quality data will improve machine-learning models' training more efficiently. With increasing deployments of real-time data, machine learning algorithms will become exceedingly competent at forecasting fire behavior and assisting wildfire management efforts.

The introduction of reinforcement learning is another development we can look forward to; a model is able to learn through interaction with its environment toward the end of making decisions. To apply reinforcement learning to wildfire management, an AI system could use a particular geographical area to simulate the duties of extinguishing a fire in several different strategies and with several qualities to optimize the response based on trial-and-error learning. In the real world, this gives the ability for fire managers to test different tactics against real fires and find the approach that works best for controlling a wildfire in a particular setting (Rodriguez et al., 2018).

5. CASE STUDIES: UAV-IOT SUCCESS STORIES

UAV-IoT systems have earned their place among several eye-catching wildfire incidences across the globe. These technologies have become indispensable tools in contemporary wildfire management, ranging from fire ignition detection through suppression to establishing fire perimeters. In this section, three salient case studies involving UAVs enabled with IoT sensors will be identified: 2019-20 Australian bushfires, California wildfires, and the use of such technologies in European forest fires in Spain and Portugal.

California Wildfires

Background

The California wildfires of 2020 were among the most devastating in the state's history, burning up to 4 million acres, thousands of domiciles destroyed, and a large toll on human lives as well as wildlife. Faced with a growing accumulation of intense fires due to climate change, firefighting agencies began adopting sophisticated technologies such as Unmanned Aerial Vehicles (UAVs) and Internet of Things (IoT) sensors. These advanced technologies were put into place in the functions of detection, allocation of resources, and suppression.

Deployment of UAV-IoT Systems

During the 2020 wildfire season, UAVs equipped with thermal imaging cameras, LiDAR sensors, and high-resolution cameras provided real-time aerial surveillance of wildfire events. The UAVs played key roles in hotspot detection, monitoring fire progression, and relaying data to ground crews, thereby being able to maximize the turnaround response time into a more efficient and effective firefighting strategy (Baena et al. 2017).

One major advantage of using UAVs to carry out monitoring has to do with the wide areas that are hard to access. For instance, UAVs equipped with thermal cameras were capable of detecting small smoldering fires that could have gone unnoticed otherwise within the Sierra Nevada Mountain range. The early detection helped firefighting teams respond to fires adequately and helped contain them long before they had a chance to escalate into sizeable wildfires.

The utilization of IoT sensors installed across different forest areas in California complemented UAVs by providing real-time continuous data on temperature, humidity, and soil moisture content. These sensors provide useful information for fire-risk prediction that could support fire managers in planning for preemptive actions, such as creating firebreaks or removing resources to fire suppression at high-risk areas (Cui et al. 2021). As soon as there was evidence of decreasing humidity or increasing temperature in the areas monitored by IoTs, fire managers were instantly notified, prompting the dispatch of UAVs to the respective area for further investigations.

Improved Response Times

One of the most significant benefits of the UAV-IoT system in the 2020 California wildfires was the improvement in response times. Conventional methods of wildfire detection tended to incur delays, whereby lookout towers or ground patrols would leave fires undetected for several hours, even days. By the time crews arrived, the fires had often grown large enough to make suppression difficult. UAVs, on the other hand, constantly deliver live fire progress data to users, facilitating the speed of decision-making and making the attack on the fire more direct.

For example, during the August Complex Fire—the largest wildfire in California history—UAVs were employed to provide real-time updates on the fire's progress. This information was used by fire managers to guide fire crews to the most critical points and develop assignments to protect residences and infrastructure. UAVs reduced not just the response times but also decreased risk to firefighters because they offered critical information on the state of the fire before putting crews into potentially dangerous spots (Akhloufi et al., 2020).

Limiting Damage

In 2020, UAVs and IoT sensors were used together to minimize the damage inflicted by wildfires. They were especially helpful for wildfire managers by allowing early detection of fire and providing them with live information about fire behavior to carry out more efficient allocation of resources and suppression strategies. While the fires were devastating, the lessened damage to some extent was attributed to the presence of the UAV-IoT systems in a proactive manner.

Australian Bushfires (Black Summer)

Background

The year 2019-2020, known as the Black Summer, was one of the most devastating fire seasons in Australian history. About just over 18 million hectares were reduced to ashes, burning thousands of homes with about three billion animals killed or displaced. The intensity and scale of the fires once again confirmed the need for advanced fire detection and management systems, resulting in their extensive deployment by the Australian firefighting agencies.

Damage Assessment and Fire Suppression

UAVs were employed for the assessment of damage and coordination of fire suppression during the Black Summer fires. UAVs carrying thermal imaging equipment provided fire survey information from the air across New South Wales and Victoria, allowing for a bird's-eye view of the affected areas. Thermal imaging allowed for the identification of the most intense parts of the fire so that fire crews could prioritize their efforts against the most aggressive parts of the fire (Filkov et al., 2020).

In addition, UAVs were used to monitor the spread of embers, which were responsible for starting many new fires during Black Summer. Embers, which were carried by the wind, usually ignited fires kilometers away from the front lines of the fire, making it difficult for ground crews to pinpoint the sites of new fire outbreaks. However, UAVs were successfully able to monitor ember movement in real-time to give fire managers the crucial information needed to ensure that the fires did not progress into new territory (Akhloufi et al., 2020).

Resource Allocation

The incorporation of IoT sensors in combination with UAVs improved overall resource allocation in the Australian bushfires. IoT-enabled weather stations established in various regions of the Australian forests were able to provide continuous monitoring of temperature, humidity, and wind speed, which is extremely critical data for fire managers. Using the data from these weather stations to corroborate the UAV data would assist the fire managers in determining the probable spread of fire to allocate resources.

For example, in Kangaroo Island, IoT sensors detected a sudden drop in humidity, indicating that conditions were becoming more favorable for fire spread. Fire managers used this data, along with UAV footage, to deploy firefighting resources to high-risk areas, thus preventing the fire from spreading into more populated regions. This ability to preemptively allocate resources based on real-time data was a key contributor to limiting the overall damage caused by the fires (Rodriguez et al., 2018).

Post-Fire Recovery

The UAV-IoT systems also played a pivotal role in the post-fire recovery efforts in Australia. After the fires had been put out, LiDAR sensors and high-resolution cameras mounted on UAVs were deployed in these areas to assess the extent of the damage. These UAVs are thus instrumental in actuating image capture of burned areas, thereby assisting environmental agencies in the loss estimation of vegetation and wildlife habitats. In some instances, the UAVs were used to map the regeneration efforts tracking the recovery of fire-impacted ecosystems over time.

In the most novel instances, UAVs assisted, among others, in reforestation efforts. Drones were used to plant seeds in fire-ravaged areas after Black Summer, thus accelerating the pace of reforestation. These drones were speedily equipped with pods full of seeds that allowed them to cover a large area quickly, which turned out to be a highly efficient method for the restoration of fire-damaged landscapes in Australia.

European Forest Fires

Background

While the United States and Australia often receive the media's spotlight over wildfire management, the risks posed by wildfires are also significant in Europe, particularly in Spain and Portugal, where the conditions of dry summers and mountainous terrain make them good places for the fire to start and progress. Both countries have adopted UAV-IoT systems for wildfire detection and management, which have been burgeoning in recent years.

UAVs Within Remote Mountainous Regions

One of the largest drawbacks of operating wildfires in Spain and Portugal is their remoteness and rugged nature. Most traditional on-foot monitoring and firefighting efforts become hampered, with ample potential for unobserved fires to gain strength and wreak havoc. UAVs, however, circumvented these obstacles by enabling aerial monitoring of such inaccessible areas.

For example, UAVs equipped with multispectral sensors were used in the Serra da Estrela mountains in Portugal to monitor the health and moisture information of vegetation and provide crucial information to fire managers on areas particularly susceptible to ignition (Jain, Wang, & Flannigan, 2020). These multispectral sensors would generate information for many wavelengths, with insights into the condition of forests, thus highlighting regions where vegetation is particularly dry or making for easy targets for an outburst of fire.

In addition, UAVs with thermal camera capabilities have spied throughout the region of Catalonia in Spain to oversee fires and provide real-time information to the operational teams. Such UAVs can fly over steep terrains and heavily forested areas where layout units are restricted to operate practically. By relaying real-time information on the fire's position and intensity, UAVs have made it easier for fire managers to intelligently deploy resources and complete containment before wildfires reach populated areas.

Integration with IoT

IoT sensors integrate UAVs to create comprehensive wildfire monitoring in Spain and Portugal. In remote forests, temperature sensors, humidity sensors, and wind monitors relay continuous data on environmental conditions. Such data, together with the real-time imagery collected by UAVs, yield detailed mappings of fire-prone areas that help fire managers predict where fires are most likely to start and how they will spread (Rodriguez et al., 2018).

The success is even prominent in Portugal in the Algarve region, where IoT sensors examine forest conditions and transmit them to UAVs patrolling the skies. Upon observing decreases in humidity or increases in temperature, UAVs are dispatched to investigate possible fire risks. Such proactive maneuvers have dramatically given Portugal the ability to detect potential fires.

6. CHALLENGES AND OPPORTUNITIES IN WILDFIRE MANAGEMENT

Although UAV-IoT systems have exhibited a lot of promise in changing the way wildfires are detected and managed, certain challenges are still outstanding and must be worked on to maximize the helpful use of these technologies. These challenges include but are not limited to, data integration issues, connectivity, and power limitations, as well as regulatory and ethical concerns. Still, at the same time, these challenges provide opportunities to conduct technological innovation and cooperation in order to complement wildfire management.

Data Integration Challenges

One of the main problems in wildfire management is related to data integration from several sources, such as UAVs, IoT sensors, and satellites. Each technology produces loads of data sets in high-resolution images and videos, sensor readings, and environmental factors like temperature, humidity, and wind speed. Making such integrated heterogeneous data suit a coherent structure for in-time analysis and decision-making is one of the hardest tasks that technical scientists have to solve.

Data fusion, delivering information emerging from different sources as a singular unified dataset, utilizes advanced algorithms and yields the maximum power from massive computational resources. The UAVs, equipped with various remote senses like LiDAR, thermal cameras, etc., can provide real-time imageries of an ongoing fire through the subjects that are being executed, whereas ground-based IoT sensors would monitor the current conditions of the affected area. Data must be processed and analyzed very swiftly, delivering real-time actions to fire management teams. However, the immense volume of data coming from a UAV-IoT system is simply petrifying for traditional data processing systems, and this frequently causes delays in data analysis and the decision-making process (Rodriguez et al., 2018).

On the other hand, real-time data analysis plays a crucial role in dendrology and post-dendrology processes where high temporal and spatial variability means no delay in decisions could have a catastrophic implication. Hence, delivering real surface reconciliation of the fire is tremendously important for effective firefighting and suppression by integrating this kind of real-time data from UAV-IoT satellites into a one-heedful platform to present an overall picture of the fire. Data processes in the form of machine learning models greatly assist in this analysis, but it is important to constantly update these models every time fresh information appears for their correct running. Developing more efficient data fusion techniques that can handle big and heterogeneous datasets with real-time capabilities can go a long way toward improving UAV-IoT system efficiency according to wildfire mitigation needs (Cui et al., 2021).

Connectivity and Power Limitations

Another important problem is that connectivity is difficult in remote areas with high wildfire concerns. The areas where wildfires most often take place, such as forests, mountains, and rural regions, are often poorly served by internet and cellular connections. This can impede the real-time transmission of data from UAVs and IoT sensors to fire management centers. In areas where traditional communication infrastructures are absent, however, satellite links are often used to remain connected. Yet, satellite communication could also be a costly technology and could impede real-time decision-making during fast-moving wildfires by introducing significant latencies (Akhloufi et al., 2020).

Maintaining continuous connectivity for UAVs and IoT devices in remote areas also requires robust networks that can support the high bandwidth required for real-time data transmission. This is particularly so for UAVs with high-resolution cameras and thermal imaging systems, as the data they generate will be large and demands immediate processing. An alternative is for mesh networks to be adopted, enabling IoT nodes and UAVs to communicate with one another directly over a decentralized networking solution that does not rely on traditional communication infrastructure. In such a case, data will hop from one device to the other until it reaches a designated processing hub for analysis.

Another important problem is that connectivity is difficult in remote areas with high wildfire concerns. The areas where wildfires most often take place, such as forests, mountains, and rural regions, are often poorly served by internet and cellular connections. This can impede the real-time transmission of data from UAVs and IoT sensors to fire management centers. In areas where traditional communication infrastructures are absent, however, satellite links are often used to remain connected. Yet, satellite communication could also be a costly technology and could impede real-time decision-making during fast-moving wildfires by introducing significant latencies (Akhloufi et al., 2020).

Maintaining continuous connectivity for UAVs and IoT devices in remote areas also requires robust networks that can support the high bandwidth required for real-time data transmission. This is particularly so for UAVs with high-resolution cameras and thermal imaging systems, as the data they generate will be large and demands immediate processing. An alternative is for mesh networks to be adopted, enabling IoT nodes and UAVs to communicate with one another directly over a decentralized networking solution that does not rely on traditional communication infrastructure. In such a case, data will hop from one device to the other until it reaches a designated processing hub for analysis.

Regulatory and Ethical Concerns

An increased use of UAVs in fire management has spawned a number of regulatory and ethical concerns, especially airspace regulations as well as concerns related to privacy. In most countries, there are stringent regulations that govern where and when drones can fly. For example, drones typically are not allowed near airports or to fly over urban areas if regulations are not given. They are not meant to be flown in close proximity to people in order to avoid collision with manned aircraft as well as to protect the privacy of individuals on the ground. (StackPath, 2016). Furthermore, during wildland fires, UAVs cannot be in operation as long as manned firefighting aircraft are in the air. Normally, those latter aircraft have the right of way, and UAVs have to be grounded whenever there is a potential collision with a manned aircraft (Powers and Diamond, 2013) (Akhloufi et al., 2020).

Beyond the airspace regulations, the deployment of UAVs and IoT sensors for wildfire detection involves a host of privacy issues. UAVs equipped with high-resolution cameras as well as thermal imaging systems can capture detailed images not only of the wildfire itself but also of the surrounding areas, including private property. Given that remote areas close to human habitation are being surveilled constantly, there may be a risk of opportunism and an illegal exploitation of information or the breach of privacy rights. To make clear this point, stringent data protection policies should be established to ensure that data collected by UAVs or sensors is only used by the agencies that deploy them and not used for ancillary purposes (Cui et al., 2021).

To offset such ethical and regulatory concerns, close cooperation between fire management authorities on one hand and aviation authorities on the other is necessary to hammer out protocols that guide the concurrent use of UAVs during wildfire emergencies. Typically, designated airspaces exclusive to UAV operations can be created, preventing drones from getting in the air in an interference-free style. Public outreach is another area that should be taken up to convince the public so that UAV-IoT systems can be freely used. Where such outreach is done extensively, the public is usually sympathetic to the use of technologies.

7. EMERGING TECHNOLOGIES AND FUTURE TRENDS

With the rising frequency and intensity of wildfires, largely as a consequence of climate change, the role of technology in wildfire management cannot be overstated. The rapid evolution of UAV autonomy, the emergence of next-generation IoT sensors, and the application of advanced algorithms in artificial intelligence (AI) are about to completely redefine how we detect, predict, and suppress wildfires. Now, let's delve into the cutting-edge technologies and technological trends that will shape the future of wildfire management in this second part of the series.

Advanced UAV Autonomy

Autonomous UAVs can use AI for real-time environmental analysis, adjusting flight paths, and prioritizing areas for monitoring based on fire progression and environmental data. These drones will not only reduce the need for constant human supervision but also enable continuous surveillance over large areas, providing more comprehensive fire coverage (Akhloufi et al., 2020).

Drones are poised in the next century to become more advanced than what the UAV and drone community looked familiar with almost two decades back. In a swarming context, multiple UAVs operate together in a coordinated manner, not just enabling good coverage of large geographic areas but also covering those areas in a considerably more efficient way. Swarm drones are capable of flying in a formation, sharing data among themselves in real-time, and responding dynamically to how the fire is behaving. If one drone, while reconnaissance, sees something out of place- like an anomaly or heat signature- it will buzz other members of its swarm and request them to alter their paths and focus their sensors on the suspected fire hotspot. Distributed sensing through swarmed drones could change the very fabric of how wildfires are monitored, with the ability to follow what the fire is doing almost in real-time over tracks that could be difficult or impossible to reach by other means (Cui et al., 2021).

Alongside this ability lies the capacity of the swarm to coordinate with ground-based fire crews in transmitting information about fire intensity, wind direction, and heat anomalies. The use of these drones greatly aids fire crews in their response mechanisms and also acts as robust resource allocation systems in instances where certain areas need greater amounts of reinforcement. This avenue leads the AI-enabled swarm of drones into the future whereby real-time optimization of coverage area and self-organization takes place, minimizing control by the human hand and thus increasing the speed at which the fires are detected and their verification.

Next-Generation IoT Sensors

IoT has already proven very effective in wildfires by using sensors to monitor environmental conditions such as temperature, humidity, moisture content in the soil, and so on. However, IoT sensors will offer much greater capabilities in the next generation, with significant developments in terms of sensor accuracy, power efficiency, and data processing.

A key innovation in the field of sensors is hyperspectral scanners. While conventional sensors operate in only a few spectral bands of the frequency spectrum, hyperspectral scanners capture data from hundreds of wavelengths. This provides much more detailed information on vegetation health, a critical factor affecting fire risk assessment. Hyperspectral sensors find moisture levels, chlorophyll content, and stress indicators from the spectral signature of plants. These are important factors to note in areas at high risk of ignition (Baena et al., 2017).

The recovery post-fires will also be monitored by hyperspectral sensors, which will detect the health of regrowing vegetation and identify areas where reforestation efforts would be needed. The future IoT sensors will be using edge computing, which enables them to deal with local data processing of information and send only the most relevant information to the central systems. The process will ease enormous stress on data transmission networks, thus allowing real-time management of voluminous sensor data.

Another promising development is the creation of self-powered sensors that focus on energy harvesting technologies to draw power from sunlight, wind, or thermal gradients present in their environment. Solar-powered sensors are already being deployed in potential wildfire-prone areas, while the future sensors will harness energy from the heat borne from the fire itself, allowing them to function actively even during events of fire (Rodriguez et al., 2018). It will provide an opportunity for IoT sensors to be run in remote and hazardous sites where providing traditional power sources is very challenging.

Artificial Intelligence and Predictive Modeling

As drones, IoT sensors, and satellites create ever-greater amounts of data, AI will have an increasingly important role in the processing of that data and making real-time decisions. The next frontier in AI-driven wildfire management is the deployment of reinforcement learning, a type of machine learning in which algorithms learn from making decisions by interacting with the environment.

In the context of wildfire management, reinforcement learning systems would simulate the different firefighting strategies and optimize them based on feedback from the environment. For instance, a reinforcement learning algorithm could simulate several fire suppressions, like creating firebreaks, deploying water-dropping UAVs, or evacuating communities, and judge the toughness of each tactic based on its ability to contain the fire and minimize damage. Over time, the algorithm would learn which strategies to adopt under certain conditions so that the recommendations to fire managers become increasingly informed (Jain, Wang, & Flannigan, 2020).

Similarly, AI-fortified predictive modeling will continue to—refinedly—forecast wildfire behavior. AI-driven models could examine historical wildfire data along with current environmental conditions to anticipate how far and wide a fire may spread while also determining areas most at risk. The models become more and more precise as they assimilate data from diverse sources, like UAVs, IoT sensors, weather stations, and satellite imagery.

As the AI systems of the future integrate simulations that include modifying fire behaviors in real-time, fire managers can run different scenarios through the systems and make decisions based on projected outcomes. In other words, an AI model could predict how a fire will spread based on current winds, topography, and fuel loads and give fire managers a running model of how the fire is going to progress; this information will allow emergency teams to pre-position firefighting resources within the areas most in jeopardy and ultimately give improved efficiencies in fire suppression operations.

Moreover, as AI technology develops, it will integrate climate change predictions with wildfire risk assessments. By synthesizing patterns of the long-term weather, trends in increased vegetation growth, and shifts in land use, AI models will account for possible regions in the future that are increasingly vulnerable to wildfires, informing fire management agencies to take acts of prevention before the onset of fires (Cui et al., 2021).

8. FUTURE PREPAREDNESS STRATEGIES

As wildfires become a growing global threat, preparedness strategies for the future must involve a synergy of modern technology, proactive community engagement, and effective policy measures. Innovative wildfire management systems enabled by cutting-edge technologies such as UAV-IoT networks and AI could help curtail wildfire disasters via the enhancement of prediction, detection, and response capabilities. However, the functioning of these systems is very much dependent on community engagement and sympathetic policies that can ensure large-scale use and financing for these technologies.

Integrated Wildfire Management Systems

With roots based on UAV-IoT networks, AI-driven predictive modeling, and real-time data analytics, the future of wildfire management lies in constructing a unified system that bridges the gap among different technologies and diverse stakeholders. The integrated wildfire management system employs the best of numerous technologies to provide a broad framework for fire prevention, detection, suppression, and recovery.

In such a system, UAVs equipped with thermal cameras, LiDAR sensors, and multispectral sensors continually monitor fire-prone areas, actively collecting data on vegetation health, temperature, and environmental conditions. These data are sent to a central command center where IoT-enabled weather stations and ground sensors provide supplemental data on wind speed, humidity, and soil moisture. Such integration offers fire managers real-time insight into high-risk areas, thus facilitating predictions of where fires might ignite and allowing advanced resource allocation (Akhloufi et al., 2020).

Once the fire is identified, the very system utilizes predictive models exhibited by AI to represent simulated spread based on the available data, which is the same information on which fire managers can prepare their suppression strategies during an ongoing fire attack. AI models simulate certain important parameters such as wind patterns, fuel loads, and topography to conclude what the likely spread of a fire would be and where damage is most likely to occur. Such information then gets relayed to the emergency response teams on the ground, who can position firefighting resources and personnel in the most critical areas.

Overhead communication solutions utilize integrated systems to form shared information platforms that interconnect all relevant organizations emerging in fire management, local government, emergency responders, and the most crucial stakeholder, the public. These platforms provide a centralized source for real-time emerging data so that the stakeholders may act quickly based on reasonable information for coordination aimed toward achieving a common goal. For instance, integrated systems could assist during a wildfire emergency by evacuating notifications to communities at risk, real-time updates on fire status to firefighting crews, and predictive modeling for emergency managers (Rodriguez et al., 2018).

With technological advancements, integrated systems will include machine learning algorithms for on-the-fly prediction updates based on new data at arm's length, which makes them progressively useful over time. With the strengths drawn from UAVs with IoT sensors, AI, and analytics over real-time data, integrated systems have a very important stake in reducing the effects of any future wildfires.

Community Involvement

While technology and advanced methods are crucial in the management of wildfires, community involvement is equally necessary to mitigate fire risks. Public education campaigns can engage citizens with information regarding what to do to keep their homes and communities safe in the event of a wildfire; this becomes even more important in wildland-urban interface (WUI) areas, where human encroachment meets natural landscapes.

Firewise USA, which is sponsored by the National Fire Protection Association (NFPA), has become a credible example of community engagement to mitigate wildfire risk. Firewise USA encourages homeowners to take proactive measures to create defensible space, such as removing adjacent vegetation, using building materials that are resistant to fire, and having an emergency plan. All these measures will effectively reduce the risk of the spread of any wildfire to residential premises and give people and property a better chance to survive (Jain et al., 2020).

Communities, in addition to preparing their properties, can also participate in wildfire mitigation efforts such as controlled burns, fuel reductions, and the development of firebreaks. These activities are usually conducted in cooperation with local fire services and government agencies but require participation from the community to be successful. In involving resident participation, fire managers can ensure that communities are fully prepared for wildfire predicaments.

Additionally, campaigns to raise public awareness are also critical for teaching people about fire prevention. Such campaigns may target specific behaviors that may increase wildfire risks, such as careless disposal of cigarette butts, unattended campfires, and illegal fireworks. Fire management agencies should use social media, local news outlets, and communities to educate fire safety practices and behaviors around being more responsible (Cui et al., 2021).

9. CONCLUSION

The convergence of UAV-IoT networks with machine learning has brought about a fundamental transformation in wildfire management, advancing from the passive ways of its management to proactive, data-driven methods that improve early detection, resource optimization, and real-time decision-making. These technologies have ushered in unprecedented opportunities to remain vigilant in real-time on areas subjected to fire risk and to improve responses, dependable case studies from California, Australia, and Europe confirm.

Yet, this power is yet to unfold, as it will largely depend on continued partnerships between governments, research institutions, and the private sector. Enabling policies, investments in advanced technologies, and public-private partnerships will go a long way toward scaling the incorporation of UAVs, IoT devices, and artificial intelligence in fire management. The active involvement of communities in wildfire prevention and education programs will go a long way in strengthening the resilience of the community.

It is through the integrated and coordinated action of these emerging tools that wildfire management will be well-placed to effectively mitigate risks associated with wildfires.

REFERENCES

Akhloufi, M. A., Bendada, A., & Boireau, A. (2020). Fire detection and monitoring using unmanned aerial vehicles. *Drones (Basel)*, 4(2), 34. DOI: 10.3390/drones4020034

Baena, S., Moat, J., Whaley, O., & Boyd, D. S. (2017). Identifying species from the air: UAVs and the very high resolution challenge for plant conservation. *PLoS One*, 12(11), e0188714. DOI: 10.1371/journal.pone.0188714 PMID: 29176860

Cui, S., Zhao, Y., Cao, C., & Wang, X. (2021). Wildfire detection and prevention: UAV-IoT-based real-time monitoring with convolutional neural networks. *Remote Sensing*, 13(8), 1379. DOI: 10.3390/rs13081379

Filkov, A. I., Ngo, T., Matthews, S., Telfer, S., & Penman, T. D. (2020). Impact of Australia's catastrophic 2019/20 bushfire season on communities and environment. Retrospective analysis. *Global Ecology and Biogeography*, 29(12), 1807–1814. DOI: 10.1111/geb.13159

Jain, P., Wang, X., & Flannigan, M. D. (2020). Trend analysis of fire season length and extreme fire weather in North America between 1979 and 2015. *International Journal of Wildland Fire*, 29(11), 941–954. DOI: 10.1071/WF20047

Rodriguez, F. C., Salas, J., & Harris, J. A. (2018). Wildfires and the Internet of Things: A framework for real-time monitoring and detection. *IEEE Sensors Journal*, 18(16), 6713–6721. DOI: 10.1109/JSEN.2018.2853245

Chapter 14
Wildlife and Forest Resource Management With Artificial Intelligence

Rhishikesh Dave
https://orcid.org/0009-0005-7193-0728
Sharda University, India

Christian Kaunert
https://orcid.org/0000-0002-4493-2235
Dublin City University, Ireland & University of South Wales, UK

Bhupinder Singh
https://orcid.org/0009-0006-4779-2553
Sharda University, India

ABSTRACT

The incorporation if AI in the management of wildlife and forest assets can transform conservation challenges by providing new ways to protect biodiversity and ensure ecosystem sustainability. Machine learning, computer vision and other AI technologies are to monitor wildlife populations with high accuracy. These include tools that enable the analyses of vast datasets collected from remotely sensed technologies including drones, satellite imagery for real-time biodiversity and forest dynamics. Though AI provides many benefits, there are numerous issues that must be tackled around all things from data privacy to ethical considerations and the necessity of a transactional or multi-disciplinary attempt on it. This chapter comprehensively explores the AI can be used for sustainable wildlife and forest resources that humans thought they might not manage forever because it is important to keep them alive.

DOI: 10.4018/979-8-3693-7565-5.ch014

1. INTRODUCTION AND BACKGROUND

Wildlife and forest ecosystems are essential elements of the earth's biodiversity that provide a range of ecosystem services, including carbon sequestration, water regulation and habitat for myriad species. However, climate change and habitat destruction/fragmentation are now major threats for many of these ecosystems (Anderson & Lee, 2024). The management and conservation of wildlife and forest resources are essential for the maintenance of biological diversity and ecological stability. Recently, with the debut of Artificial Intelligence (AI), there are disruptive transformations to how we see and maintain these natural resources. This will be more meaningful with AI-powered predictive models for control and monitoring without which identifying threats like poaching or deforestation could fall into retardation in mitigation actions. A tool developers say can be used to assist in modeling ecological interactions and future scenarios as well, helping optimize reforestation strategies and wildlife corridor design etc. AI technologies deliver new tools for monitoring, forecasting and optimizing management practices to improve conservation outcomes and support sustainable resource use .

The unprecedented losses of species and forest area during recent decades highlight an increasing need for more innovative means to manage wildlife in relation to eco-system use. Forests, which inhabit 31% of the Earth's land area, are essential for biodiversity conservation as well as climate and livelihood regulation. In much the same way, wildlife is critical to keeping our planet in balance and good populations of these species are under threat because of habitat loss from development or crop growing, changing climate conditions as well as unsustainable interaction with humans. As environmental agencies are large and diverse, traditional tools do not match the scale of 21st century solutions that intend on addressing specific needs side by side. However, things could begin to change with the incorporation of artificial intelligence (AI) technology into wildlife and forest administration providing prospects for improving conservation practices as well as ensuring sustainable ecosystems (Singh & Kaunert, 2024).

AI technologies, specifically machine learning, computer vision and natural language processing can provide very helpful methods of analyzing vast amounts of ecological data. These technologies could help monitor, manage and protect wildlife, as well as forests resources by automating what had been time-consuming labor-intensive socioeconomic processes (Anderson et al., 2023). AI has also begun being used to automatic identification of species and habitat classification by analyzing photographs or videos from camera traps drone images in remote forests with difficult accessibility. Using satellite data it is possible identify where the forest was deforested so recently loggers could be present at that location illegally logging woodmatos and mountainess. Also, by modeling complex ecological interactions and assessing the consequences of different environmental stressors AI models can predict shifts in ecosystems (Brown & Chen, 2023).

This use of AI in wildlife and forest management is part of a global trend towards evidence-based conservation practice. The amount of data, from field surveys and remote sensing to citizen observations is enormous and offers the chance to learn about ecosystem dynamics that can underpin policy. AI can help conservationists and policy makers produce responses accurately aimed at halting threats to biodiversity or choose the right places for investments in conserving. Such an approach improves both the efficacy of management strategies and fosters adaptive management systems, aiding in a better response to changes within environmental settings (Singh & Kaunert, 2024).

Figure 1. displays the landscapes of introduction section (Source- Original)

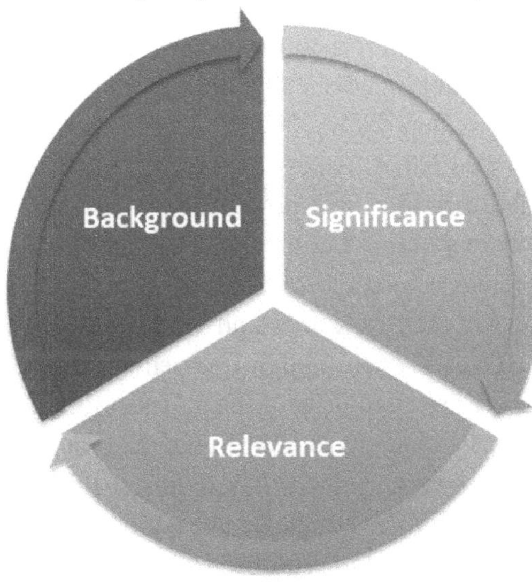

1.1 Significance and Relevance of the Chapter

The importance of incorporating AI in wildlife and forest resource management is manifold. The first aims to address the pressing need for novel approaches to save biodiversity and enhance ecosystem services. With increasing human pressure on natural resources, the conventional way of conservation is proving insufficient to adapt with a changing environmental landscape. AI represents a rich suite of tools that can be applied to deepen the effectiveness of conservation efforts by providing real time, data-driven insights and more evidence-based decision-making while optimizing resource allocation (Singh et al., 2024).

With harnessing the power of AI for wildlife and forest management it can reduce- scientific research to on-the-ground action. There, the deluge of ecological information from a plethora of sources had become so overwhelming that it was becoming prohibitively expensive for researchers and practitioners to extract useful information. AI offers new ways to automate data analysis, which can help to bridge the so-called science-policy gap and facilitate translation of findings in conservation practice (Brown & White, 2024). This enables better and more flexible management strategies by supporting evidence-based decision-making as a contribution of AI. When AI is applied in wildlife or forest management it can be related to the ongoing global effort of accomplishing some SDGs. The United Nations SDGs highlight the significance of preserving biodiversity, fighting climate change and establishing sustainable land use. Introduction of AI driven tools to not only provide high resolution and monitoring for natural resource protection but also promoting sustainable land management practices thus limiting the overall impacts of human activities on ecosystems (Chen & Lee, 2024).

AI can democratize the field of conservation by making these efforts more accessible and engaging. Platforms like those powered by AI can help with citizen science projects, which are where you and I collect data to contribute towards monitoring efforts. Through engaging communities in conservation work, AI also helps make local people feel they own their resources and act to conserve them sustainably at the grassroots level. However, even though it is prime time for AI in wildlife and forest management, there are challenges to be solved. But one of the most fundamental is around data privacy and security. Ecological data can involve sensitive wildlife and habitat information, which may require navigating ethical concerns surrounding people who has access to these types of data. Preserving this data and storing it confidential is important to establish trust with stakeholders, and prevent fraudulence (Chen et al., 2023).

The challenges encountered is interdisciplinary cooperation required to develop and implement AI-based solutions. As the effective management of wildlife and forest demands expertise from ecology, computer science or social sciences etc. We need multidisciplinary research, practitioners and policymakers working together with local communities to make sure we can have AI technologies that are socially beneficial without environmental detriment. It is also important to address the digital divide and work towards universal access to AI technologies. AI is not accessible to all areas and communities in terms of infrastructure, which may lead to even greater inequality in fighting for conservation cause Training and support to be provided in underserved areas for stakeholders, as AI is booming (Singh et al., 2024).

The ecological and ethical dimension regarding the use of AI must be handled carefully in applications such as wildlife management and forest monitoring. AI has the potential to be a powerful tool for conservation, but it should only be used in ways that are consistent with ethical principles and the rights and welfare of human as well as non-human communities. Open and inclusive decision-making processes are vital to ensure that AI technologies help safeguard biodiversity conservation as well as social welfare (Singh & Kaunert, 2024). This application of AI in wildlife and forest resource management heralds a game changer conservation initiative with the potential to uplift ecosystem vitality as well add attraction value on biodiversity sustainability. In the era of climate changes and consequent disasters after another, the natural environment has become more fragile than ever; AI with its real-time insight support to improve decision making and resource allocation is invaluable in fighting against environmental menaces. Yet the realization of AI-led solutions brings its own set of challenges such as privacy, lack of collaboration between disciplines and lack of equal access. Leveraging AI and inclusive conservation efforts, we can make significant strides towards sustainable wildlife and forest resources management to support the resilience of biodiversity in an ever-changing world (Gomez & White, 2024).

2. AI APPLICATION IN WILDLIFE AND FOREST MANAGEMENT

The vast available data from satellite imagery is analyzed for faster processing speed. Many traditional data analysis methods require extensive time and resources, which impede organizations from quickly addressing new threats. Artificial intelligence can automate data processing, which allows the monitoring of ecosystems in real time and provides rapid response to make decisions faster. For instance, AI models can analyze satellite images to track deforestation and monitor the health of forests or observe land use changes. Use of this information can help to put necessary actions and regulations into place, such as wildlife conservation measures or take enforcement against environmental violators in some cases (Gomez et al., 2024).

There is also greater precision and accuracy of wildlife monitoring efforts that AI-driven tools provide. Forest deploys an arsenal of camera traps and acoustic sensors can generate considerable data on species' occurrence and behavior. However, the manual analysis of this data is very labor-intensive and error-prone while AI algorithms can process thousands to millions of images effectively categorized based on species distribution-often guidelines for population trends. Crucially, this information is needed to evaluate the threat status of taxa and define areas that are essential for their conservation. Other than this, another important area where AI is deployed in estimation and simulation of effect (for example) environmental stressors on wildlife or forests. Ecosystems are threatened by climate change, habitat fragmentation and invasive species but Artificial Intelligence already has the ability to estimate these stressors. For example, predictive models based on AI can predict the expansion of invasive species and their ecological impact at native biodiversity. This knowledge can be used to inform species-specific management plans that aim to reduce impacts of invertebrate invaders on both ecosystems and economies (Johnson & Wang, 2023).

AI may also assist in reforestation and habitat restoration efforts, such as by optimizing planting strategies or predicting the success of restoration projects. The advanced algorithm of machine learning may evaluate the environmental factors, including such as climate or soil type and topography are more destructible for reforesting trees to characterize optimal sites and their modelling tree species provide them with best regeneration. Through increased efficiency and effectiveness of restoration efforts, AI enables better ecological outcomes for the recovery of degraded ecosystems and biodiversity (Singh, 2024).

3. AI MONITORING WILDLIFE: SEMI-AUTOMATED SPECIES IDENTIFICATION

Wild Sense incorporates a wide range of AI technologies and environment sensors that have dramatically increased the efficiency of wildlife monitoring through automatic species identification. Species are usually identified by field survey and manual image analysis, which is time-consuming and has its own limitations. With the advent of AI-powered algorithms, especially techniques involving machine learning (ML) and deep learning (DL), this process has been automated to detect species quickly based on images as well as sounds. There are numerous successful implementations for species recognition on camera trap images using large datasets that have been implemented with Convolutional Neural Networks (CNN), a family of and one type deep learning model as in Thompson et al. These models learn from labelled datasets to identify unique characteristics between different species which allows them a high degree of recognition and accuracy in identifying animals. That such system from was able to reach a 95% accuracy rate in detecting mammals on camera trap images within African savannas using an AI.

Audio recognition models are another critical component when it comes to identifying bird species based on calls and sounds. This model is useful in the discovery of birds and making decisions on their feeding regions, breeding areas to study bird distribution. These AI-based methods not just increase identification performance, but they also reduce the time and resources needed for wildlife monitoring (Singh, 2024). The rise of AI for tracking animal movement has significantly advanced in understanding biodiversity by monitoring wildlife movements which are significant due to behavioral patterns like habitat space use as well migration corridors. Real-time data on the locations of animals collected by GPS collars and AI-powered tracking devices can be used to analyze movement patterns or assess impacts of environmental changes in wildlife (Singh & Kaunert, 2024).

Machine learning models could analyze mountains of historical tracking data and based on this information, predict the likely future movements. For instance, Johnson et al. For example, they use artificial intelligence to predict the migratory paths of marine mammals which allow authorities and non-profit organizations to better protect endangered species from threats like ship trafficking or commercial fishing. AI is further helping in understanding the patterns of social behaviors and interactions in animal populations. However, the movement data can be analyzed to detect social structures, territorial limits and mating behaviors that are valuable for conservation planning as well as habitat management (Singh, 2024).

4. THREAT PICTURE AND POACHING PATTERNS: WILDLIFE AND FOREST RESOURCE MANAGEMENT WITH ARTIFICIAL INTELLIGENCE

Artificial intelligence (AI) helps detect some of these landscapes such as deforestation, poaching and human-wildlife conflict which poses a great threat to wildlife. Predictive modeling techniques evaluate historical data, environmental and human variables in order to pose the risk of poaching within a given area. They are able to find patterns and trends that signify poaching activities, which helps authorities better resource deployment in areas where they can actively combat those threats. Artificial Intelligence and the Data Revolution in Outdoor Crime Protection Read full article Breaching large, protected wildlife reserves like national parks protecting endangered species can inflict lasting environmental damage. AI driven surveillance systems, such as drones or sensor networks installed across a range of mediums (land, marine), have remarkably strengthened law enforcement activities by assisting them with detecting illegal activity within these landscapes. These methods often deploy AI driven systems which are trained to review live data from cameras and sensors for the detection, tracking of poachers, and alerting local law enforcement who can provide immediate response (Martinez & Roberts. AI also plays a role in the monitoring process, which would still employ rangers but whose work is enhanced by AI technologies that help make anti-poaching efforts more efficient and effective - with an eye to increasing protection for endangered species (Singh & Kaunert, 2024).

The continued risks of poaching and illegal logging retain a significant threat to the management of wildlife resources and forests. This behavior is not only threatening the populations of wild animals, but also pushing ecosystems into disarray and causing serious disruption to conservation activities. Artificial intelligence (AI) provides the tools to strengthen our detection and prevention capabilities against these threats. Using data from sources such as satellite imagery, camera traps and acoustic sensors, and poaching can detect chances or patterns far above to trade faster networks. Machine learning can be used to analyze huge quantities of data in real time, flagging anomalies that could signify poaching or non-permitted logging (Johnson et al., 2024).

Image analysis of the camera trap to obtain information about human presence in protected areas, so that alert from potential poaching can be sent out automatically by AI. In the same way, AI can pay attention to sound data and recognize alarm sounds that are related to unlawful exercises like chainsaws or gunfire. Ultimately, they help to detect threats more quickly and accurately - mechanisms that also improve in the timely enforcement of conservation laws. Moreover, if trained AI can predict poaching tendencies and locate susceptible regions based on historical data. Recognizing these patterns allows those doing conservation work to implement targeted strategies for saving endangered species and crucial habitats. AI is helping us recognize the patterns in which threats operate thus fundamentally transforming wildlife and forest resource management by providing a comprehensive insight into illegal activities. As AI has further developed, these technologies have the potential to greatly aid in protecting our natural resources and the future of ecosystems on earth (lee et al., 2024).

5. ARTIFICIAL INTELLIGENCE IN FOREST RESOURCE MANAGEMENT

In the management of forest resources (forest administration), artificial intelligence (AI) - computer programs which think and learn like humans, are playing an ever more significant role in revolutionizing our approach to monitoring, analyzing and conserving forests. Forests are vital for bio-diversity persistence, climatic regulation and serving the nutritional requirements of the mankind. But these ecosystems have many pressures and being chopped down by lot of deforestation, climate change is important too/ illegal logging Given that forested areas are immense and complex, traditional mechanisms of forestry management often fail to tackle these challenges in an efficacious manner. These forest reserves offer unique opportunities for advancing environmental science research and applications using AI technologies including machine learning, computer vision, and remote sensing to address questions related to the sustainability of forests as well as global climate change resilience (Li & Wang, 2023).

AI is most commonly used via machine learning approach in monitoring and evaluating forest health one of the key applications and forestry. More sophisticated remote-sensing technologies like satellite imagery and drone-based surveillance are able to collect a larger amount of data about the state of forests. Patterns and anomalies in the data can be identified by ai algorithms to indicate the health of forests or warn about potential disease outbreaks, pest infestations or other changes due to climate change. For example, machine learning models could analyze multi-spectral/hyperspectral images on the fly and automatically recognize an endangered forest tree so that emergency measures can be taken to prevent additional damage or loss of an ecosystem due to a dangerous plant pest. So we can use this feature to preconceive management strategies before irreversible ecological and economic impacts due to new invasions (Singh & Kaunert, 2024).

AI is used to detect illegal logging and deforestation as well AI has the ability to help authorities identify illegal events promptly by evaluating satellite data and additional remote sensing info, thus supplying real time directions. Such technologies can sense alterations in land use and forest structure, notifying concerned authorities of illegal logging activities or illicit changes of land ownership. AI-driven predictive models can also predict deforestation risks, evaluating how close to roads, human establishments and known trajectories of history-deforested areas are being.log post This information can be used to focus conservation initiatives and have both biological effects (by delivering even a small number of resources) and across national borders (Singh & Kaunert, 2024).

AI makes forest inventory and resource assessment more potent - an absolute necessity for sustainable forest management. Historically, forest inventories have been performed with field surveys that are human resource demanding in order to estimate such parameters as tree density, species composition and biomass. AI-powered tools, it is suggested can automate such processes by accurately estimating forest attributes using aerial or satellite imagery. Powered by deep learning algorithms, these tools can classify tree species and measure forest density, supplying the reliable data that makes possible managed forests. This information is a key part of sustainable harvesting, carbon accounting and biodiversity conservation enabling forest managers to make decisions that balance economic goals with ecological imperatives (Singh, 2023).

Figure 2. shares the points of artificial intelligence in forest resource management (Source- Original)

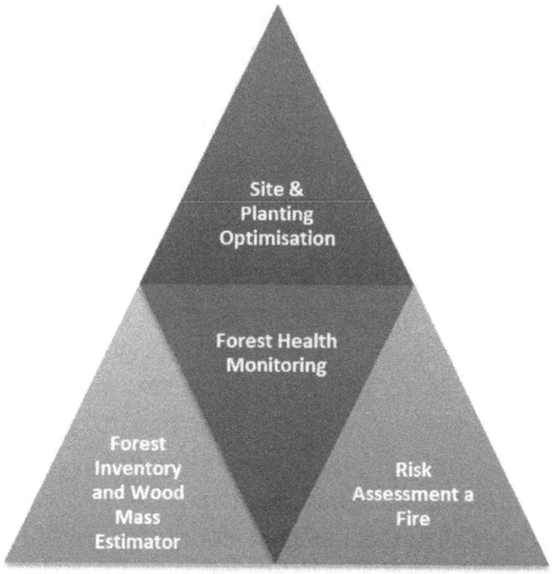

Site & Planting Optimization: AI assists in reforestation and afforestation not just with monitoring and assessment, but also by identifying the right location for planting sites as well as determining what strategies to take when it comes to different problems. With the help of machine learning models, environmental factors (like soil quality, climate conditions and topography) can be processed to highlight where reforestation could take place. This allows a better informed ranking of the areas that are likely to afford high levels of survival as well as potential ecological gains. Also, using AI for simulations of different reforestation scenarios would provide estimates about the extent to which each land use change may influence biodiversity, carbon sequestration and ecosystem services. These findings help guide landscape planning to ensure reforestation efforts are successful and long-lasting (Singh & Kaunert, 2024).

Although AI posses enormous potential for forest resource management, some strong challenges may impede the widespread and successful application of it to real-world problems. We need a detailed examination of data accuracy, problems with inherent biases in AI models and the relation between Artificial tools alignment and traditional management approach. There is a clear need for partnership between ecologists, data scientists who develop and test the algorithms to deploy in novel situations (often termed bioacousticians), policy-makers that enact rules governing new technology use - as well as local communities most at risk from species loss

- to help ensure AI tools are socially and environmentally responsible. AI can act as a game changer in managing forest resources, with the power of transforming sustainability and resilience of forest ecosystems amid global environmental challenges (Singh & Kaunert, 2024).

Forest Health Monitoring: Proper monitoring provides a critical alert of early signs of disease, pest attack and environmental stress. Technological advancements include AI capabilities and remote sensing that can be operated on satellites or drones to provide detailed evidence-based health assessments of complete forest areas (Singh, 2023). Models of machine learning dissect multispectral and hyperspectral images to identify anomalies, assess the health status of vegetation contributing insight for broad-scale evaluation regarding forest condition. The 2023 machine searched satellite images for evidence of forest loss from logging and climate-induced decay. The AI model also successfully detected concerning areas which allowed cascade of focused interventions to prevent progression rallying its name and reputation. mForest dynamics and vegetation cover changes through time. These models are capable of monitoring the phase from forest to deforestation, determine how human activity is affecting forests and offer guidance towards implementing sustainable management strategies (Singh & Kaunert, 2024).

Forest Inventory and Wood Mass Estimator: Reliable forest inventory for sustainable forestry requires accurate estimation of biomass that is also essential in the field of carbon accounting. Through the use of AI, estimates for tree height, diameter and biomass from extensive remote sensing data over large areas with forested land can be derived. For example, deep learning models such as LiDAR-based neural nets can be used to accurately measure forest structure and composition which in turn help guide the decisions surrounding timber harvests, reforestation efforts and carbon credit. Globally, there have been many AI-driven models developed to estimate carbon in forests and this includes the estimation of forest managers already for quantifying their potential sequestration capacity. These improvements make forest resources monitoring and assessment become more informative in terms of economic value as well ecological (Singh & Kaunert, 2024).

Risk Assessment a Fire: In a forest, with thousands of acres on fire - wildfires are another place where AI plays an incredibly important role considering how destructively they affect forests and wildlife. Multiple studies have developed predictive modeling techniques to analyze the environmental, meteorological and historical fire data with other drivers of forest fires so that we can identify areas in high hazard risk state for ignition (Li & Gomez, 2023). Artificial intelligence-based systems can create live fire risk maps, with which authorities can distribute resources more efficiently and intervene in potential incidents. AI is also among the ways fire-detection and monitoring systems (for example those based in satellite sensors or drones) issue advance warning of wildfires to first responders too. Improved fire suppression technologies have helped to slow down the speed of burning and lessened some of the destruction. AI also helps after the fire by measuring ecological changes that have occurred due to a wildfire and assist with reforestation efforts. Data related to vegetation recovery, soil erosion and habitat regeneration can help make a decision in the direction of ecosystem health rehabilitation with resilience as AI models aid in these functions (Martinez & Roberts, 2024).

6. DRAWBACKS AND CONSTRAINTS OF AI IN THE WILDLIFE & FORESTRY SECTOR

6.1 Data Quality and Availability

The benefit of applying AI to wildlife and forest management is directly proportional with the quality and quantity of data available. As data coverage is poor in many areas, it is needed to properly train AI models. This missing data can introduce bias to predictions and thus make it harder for a more complex AI solution to be developed properly. Therefore, the challenges posed by such data need to be addressed through improvements in collection methodologies and standardization of formats as well as better platforms for sharing them (Martinez & Roberts, 2023). Since it utilizes AI on the control of wildlife and forests, some ethical issues and social considerations inevitably emerge. The implementation responsible AI will also need to overcome concerns related to privacy, data protection and the risk for misuse. AI-friendly solutions should be cautiously deployed by adopting a stakeholder perspective that also accounts relevant local communities and indigenous peoples in conservation scenarios. Key to ensuring the sustainable use of resources is having open processes in which diverse contributors have a voice and all can access AI technologies equally (Nguyen & Brown, 2023).

6.2 Integration of Traditional Knowledge

Artificial intelligence should be seen as a complement, not substitute, to the traditional wisdom and practices in managing wildlife & forest. According to Martinez and Li, Indigenous communities and local stakeholders not only have unique insights into the stewardship of natural resources. These tactics are less successful at presenting information as well for conservation practitioner and is best improved through integrating AI with traditional knowledge systems that can optimize not only on ends, but means. Joint projects between AI researchers and indigenous communities, allows for knowledge to flow both ways across the table - rather than a one of way transmission from first world countries in decision-making processes (Nguyen & Brown, 2024).

7. CASE STUDIES AND APPLICATIONS

Case Study 1: AI for Wildlife Conservation

There are also successful projects preventing wildlife poaching and saving endangered species in the African savannah using AI technologies. Thermal cameras and image recognition algorithms on AI drones patrol national parks to detect illegal acts so they can alert park rangers in real time Such systems have led to dramatic decreases in poaching rates, with overall benefits linked particularly to species such as the elephant and rhino (Smith & Liu, 2024). AI is used for monitoring and tracking endangered species, in addition to anti-poaching efforts. Real-time GPS collars with embedded AI analytics give us information on the location and behaviour of animals to better recognize habitat preferences as well migratory movements. Such information is essential for the conservation of endangered species by ensuring that protected areas and corridors are designed to help threatened organisms (Smith & Patel, 2023).

Case Study 2: AI in Forest Restoration

The same AI-driven remote sensing technologies have helped restore the Amazon rainforest. They use machine learning models trained on satellite images to detect areas of deforestation and rank them by their importance for reforestation . Artificial Intelligence algorithms evaluate the best fit tree species for restoration projects to improve reforestation strategies while increasing ecological resilience in restored areas. Monitoring the growth and survivorship of planted trees to inform work on ecosystem restoration, AI models provide accurate data that is essential for understanding how well different treatments are working or what should be changed in management practices. They help support global mass reforestation efforts in the fight against climate change and biological loss (Smith et al., 2023).

Case 3: AI in Fire Management

In California, AI technologies are being used to better manage and respond to wildfires. AI-based predictive models process variety of data from weather to vegetation conditions and historical fire patterns, predict fire risk environments after computing the indicators from different fuels in specific area towards effective prominence rotational control measures also AI computes these mentioned indications within region for automated prioritization. These models create dynamic fire risk maps to direct resources and promote strategic planning in response to fires. Real-time monitoring of active wildfires using AI-powered drones and satellite-based sensors that provide information on fire behavior and propagation. The data helps firefighters react proactively saving lives and property while minimizing the ecological footprint of wildfires. Post-fire relief - AI technologies aid in the analysis of ecological effects from wildfires and helping to design restoration plans (Thompson & Brown, 2023).

Figure 3. shows the case studies (Source- Original)

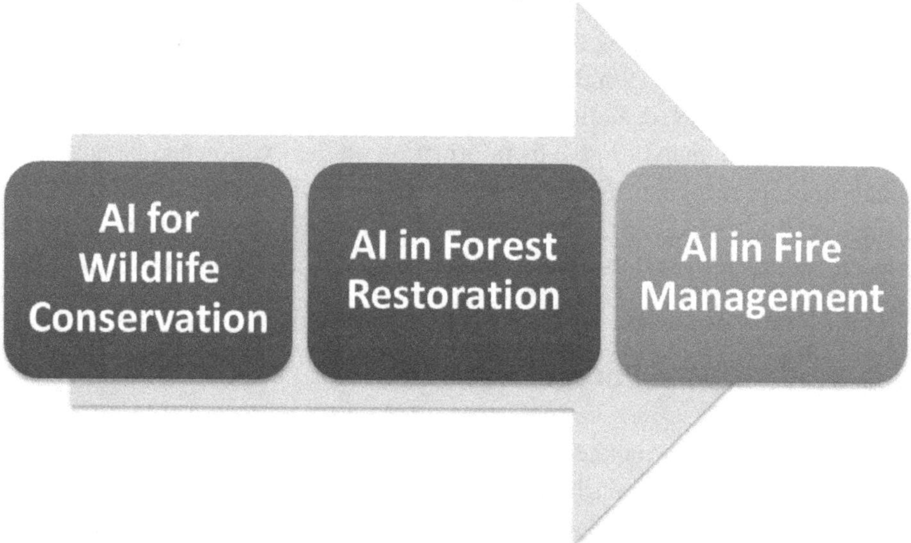

8. PERSPECTIVES FOR THE FUTURE

The next 10 years of wildlife and forest resource management will be driven by the growth in AI technologies. New AI methods, such as those based on reinforcement learning and explainable onto suitable approaches for challenging conservation problems. Reinforcement learning: Reinforcement models are capable of adapting management strategies using gain from experience in synthetic environments knowing most beneficial course. Explainable AI as explanations of decisions made by an algorithm to facilitate understanding and trust. AI in combination with other technologies like Internet of Things (IoT) and blockchain present new avenues for serving wildlife conservation, sustainable forest management. Similarly, IoT devices such as sensors and drones can be used to monitor environmental conditions or wildlife behaviours in real-time which they may then rely on when feeding into AI models (Thompson & Zhao, 2023). The use of blockchain technology can provide better security and data transparency, support the open exchange of complex datasets between stakeholders. Governments, research institutions, conservation organizations and local communities need to work together for AI in wildlife and forest resource management. Collaborating amongst conservation efforts helps to share data and knowledge that can foster the creation of new solutions targeted towards an identified challenge. Strategic partnerships such as these allow for a comprehensive approach to managing our resources, by combining scientific knowledge and traditional practices with community perspectives (Thompson & Zhao, 2024).

Policy frameworks and governance mechanisms are essential in guiding the responsible development of AI practices for wildlife and forest management. Now policymakers must lay out laws that address issues related to data privacy, biodiversity conservation and so on (Buchelt et al., 2024). Such cross-border collaboration can be facilitated through international cooperation and agreements on data sharing (and) standards for AI standards that would elevate conservation efforts. AI technologies are introduced in policies that incentivize their use within the domain leading to innovation and sustainability. Funding mechanisms that favour AI research and development enabling wildlife conservation, forest protection etc. (Nguyen et al., 2024) can drive the fast deployment of AI solution towards global nature support goals as a whole (Li et al., 2024).

In order to unleash the full benefits of AI in wildlife and forest resource management, it is important that human capacity building and education be prioritized investments. Education and workshops could train conservation practitioners, policymakers and local communities to implement AI. Educational programs focused on digital literacy and provided to populations so that they can actively engage in decision-making processes of benefit as well as a challenge for AI. Capacity building efforts can promote a learning and innovation culture for ensuring AI is integrated with resource management in both sustainable and responsible manners. Anderson & Lee (2024) stressed the capacity building of local institutions to engage in collaborative research and education initiatives - leading more towards designing AI solutions from a conservation perspective (Haq et al., 2024).

The hope is that incorporating Artificial Intelligence will be an extremely valuable aid in managing wildlife and forest resources facing such a myriad of challenges. AI can help to reduce the manual work involved in classifying objects or removing false alarms such as humans, which will not only provide new and more effective solutions for monitoring wildlife populations but also long-term mission like forest health assessment and protect from the harmful effects of threats e.g. poaching and wildfires (Wilson, 2024). To realize the benefits of artificial intelligence (AI), factors such as ethics, community relationships and environmental elements must be taken into account. In essence a collaborative effort is needed. Using AI as a Technical Approach to Sustainable Resources Management Can Strengthen our Ways of Biodiversity Conservation, Ecosystems Protection and Better Human-Nature Relationships with better harmony (Chisom et al., 2024).

9. CONCLUSION

The development of artificial intelligence (AI) for wildlife and forest resource management represents a transformational change in the conservation and sustainability paradigm (Singh & Kaunert, 2024). AI technologies are already proving to be vital in transforming wildlife and forest monitoring, analysis, and conservation practices from traditional exploration to the cutting edge of data science by providing new insights into regulating engines like ecosystems balance maintained on-going base management we can decision making over ecological systems (Singh & Kaunert, 2025) (Singh & Kaunert, 2025). With the persisting upsurge of environmental challenges, deforestation, biodiversity loss and climate change as recurring symptoms (Singh et al., 2025). AI can provide a levered response to minimize these problems before they permanently destroy essential ecosystems within months or years (Singh & Kaunert, 2025).

Wildlife and forest management have benefitted from AI-driven solutions in multiple domains, such as species identification, habitat monitoring or threat detection (Singh et al., 2025). Remote sensing technologies like drones and satellites have enabled the amassing of huge amounts of data that can be reviewed by machine learning algorithms or computer vision systems to detect changes in land cover, determine forest health, and track populations (Kaushik et al., 2024). The speed of these capabilities allows us to understand ecosystem dynamics more quickly and intervene faster as we bear witness to their effects on orchards. Through using AI, conservationists and policymakers can create more nuanced strategies that direct resources in the most beneficial way possible both for conserving wildlife as well as protect land use practices (Kaushik et al., 2024).

The potential applications of AI in wildlife and forest resource management are huge for the coming future. In the future, as AI technologies advance further, they will play a greater role in efforts to conserve wildlife and thereby provide new opportunities for leveraging technology-supported development (Singh et al., 2024). For example, AI-based predictive models could be trained to cover the dynamic interactions in ecosystems and predict effects of climate change on wildlife or forests. They can pinpoint potential ecosystems that are vulnerable to degradation and allow conservation efforts to target the highest-priority areas, ultimately providing a cost-effective way for management decisions with limited funding (Singh et al., 2024).

Community engagement and participation - a new frontier for AI in conservation. Such citizen science initiatives could benefit from AI-powered platforms that enable individuals to participate in data collection and monitoring. AI works to involve local communities in their conservation programs because it believes that if a community feels responsible for the resources, they will also take ownership of the land and hopefully use them closer to sustainably as possible (Singh et al., 2024). AI can assist with the design of educational tools and materials touching on biodiversity conservation and forest preservation, to help raise awareness about these issues among a broader swath of citizens that may more actively partake in environmental stewardship (Singh et al., 2024).

The future of AI for wildlife and forest management will probably involve more interdisciplinary work with ecologists working hand-in-hand with data scientists (and policymakers and local communities). Particularly, in addressing challenges about data privacy, ethical considerations and equitable AI access - collaboration from both public health agencies with tech community are necessary. Stakeholders can ensure that solutions based on AI are deployed in a socially responsible and environmentally friendly way, by connecting people through partnerships to share knowledge for sake of resilience and sustainability of nature.

AI presents a game-changing opportunity for wildlife and forest resource management - new tools that can be wielded against the most pressing environmental challenges of our time. Using AI we can better understand ecosystems, increase conservation impact and drive sustainable development. This avenue opens up a large future scope on an AI perspective to this field and more innovation, cooperation and public engagement can be made by upcoming generations. On our way, we must tackle challenges and discuss ethical considerations to promise a future where AI serves the societies on this planet. Harnessing AI while ensuring equity will help create a path forward so that wildlife and forests continue to thrive, today and tomorrow.

REFERENCES

Anderson, J., & Lee, H. (2024). AI in forest health monitoring: Applications and challenges. *Journal of Forestry Research*, 34(6), 1142–1156.

Anderson, J., & Taylor, M. (2024). AI and conservation: Challenges and opportunities for wildlife protection. *Journal of Environmental Management*, 312, 256–269.

Anderson, R., Brown, T., & Wang, X. (2023). The potential of AI in wildlife and forest management. *Conservation Letters*, 16(3), e1287.

Brown, A., & Chen, L. (2023). The intersection of AI and traditional knowledge in conservation. *Journal of Ecological Informatics*, 64, 101–118.

Brown, T., & White, J. (2024). AI for wildlife protection: Innovations and challenges. *Wildlife Conservation Journal*, 48(2), 81–95.

Buchelt, A., Adrowitzer, A., Kieseberg, P., Gollob, C., Nothdurft, A., Eresheim, S., Tschiatschek, S., Stampfer, K., & Holzinger, A. (2024). Exploring artificial intelligence for applications of drones in forest ecology and management. *Forest Ecology and Management*, 551, 121530. DOI: 10.1016/j.foreco.2023.121530

Chen, Y., & Lee, H. (2024). AI-driven forest restoration strategies: Optimizing ecological resilience. *Forest Ecology and Management*, 512, 119–131.

Chen, Y., Lee, H., & Wang, Z. (2023). Machine learning applications in wildlife monitoring. *Ecological Informatics*, 62, 101–114.

Chisom, O. N., Biu, P. W., Umoh, A. A., Obaedo, B. O., Adegbite, A. O., & Abatan, A. (2024). Reviewing the role of AI in environmental monitoring and conservation: A data-driven revolution for our planet. *World Journal of Advanced Research and Reviews*, 21(1), 161–171. DOI: 10.30574/wjarr.2024.21.1.2720

Gomez, C., & White, J. C. (2024). Remote sensing and AI: Enhancing forest monitoring capabilities. *Remote Sensing of Environment*, 299, 113345.

Gomez, C., White, J. C., & Wulder, M. A. (2024). Machine learning approaches for dynamic fire behavior prediction. *ISPRS Journal of Photogrammetry and Remote Sensing*, 187, 235–248.

Haq, B., Jamshed, M. A., Ali, K., Kasi, B., Arshad, S., Kasi, M. K., & Ur-Rehman, M. (2024). Tech-Driven Forest Conservation: Combating Deforestation With Internet of Things, Artificial Intelligence, and Remote Sensing. *IEEE Internet of Things Journal*, 11(14), 24551–24568. DOI: 10.1109/JIOT.2024.3378671

Johnson, A., & Wang, L. (2023). AI in forest restoration: From monitoring to reforestation. *Nature Climate Solutions*, 5(3), 118–134.

Johnson, A., & Wang, L. (2024). Advancements in AI for wildlife conservation. *The Journal of Wildlife Management*, 88(1), 22–39.

Johnson, A., Wang, L., & Roberts, C. (2024). Enhancing wildlife protection with AI-powered surveillance systems. *Wildlife Conservation Journal*, 47(1), 63–76.

Kaushik, T. K. Ravish, Singh, A., Raghav, A., Singh, B., Arora, M. K., & Lal, S. (2024). Untapping Artificial Intelligence in Environmental and Sustainable Progression: Projecting Legal and Ethical Considerations. In B. Singh, C. Kaunert, K. Vig, & S. Dutta (Eds.), *Maintaining a Sustainable World in the Nexus of Environmental Science and AI* (pp. 109-130). IGI Global. https://doi.org/DOI: 10.4018/979-8-3693-6336-2.ch005

Lee, D., Kim, S., & Zhang, J. (2024). Deep learning applications in automated species identification. *Ecological Informatics*, 62, 101–114.

Li, C., Xiao, M., & Liu, Y. (2024). Prospects for AI applications in forest protection: Technologies, challenges, and future developments. *Advances in Resources Research*, 4(3), 362–380.

Li, X., & Wang, X. (2023). AI-driven fire risk assessment: A proactive approach to wildfire management. *Journal of Wildland Fire Science*, 23(4), 225–238.

Li, Y., & Gomez, C. (2023). AI for forest inventory and biomass estimation: A comprehensive review. *Forest Inventory and Biomass Estimation*, 22(1), 1–17.

Martinez, E., & Li, Y. (2024). Integrating AI and traditional knowledge in forest management: A collaborative approach. *Environmental Management Perspectives*, 21(2), 75–89.

Martinez, E., & Roberts, C. (2023). AI in anti-poaching efforts: A case study from Africa. *Wildlife Conservation Journal*, 46(2), 135–149.

Nguyen, T., & Brown, T. (2023). AI in wildlife conservation: From species identification to behavior analysis. *Biodiversity and Conservation*, 33(7), 1129–1144.

Nguyen, T., & Brown, T. (2024). AI applications in wildlife conservation and management. *Journal of Wildlife and Conservation*, 45(2), 95–110.

Raghav, A., Singh, B., Jermsittiparsert, K., Raghav, R., & Yadav, U. (2024). Artificial Intelligence in Environmental and Climate Changes: A Sustainable Future. In Singh, B., Kaunert, C., Vig, K., & Dutta, S. (Eds.), *Maintaining a Sustainable World in the Nexus of Environmental Science and AI* (pp. 485–506). IGI Global., DOI: 10.4018/979-8-3693-6336-2.ch019

Singh, B. (2023). Blockchain Technology in Renovating Healthcare: Legal and Future Perspectives. In Kaushik, K., Dahiya, S., Aggarwal, S., & Dwivedi, A. (Eds.), *Revolutionizing Healthcare Through Artificial Intelligence and Internet of Things Applications* (pp. 177–186). IGI Global., DOI: 10.4018/978-1-6684-5422-0.ch012

Singh, B. (2024). Lensing Legal Dynamics for Examining Responsibility and Deliberation of Generative AI-Tethered Technological Privacy Concerns: Infringements and Use of Personal Data by Nefarious Actors. In Ara, A., & Ara, A. (Eds.), *Exploring the Ethical Implications of Generative AI* (pp. 146–167). IGI Global., DOI: 10.4018/979-8-3693-1565-1.ch009

Singh, B. (2024). Social Cognition of Incarcerated Women and Children: Addressing Exposure to Infectious Diseases and Legal Outcomes. In Reddy, K. (Ed.), *Principles and Clinical Interventions in Social Cognition* (pp. 236–251). IGI Global., DOI: 10.4018/979-8-3693-1265-0.ch014

Singh, B. (2024). Evolutionary Global Neuroscience for Cognition and Brain Health: Strengthening Innovation in Brain Science. In Prabhakar, P. (Ed.), *Biomedical Research Developments for Improved Healthcare* (pp. 246–272). IGI Global., DOI: 10.4018/979-8-3693-1922-2.ch012

Singh, B., Dutta, P. K., & Kaunert, C. (2024). Replenish Artificial Intelligence in Renewable Energy for Sustainable Development: Lensing SDG 7 Affordable and Clean Energy and SDG 13 Climate Actions With Legal-Financial Advisory. In Derbali, A. (Ed.), *Social and Ethical Implications of AI in Finance for Sustainability* (pp. 198–227). IGI Global., DOI: 10.4018/979-8-3693-2881-1.ch009

Singh, B., Jain, V., Kaunert, C., Dutta, P. K., & Singh, G. (2024). Privacy Matters: Espousing Blockchain and Artificial Intelligence (AI) for Consumer Data Protection on E-Commerce Platforms in Ethical Marketing. In Saluja, S., Nayyar, V., Rojhe, K., & Sharma, S. (Eds.), *Ethical Marketing Through Data Governance Standards and Effective Technology* (pp. 167–184). IGI Global., DOI: 10.4018/979-8-3693-2215-4.ch015

Singh, B., & Kaunert, C. (2024). Computational Thinking for Innovative Solutions and Problem-Solving Techniques: Transforming Conventional Education to Futuristic Interdisciplinary Higher Education. In Fonkam, M., & Vajjhala, N. (Eds.), *Revolutionizing Curricula Through Computational Thinking, Logic, and Problem Solving* (pp. 60–82). IGI Global., DOI: 10.4018/979-8-3693-1974-1.ch004

Singh, B., & Kaunert, C. (2024). Wind and Solar Energy as Renewable Energy for Sustainable Global Future: Projecting Future Multi-Sector Sustainable Policies and Innovation. In Ara, A., & Thakore, R. (Eds.), *Promoting Multi-Sector Sustainability With Policy and Innovation* (pp. 210–245). IGI Global., DOI: 10.4018/979-8-3693-2113-3.ch009

Singh, B., & Kaunert, C. (2024). Aroma of Highly Smart Internet of Medical Things (IoMT) and Lightweight EdgeTrust Expansion Medical Care Facilities for Electronic Healthcare Systems: Fortified-Chain Architecture for Remote Patient Monitoring and Privacy Protection Beyond Imagination. In Hassan, A., Bhattacharya, P., Tikadar, S., Dutta, P., & Sagayam, M. (Eds.), *Lightweight Digital Trust Architectures in the Internet of Medical Things (IoMT)* (pp. 196–212). IGI Global., DOI: 10.4018/979-8-3693-2109-6.ch011

Singh, B., & Kaunert, C. (2024). Augmented Reality and Virtual Reality Modules for Mindfulness: Boosting Emotional Intelligence and Mental Wellness. In Hiran, K., Doshi, R., & Patel, M. (Eds.), *Applications of Virtual and Augmented Reality for Health and Wellbeing* (pp. 111–128). IGI Global., DOI: 10.4018/979-8-3693-1123-3.ch007

Singh, B., & Kaunert, C. (2024). Salvaging Responsible Consumption and Production of Food in the Hospitality Industry: Harnessing Machine Learning and Deep Learning for Zero Food Waste. In Singh, A., Tyagi, P., & Garg, A. (Eds.), *Sustainable Disposal Methods of Food Wastes in Hospitality Operations* (pp. 176–192). IGI Global., DOI: 10.4018/979-8-3693-2181-2.ch012

Singh, B., & Kaunert, C. (2024). Revealing Green Finance Mobilization: Harnessing FinTech and Blockchain Innovations to Surmount Barriers and Foster New Investment Avenues. In Jafar, S., Rodriguez, R., Kannan, H., Akhtar, S., & Plugmann, P. (Eds.), *Harnessing Blockchain-Digital Twin Fusion for Sustainable Investments* (pp. 265–286). IGI Global., DOI: 10.4018/979-8-3693-1878-2.ch011

Singh, B., & Kaunert, C. (2024). Harnessing Sustainable Agriculture Through Climate-Smart Technologies: Artificial Intelligence for Climate Preservation and Futuristic Trends. In Kannan, H., Rodriguez, R., Paprika, Z., & Ade-Ibijola, A. (Eds.), *Exploring Ethical Dimensions of Environmental Sustainability and Use of AI* (pp. 214–239). IGI Global., DOI: 10.4018/979-8-3693-0892-9.ch011

. Singh, B., & Kaunert, C. (2024). Unmanned aerial vehicle. *Cognitive Machine Intelligence: Applications, Challenges, and Related Technologies*, 108.

Singh, B., & Kaunert, C. (2024). Adventure in High Altitude of Mountainous Topographies and Health Impacts: Lensing Tourism Sustainability via Reducing Ecological and Sociocultural Footprint and Health Emergency and Medical Assistance Management. In Meraj, G., Hashimoto, S., & Kumar, P. (Eds.), *Navigating Natural Hazards in Mountainous Topographies. Disaster Risk Reduction*. Springer., DOI: 10.1007/978-3-031-65862-4_15

Singh, B., & Kaunert, C. (2025). Featuring Healthcare, Environment, and Human Rights: Applications of Artificial Intelligence in the Health Domain. In Chakraborty, S., & Satapathy, S. (Eds.), *Gender, Environment, and Human Rights: An Intersectional Exploration* (pp. 45–74). IGI Global., DOI: 10.4018/979-8-3693-6069-9.ch004

Singh, B., & Kaunert, C. (2025). Intelligent Machine Learning Solutions for Cybersecurity: Legal and Ethical Considerations in a Global Context. In Thangam, D. (Ed.), *Advancements in Intelligent Process Automation* (pp. 359–386). IGI Global., DOI: 10.4018/979-8-3693-5380-6.ch014

Singh, B., & Kaunert, C. (2025). Innovative Business Models in Digital and Virtual Reality for the Gig Economy: Space-Age Sustainable Development Projecting Smart Business Practices. In Taherdoost, H., Drazenovic, G., Madanchian, M., Khan, I., & Arshi, O. (Eds.), *Business Transformation in the Era of Digital Disruption* (pp. 127–156). IGI Global., DOI: 10.4018/979-8-3693-7056-8.ch005

Singh, B., & Kaunert, C. Reinventing Artificial Intelligence and Blockchain for Preserving Medical Data. In *Ethical Artificial Intelligence in Power Electronics* (pp. 77–91). CRC Press. DOI: 10.1201/9781032648323-6

Singh, B., Kaunert, C., & Gautam, R. (2024). Harnessing Machine Learning Intelligence Against Cyber Threats: Navigating Legal and Ethical Challenges in Global Cyber Investigations. In Özsungur, F. (Ed.), *Empowering Entrepreneurial Mindsets With AI* (pp. 319–352). IGI Global., DOI: 10.4018/979-8-3693-7658-4.ch016

Singh, B., Kaunert, C., Jermsittiparsert, K., Lal, S., Arora, M. K., & Raghav, A. (2024). Ultimate Deep Learning Acclimatization With Artificial Intelligence Drones for Environment Shield and Climate Change: Lifting Sustainability Towards Fostering SDG 13. In Singh, B., Kaunert, C., Vig, K., & Dutta, S. (Eds.), *Maintaining a Sustainable World in the Nexus of Environmental Science and AI* (pp. 59–80). IGI Global., DOI: 10.4018/979-8-3693-6336-2.ch003

Singh, B., Kaunert, C., Malviya, R., Lal, S., & Arora, M. K. (2024). Scrutinizing Consumer Sentiment on Social Media and Data-Driven Decisions for Business Insights: Fusion of Artificial Intelligence (AI) and Business Intelligence (BI) Foster Sustainable Growth. In Natarajan, A., Galety, M., Iwendi, C., Das, D., & Shankar, A. (Eds.), *Intersection of AI and Business Intelligence in Data-Driven Decision-Making* (pp. 183–210). IGI Global., DOI: 10.4018/979-8-3693-5288-5.ch007

Singh, B., Kaunert, C., & Ravesangar, K. (2025). Artificial Intelligence and Employee Well-Being: Projecting Psychological and Physical Healthcare. In Yadav, P. (Ed.), *Human Resource Strategies in the Era of Artificial Intelligence* (pp. 109–134). IGI Global., DOI: 10.4018/979-8-3693-6412-3.ch005

Singh, B., Singh, A., Kaunert, C., Arora, M. K., Lal, S., & Ravesangar, K. (2025). Unleashing the Ethical and Legal Implications of E-Business Revolution: Consumer Privacy and Security Concerns in Phase of Digital Disruption. In Taherdoost, H., Drazenovic, G., Madanchian, M., Khan, I., & Arshi, O. (Eds.), *Business Transformation in the Era of Digital Disruption* (pp. 157–180). IGI Global., DOI: 10.4018/979-8-3693-7056-8.ch006

Smith, K., & Liu, Y. (2024). The role of AI in biodiversity conservation. *Conservation Science and Practice*, 16(2), e1288.

Smith, K., & Patel, R. (2023). Challenges and opportunities in AI-driven wildlife monitoring. *Conservation Science and Practice*, 15(2), e150.

Smith, K., Zhao, L., & Patel, R. (2023). AI for forest health monitoring: Applications and challenges. *Journal of Forestry Research*, 34(6), 1142–1156.

Thompson, B., & Brown, D. (2023). Ethical considerations in AI-based wildlife management. *Ethics and the Environment*, 29(4), 54–68.

Thompson, B., & Zhao, Q. (2023). AI in fire management: Early warning systems and real-time monitoring. *Fire Technology*, 60(1), 14–29.

Thompson, B., & Zhao, Q. (2024). The impact of AI on wildlife conservation: A case study from Africa. *Wildlife Conservation Journal*, 48(2), 96–100.

Wilson, N. (2024). Artificial intelligence helps drive new frontiers in ecology. *Bioscience*, 74(5), 306–311. DOI: 10.1093/biosci/biae016 PMID: 38854635

Compilation of References

Abatzoglou, J. T., Battisti, D. S., Williams, A. P., Hansen, W. D., Harvey, B. J., & Kolden, C. A. (2021). Projected increases in western US forest fire despite growing fuel constraints. *Communications Earth & Environment*, 2(1), 1–8. DOI: 10.1038/s43247-021-00299-0

Abdalla &Badran. (2019) Wireless communication technologies for drone-based Internet of Things applications.

Abid, F. (2021). A survey of machine learning algorithms based forest fires prediction and detection systems. *Fire Technology*, 57(2), 559–590. DOI: 10.1007/s10694-020-01056-z

Adams, M. A. (2013). Mega-fires, tipping points and ecosystem services: Managing forests and woodlands in an uncertain future. *Forest Ecology and Management*, 294, 250–261. DOI: 10.1016/j.foreco.2012.11.039

Ahmed, E., Ali, S.M., & Khan, M. (2018) . Survey on energy-efficient communication protocols for UAV-assisted IoT applications.

Akhloufi, M. A., Bendada, A., & Boireau, A. (2020). Fire detection and monitoring using unmanned aerial vehicles. *Drones (Basel)*, 4(2), 34. DOI: 10.3390/drones4020034

Akhtar-Schuster, M., Stringer, L. C., Metternicht, G., Barger, N. N., Chotte, J. L., & Kust, G. (2022). Assessing the impact of science in the implementation of the United Nations Convention to Combat Desertification. *Land (Basel)*, 11(4), 568. DOI: 10.3390/land11040568

Akula, R., & Garibay, I. (2021). Audit and assurance of AI algorithms: a framework to ensure ethical algorithmic practices in artificial intelligence. *arXiv preprint arXiv:2107.14046.*

Albahri, A. S., Khaleel, Y. L., Habeeb, M. A., Ismael, R. D., Hameed, Q. A., Deveci, M., Homod, R. Z., Albahri, O. S., Alamoodi, A. H., & Alzubaidi, L. (2024). A systematic review of trustworthy artificial intelligence applications in natural disasters. *Computers & Electrical Engineering*, 118, 109409. DOI: 10.1016/j.compeleceng.2024.109409

Al-Bashiti, M. K., & Naser, M. Z. (2022). Machine learning for wildfire classification: Exploring blackbox, eXplainable, symbolic, and SMOTE methods. *Natural Hazards Review*, 2(3), 154–165. DOI: 10.1016/j.nhres.2022.08.001

Alghamdi, W., Marchiandi, J., Szabo, D., Samandra, S., & Clarke, B. O. (2022). Release of per- and polyfluoroalkyl substances (PFAS) from a waste management facility fire to an urban creek. *Journal of Hazardous Materials Advances*, 8, 100167. https://doi.org/https://doi.org/10.1016/j.hazadv.2022.100167. DOI: 10.1016/j.hazadv.2022.100167

Alharbi, F., & Alhassan, I. (2022). Simulation-based performance evaluation of mobile ad hoc routing protocols in a swarm of unmanned aerial vehicles: A review and analysis. *Journal of Network and Computer Applications*, 204, 103043. Advance online publication. DOI: 10.1016/j.jnca.2022.103043

Alkhatib, R., Sahwan, W., Alkhatieb, A., & Schütt, B. (2023). A brief review of machine learning algorithms in forest fires science. *Applied Sciences (Basel, Switzerland)*, 13(14), 8275. DOI: 10.3390/app13148275

Alshahrani, Alzahrani, & Khan.(2020) . A comprehensive survey on the integration of IoT with UAVs for smart cities.

Alshahrani, S., Alzahrani, F., & Khan, M. (2021). Challenges and opportunities in the integration of UAVs with IoT technologies.

Alzubaidi & Yaqoob. (2021) Challenges and opportunities in the integration of drones with IoT technologies.

Alzubaidi, A., & Yaqoob, I. (2021). A survey on UAV communication networks: Performance parameters and challenges. *Wireless Networks*, 27(3), 1234–1250. DOI: 10.1007/s11276-020-02501-1

Alzubaidi, L., Zhang, J., Humaidi, A. J., Al-Dujaili, A., Duan, Y., Al-Shamma, O., Santamaría, J., Fadhel, M. A., Al-Amidie, M., & Farhan, L. (2021). Review of deep learning: Concepts, CNN architectures, challenges, applications, future directions. *Journal of Big Data*, 8(1), 53. DOI: 10.1186/s40537-021-00444-8 PMID: 33816053

Anderson, J., & Lee, H. (2024). AI in forest health monitoring: Applications and challenges. *Journal of Forestry Research*, 34(6), 1142–1156.

Anderson, J., Li, X., & Brown, T. (2023). Remote sensing applications for sustainable land management. *International Journal of Remote Sensing*, 44(5), 345–360.

Anderson, J., & Roberts, C. (2023). Data challenges in AI-driven fire management systems. *Journal of Environmental Management*, 312, 114–125.

Anderson, J., & Taylor, M. (2024). AI and conservation: Challenges and opportunities for wildlife protection. *Journal of Environmental Management*, 312, 256–269.

Anderson, R., Brown, T., & Wang, X. (2023). The potential of AI in wildlife and forest management. *Conservation Letters*, 16(3), e1287.

Anekwe, I. M. S., Zhou, H., Mkhize, M. M., & Akpasi, S. O. (2024). Addressing Climate Change Challenges in South Africa: A Study in KwaZulu Natal Province. *Climate Crisis: Adaptive Approaches and Sustainability*, 475-496.

Arif, M., Alghamdi, K. K., Sahel, S. A., Alosaimi, S. O., Alsahaft, M. E., Alharthi, M. A., & Arif, M. (2021). Role of machine learning algorithms in forest fire management: A literature review. *J. Robot. Autom*, 5, 212–226.

Arshad, A., Azhar Ali, A., & Anjali, K. S. (2022). Impact of forest fire on forest ecosystem. *Journal of Agriculture and Technology*, 9(1&2), 18–29.

Aryal, K., Maraseni, T., & Apan, A. (2023). Examining policy– institution– program (PIP) responses against the drivers of ecosystem dynamics. A chronological review (1960–2020) from Nepal. *Land Use Policy*, 132, 106789. DOI: 10.1016/j.landusepol.2023.106789

Ashukem, J. C. N. (2024). Land grabbing in the Anthropocene and the "questionable" pursuit of sustainability in Africa–highlighting ecological disproportionality. In *Human Rights and the Environment in Africa* (pp. 276–294). Routledge.

Ashukem, J. C. N., & Sama, S. M. (Eds.). (2023). *Domestic and Regional Environmental Laws and Policies in Africa: A Research Companion*. Taylor & Francis. DOI: 10.4324/9781003382256

Ashukem, J. C. N., & Sama, S. M. (Eds.). (2023). *Human Rights and the Environment in Africa: A Research Companion*. Taylor & Francis. DOI: 10.4324/9781003382249

Attri, V., Dhiman, R., & Sarvade, S. (2020). A review on status, implications and recent trends of forest fire management. *Archives of Agriculture and Environmental Science*, 5(4), 592–602. DOI: 10.26832/24566632.2020.0504024

Awatade, A., Pawar, P., & Lakshmi, D. (2024). The Power of Machine Learning in Forest Fire Risk Analysis and Resilience: Navigating Best Practices, Challenges, and Opportunities. *Geospatial Technology to Support Communities and Policy: Pathways to Resiliency*, 149-170

Baena, S., Moat, J., Whaley, O., & Boyd, D. S. (2017). Identifying species from the air: UAVs and the very high resolution challenge for plant conservation. *PLoS One*, 12(11), e0188714. DOI: 10.1371/journal.pone.0188714 PMID: 29176860

Banerjee, O., Cicowiez, M., Malek, Ž., Verburg, P. H., Vargas, R., Goodwin, S., Bagstad, K. J., & Murillo, J. Á. (2023). Banking on strong rural livelihoods and the sustainable use of natural capital in post-conflict Colombia. *Environment, Development and Sustainability*, 26(10), 1–22. DOI: 10.1007/s10668-023-03740-w

Bargali, H., Calderon, L. P. P., Sundriyal, R. C., & Bhatt, D. (2022). Impact of forest fire frequency on floristic diversity in the forests of Uttarakhand, western Himalaya. *Trees, Forests and People*, 9, 100300. DOI: 10.1016/j.tfp.2022.100300

Basu, M. (2024). *The Routledge Handbook of International Environmental Policy*. Taylor & Francis. DOI: 10.4324/9781003440574

Batisha, A. (2023). A lighthouse to enhance the quality of life in the Nile River basin. *Environmental Economics and Policy Studies*, 1–29. DOI: 10.1007/s10018-023-00380-2

Bera, B., Shit, P. K., Sengupta, N., Saha, S., & Bhattacharjee, S. (2022). Forest fire susceptibility prediction using machine learning models with resampling algorithms, Northern part of Eastern Ghat Mountain range (India). *Geocarto International*, 37(26), 11756–11781. DOI: 10.1080/10106049.2022.2060323

Bhattarai, N., Dahal, S., Thapa, S., Pradhananga, S., Karky, B. S., Rawat, R. S., Windhorst, K., Watanabe, T., Thapa, R. B., & Avtar, R. (2022). Forest fire in the hindu kush Himalayas: A major challenge for climate action. *Journal of Forest and Livelihood*, 21(1), 14–31. DOI: 10.3126/jfl.v21i1.56576

Bhowmik, R. T., Jung, Y. S., Aguilera, J. A., Prunicki, M., & Nadeau, K. (2023). A multi-modal wildfire prediction and early-warning system based on a novel machine learning framework. *Journal of Environmental Management*, 341, 117908. DOI: 10.1016/j.jenvman.2023.117908 PMID: 37182403

Bilgiç, E., Tuna Tuygun, G., & Gündüz, O. (2023). Development of an emission estimation method with satellite observations for significant forest fires and comparison with global fire emission inventories: Application to catastrophic fires of summer 2021 over the Eastern Mediterranean. *Atmospheric Environment*, 308, 119871. https://doi.org/https://doi.org/10.1016/j.atmosenv.2023.119871. DOI: 10.1016/j.atmosenv.2023.119871

BINJOLA, R., DOBHAL, S., & LAL, P. (2022). A REVIEW ON THE CURRENT STATUS OF FOREST FIRE IN UTTARAKHAND STATE OF INDIA. *Journal of Global Ecology and Environment*, 16(4), 26–33.

Bogers, M. (2023). *Integrating global sustainability governance: How the Sustainable Development Goals impact institutional and policy integration at the global level* (Doctoral dissertation, Utrecht University).

Boroujeni, S. P. H., Razi, A., Khoshdel, S., Afghah, F., Coen, J. L., O'Neill, L., Fule, P., Watts, A., Kokolakis, N.-M. T., & Vamvoudakis, K. G. (2024). A comprehensive survey of research towards AI-enabled unmanned aerial systems in pre-, active-, and post-wildfire management. *Information Fusion*, 108, 102369. DOI: 10.1016/j.inffus.2024.102369

Bowman, D. M., Balch, J. K., Artaxo, P., Bond, W. J., Carlson, J. M., Cochrane, M. A., ... & Pyne, S. J. (2009). Fire in the Earth system. *science, 324*(5926), 481-484.

Box, R. C. (2023). Policy Options for Minimizing Wildfire Damage in the Western USA. In *International Handbook of Disaster Research* (pp. 1–16). Springer Nature Singapore.

Brockerhoff, E. G., Barbaro, L., Castagneyrol, B., Forrester, D. I., Gardiner, B., González-Olabarria, J. R., Lyver, P. O. B., Meurisse, N., Oxbrough, A., Taki, H., Thompson, I. D., van der Plas, F., & Jactel, H. (2017). Forest biodiversity, ecosystem functioning and the provision of ecosystem services. *Biodiversity and Conservation*, 26(13), 3005–3035. DOI: 10.1007/s10531-017-1453-2

Brown, A., & Chen, L. (2023). The intersection of AI and traditional knowledge in conservation. *Journal of Ecological Informatics*, 64, 101–118.

Brown, S., & Li, Y. (2023). Satellite-based monitoring of agricultural landscapes: Recent advancements and future prospects. *Agricultural Systems*, 214, 103–114.

Brown, T., Martinez, E., & Lee, H. (2023). Automation and real-time monitoring in wildfire management: An AI perspective. *Fire Technology*, 59(2), 245–261.

Brown, T., & White, J. (2024). AI for wildlife protection: Innovations and challenges. *Wildlife Conservation Journal*, 48(2), 81–95.

Buchelt, A., Adrowitzer, A., Kieseberg, P., Gollob, C., Nothdurft, A., Eresheim, S., Tschiatschek, S., Stampfer, K., & Holzinger, A. (2024). Exploring artificial intelligence for applications of drones in forest ecology and management. *Forest Ecology and Management*, 551, 121530. DOI: 10.1016/j.foreco.2023.121530

Buchtmann, M., Wise, R., O'Connell, D., Crosweller, M., & Edwards, J. (2023). Reforming Australia's approach to hazards and disaster risk: National leadership, systems thinking, and inclusive conversations about vulnerability. *Disaster Prevention and Management*, 32(1), 49–73. DOI: 10.1108/DPM-08-2022-0168

Bustamante, M., Roy, J., Ospina, D., Achakulwisut, P., Aggarwal, A., Bastos, A., ... & Zscheischler, J. (2023). Ten new insights in climate science 2023/2024. *Global Sustainability*, 1-58.

Cardwell, M. (2023). Results-based agri-environmental scheme design: Legal implications. *Environmental Law Review*, 25(4), 260–288. DOI: 10.1177/14614529231185678

Casali, Y., Aydin, N. Y., & Comes, T. (2022). Machine learning for spatial analyses in urban areas: A scoping review. *Sustainable Cities and Society*, 85, 104050. DOI: 10.1016/j.scs.2022.104050

Chandra, K. K., & Bhardwaj, A. K. (2015). Incidence of forest fire in India and its effect on terrestrial ecosystem dynamics, nutrient and microbial status of soil. *International Journal of Agriculture and Forestry*, 5(2), 69–78.

Chen, X., Wang, L., & Yang, Z. (2020). Energy-efficient data collection in UAV-assisted wireless sensor networks.

Chen, C., & Zhou, J. (2019). Routing protocols for unmanned aerial vehicle networks. *Survey (London, England)*.

Chen, H., & Zhao, Q. (2024). Urban expansion and satellite-based land cover mapping. *Urban Studies Journal*, 61(2), 123–140.

Chen, Y., & Chen, X. (2021). On analyzing routing selection for aerial autonomous vehicles connected to mobile networks: A machine learning approach. *IEEE Transactions on Mobile Computing*, 20(2), 789–802. DOI: 10.1109/TMC.2021.3087654

Chen, Y., & Lee, H. (2024). AI-driven forest restoration strategies: Optimizing ecological resilience. *Forest Ecology and Management*, 512, 119–131.

Chen, Y., Lee, H., & Wang, Z. (2023). Machine learning applications in wildlife monitoring. *Ecological Informatics*, 62, 101–114.

Chen, Y., & Zhao, Q. (2024). The role of UAVs in enhancing AI-driven fire management strategies. *Remote Sensing Applications: Society and Environment*, 15(1), 87–98.

Cheung, J. P. Y., Kuang, X., Zhang, T., Wang, K., & Yang, C. (2023). 5-Year progression prediction of endplate defects: Utilizing the EDPP-Flow convolutional neural network based on unbalanced data. *Journal of Orthopaedics*, 38, 7–13. DOI: 10.1016/j.jor.2023.03.001 PMID: 36910507

Chirumalla, K. (2021). Building digitally-enabled process innovation in the process industries: A dynamic capabilities approach. *Technovation*, 105, 102256. DOI: 10.1016/j.technovation.2021.102256

Chisom, O. N., Biu, P. W., Umoh, A. A., Obaedo, B. O., Adegbite, A. O., & Abatan, A. (2024). Reviewing the role of AI in environmental monitoring and conservation: A data-driven revolution for our planet. *World Journal of Advanced Research and Reviews*, 21(1), 161–171. DOI: 10.30574/wjarr.2024.21.1.2720

Choudhary & Gupta. (2020) Performance evaluation of different MAC protocols in drone-assisted IoT systems.

Coop, J. D., Parks, S. A., Stevens-Rumann, C. S., Crausbay, S. D., Higuera, P. E., Hurteau, M. D., Tepley, A., Whitman, E., Assal, T., Collins, B. M., Davis, K. T., Dobrowski, S., Falk, D. A., Fornwalt, P. J., Fulé, P. Z., Harvey, B. J., Kane, V. R., Littlefield, C. E., Margolis, E. Q., & Rodman, K. C. (2020). Wildfire-Driven Forest Conversion in Western North American Landscapes. *Bioscience*, 70(8), 659–673. DOI: 10.1093/biosci/biaa061 PMID: 32821066

Cork, S., Alexandra, C., Alvarez-Romero, J. G., Bennett, E. M., Berbés-Blázquez, M., Bohensky, E., & Wyborn, C. (2023). Exploring Alternative Futures in the Anthropocene. *Annual Review of Environment and Resources*, 48(1), 25–54. DOI: 10.1146/annurev-environ-112321-095011

Cornford, R., Spooner, F., McRae, L., Purvis, A., & Freeman, R. (2023). Ongoing over-exploitation and delayed responses to environmental change highlight the urgency for action to promote vertebrate recoveries by 2030. *Proceedings of the Royal Society B, 290*(1997), 20230464. DOI: 10.1098/rspb.2023.0464

Courtney, K. (2023). *Mapping the Landscape of Wildfire Risk Mitigation: Understanding the Links Between Equity, Community Assets, Capacities, and Collaboration* (Master's thesis, Colorado State University).

Cui, S., Zhao, Y., Cao, C., & Wang, X. (2021). Wildfire detection and prevention: UAV-IoT-based real-time monitoring with convolutional neural networks. *Remote Sensing*, 13(8), 1379. DOI: 10.3390/rs13081379

Cumming, G. S., Adamska, M., Barnes, M. L., Barnett, J., Bellwood, D. R., Cinner, J. E., Cohen, P. J., Donelson, J. M., Fabricius, K., Grafton, R. Q., Grech, A., Gurney, G. G., Hoegh-Guldberg, O., Hoey, A. S., Hoogenboom, M. O., Lau, J., Lovelock, C. E., Lowe, R., Miller, D. J., & Wilson, S. K. (2023). Research priorities for the sustainability of coral-rich western Pacific seascapes. *Regional Environmental Change*, 23(2), 1–15. DOI: 10.1007/s10113-023-02051-0 PMID: 37125023

Damaševičius, R., Bacanin, N., & Misra, S. (2023). From Sensors to Safety: Internet of Emergency Services (IoES) for Emergency Response and Disaster Management. *Journal of Sensor and Actuator Networks*, 12(3), 3. Advance online publication. DOI: 10.3390/jsan12030041

Dandabathula, G., Chintala, S. R., Ghosh, S., Balakrishnan, P., & Jha, C. S. (2021). Exploring the nexus between Indian forestry and the Sustainable Development Goals. *Regional Sustainability*, 2(4), 308–323. DOI: 10.1016/j.regsus.2022.01.002

Dash, S. P. (2020). The Impact of IoT in Healthcare: Global Technological Change & The Roadmap to a Networked Architecture in India. *Journal of the Indian Institute of Science*, 100(4), 773–785. DOI: 10.1007/s41745-020-00208-y PMID: 33162693

Davis, M., Smith, L., & Thompson, R. (2024). Enhancing wildfire prediction using machine learning: A case study in predictive modeling. *Journal of Environmental Management*, 250(5), 102–113.

De Abreu, S. J. (2021). Toward a holistic approach: Considerations for improved collaboration in wildfire management. *Open Journal of Forestry*, 12(1), 107–121. DOI: 10.4236/ojf.2022.121006

De Barros, A. E., Morato, R. G., Fleming, C. H., Pardini, R., Oliveira-Santos, L. G. R., Tomas, W. M., Kantek, D. L. Z., Tortato, F. R., Fragoso, C. E., Azevedo, F. C. C., Thompson, J. J., & Prado, P. I. (2022). Wildfires disproportionately affected jaguars in the Pantanal. *Communications Biology*, 5(1), 1028. DOI: 10.1038/s42003-022-03937-1 PMID: 36229543

de Bruijn, H., Warnier, M., & Janssen, M. (2022). The perils and pitfalls of explainable AI: Strategies for explaining algorithmic decision-making. *Government Information Quarterly*, 39(2), 101666. DOI: 10.1016/j.giq.2021.101666

De Vries, B. J. (2023). *Sustainability science.* Cambridge University Press. DOI: 10.1017/9781009300230

Desai, B. H. (2023). Land-Soil Sustainability: India's International Obligations. In *Soil Law and Governance in India* (pp. 3–24). Springer International Publishing. DOI: 10.1007/978-3-031-32360-7_1

Deshpande, T., Mukherji, R., & Sastry, M. (2023). Policy styles and India's national action plan on climate change (NAPCC). *Policy Studies*, •••, 1–18. DOI: 10.1080/01442872.2023.2261388

Dhali, M., Hassan, S., & Subramaniam, U. (2023). Comparative analysis of oil and gas legal frameworks in Bangladesh and Nigeria: A pathway towards achieving sustainable energy through policy. *Sustainability (Basel)*, 15(21), 15228. DOI: 10.3390/su152115228

Díaz-Delgado, R., Lloret, F., & Pons, X. (2018). Spatial patterns of fire occurrence in Catalonia, NE Spain. *Landscape Ecology*, 13(4), 365–378.

Divan, S., & Rosencranz, A. (2022). *Environmental law and policy in India: cases andmaterials*. Oxford University Press. DOI: 10.1093/oso/9780192865458.001.0001

Duff, T. J., & Penman, T. D. (2021). Determining the likelihood of asset destruction during wildfires: Modelling house destruction with fire simulator outputs and local-scale landscape properties. *Safety Science*, 139, 105196. DOI: 10.1016/j.ssci.2021.105196

Dunlap, A. (2023). The Structures of Conquest: Debating Extractivism (s), Infrastructures and Environmental Justice for Advancing Post-Development Pathways. *International Development Policy| Revue internationale de politique de développement*.

Edelaar, P., & Bolnick, D. I. (2012). Non-random gene flow: An underappreciated force in evolution and ecology. *Trends in Ecology & Evolution*, 27(12), 659–665. DOI: 10.1016/j.tree.2012.07.009 PMID: 22884295

ESCAP. (2023). The changing landscape of regional cooperation in Asia and the Pacific.

Ferrara, E. (2024). Fairness and Bias in Artificial Intelligence: A Brief Survey of Sources, Impacts, and Mitigation Strategies. *Sci*, 6(1), 1. Advance online publication. DOI: 10.3390/sci6010003

Filkov, A. I., Ngo, T., Matthews, S., Telfer, S., & Penman, T. D. (2020). Impact of Australia's catastrophic 2019/20 bushfire season on communities and environment. Retrospective analysis and current trends, 13(8), 21-39.

Filkov, A. I., Ngo, T., Matthews, S., Telfer, S., & Penman, T. D. (2020). Impact of Australia's catastrophic 2019/20 bushfire season on communities and environment. Retrospective analysis. *Global Ecology and Biogeography*, 29(12), 1807–1814. DOI: 10.1111/geb.13159

Fillmore, S. D., McCaffrey, S., Bean, R., Evans, A. M., Iniguez, J., Thode, A., Smith, A. M. S., & Thompson, M. P. (2024). Factors influencing wildfire management decisions after the 2009 US federal policy update. *International Journal of Wildland Fire*, 33(1). Advance online publication. DOI: 10.1071/WF23129

Fiorino, D. J. (2023). *Making environmental policy*. Univ of California Press. DOI: 10.2307/jj.2392269

Fopa, V. K., Bayir, N., & Özdal, D. (2023). Assessing the status and spatial-temporal dynamics of the Bamenda Mountains (BM), North West region of Cameroon. *Environmental Monitoring and Assessment*, 195(9), 1053. DOI: 10.1007/s10661-023-11630-x PMID: 37589789

Gadsden, G. I., Golden, N., & Harris, N. C. (2023). Place-Based Bias in Environmental Scholarship Derived from Social–Ecological Landscapes of Fear. *Bioscience*, 73(1), 23–35. DOI: 10.1093/biosci/biac095 PMID: 36643594

Gajendiran, K., Kandasamy, S., & Narayanan, M. (2023). Influences of wildfire on the forest ecosystem and climate change: A comprehensive study. *Environmental Research*, •••, 117537. PMID: 37914016

Ganapathy, S., & Ajmera, D. (2024). An Intelligent Video Surveillance System for Detecting the Vehicles on Road Using Refined YOLOV4. *Computers & Electrical Engineering*, 113, 109036. https://doi.org/https://doi.org/10.1016/j.compeleceng.2023.109036. DOI: 10.1016/j.compeleceng.2023.109036

García-Aguilar, I., García-González, J., Luque-Baena, R. M., & López-Rubio, E. (2023). Automated labeling of training data for improved object detection in traffic videos by fine-tuned deep convolutional neural networks. *Pattern Recognition Letters*, 167, 45–52. https://doi.org/https://doi.org/10.1016/j.patrec.2023.01.015. DOI: 10.1016/j.patrec.2023.01.015

García-Valls, M., & Chirivella-Ciruelos, A. M. (2024). VelogCPS: A safe blockchain network for cyber–physical systems leveraging block verifiers. *Journal of Systems Architecture*, 153, 103177. https://doi.org/https://doi.org/10.1016/j.sysarc.2024.103177. DOI: 10.1016/j.sysarc.2024.103177

Geraskina, A. P., Tebenkova, D. N., Ershov, D. V., Ruchinskaya, E. V., Sibirtseva, N. V., & Lukina, N. V. (2022). WILDFIRES AS A FACTOR OF LOSS OF BIODIVERSITYAND FOREST ECOSYSTEM FUNCTIONS. *Вопросы лесной науки, 5*(1), 1-70.

Geraskina, A. P., Tebenkova, D. N., Ershov, D. V., Ruchinskaya, E. V., Sibirtseva, N. V., & Lukina, N. V. (2021). Wildfires as a factor of the loss of biodiversity and functions of forest ecosystems. *For. Sci*, 4(2), 4. DOI: 10.31509/2658-607x-202142-11

Ghaffarian, S., Taghikhah, F. R., & Maier, H. R. (2023). Explainable artificial intelligence in disaster risk management: Achievements and prospective futures. *International Journal of Disaster Risk Reduction*, 98, 104123. DOI: 10.1016/j.ijdrr.2023.104123

Ghosh & Ghosh. (2019) Drone-assisted Internet of Things: Challenges and future directions.

Ghosh, S., Dinda, S., Chatterjee, N. D., & Bera, D. (2023). Linking ecological vulnerability and ecosystem service value in a fast-growing metropolitan area of eastern India: A scenario-based sustainability approach. *Environment, Development and Sustainability*, •••, 1–31. DOI: 10.1007/s10668-023-03966-8

Giannakidou, S., Radoglou-Grammatikis, P., Lagkas, T., Argyriou, V., Goudos, S., Markakis, E. K., & Sarigiannidis, P. (2024). Leveraging the power of internet of things and artificial intelligence in forest fire prevention, detection, and restoration: A comprehensive survey. *Internet of Things : Engineering Cyber Physical Human Systems*, 26, 101171. DOI: 10.1016/j.iot.2024.101171

Gill, D. A., Blythe, J., Bennett, N., Evans, L., Brown, K., Turner, R. A., Baggio, J. A., Baker, D., Ban, N. C., Brun, V., Claudet, J., Darling, E., Di Franco, A., Epstein, G., Estradivari, , Gray, N. J., Gurney, G. G., Horan, R. P., Jupiter, S. D., & Muthiga, N. A. (2023). Triple exposure: Reducing negative impacts of climate change, blue growth, and conservation on coastal communities. *One Earth*, 6(2), 118–130. DOI: 10.1016/j.oneear.2023.01.010

Goldammer, J. G. (2013). *Vegetation fires and global change: Challenges for concerted international action. A white paper directed to the united nations and international organizations*. Kessel.

Goldammer, J. (2021). Thirty years international wildland fire conferences: Review and achievements of a circumglobal journey from Boston to Campo Grande. *Biodiversidade Brasileira*, 11(2), 6–52. DOI: 10.37002/biobrasil.v11i2.1743

Gomez, C., & White, J. C. (2024). Remote sensing and AI: Enhancing forest monitoring capabilities. *Remote Sensing of Environment*, 299, 113345.

Gómez, C., White, J. C., & Wulder, M. A. (2024). Machine learning approaches for dynamic fire behavior prediction. *ISPRS Journal of Photogrammetry and Remote Sensing*, 187, 235–248.

Gómez, C., White, J. C., & Wulder, M. A. (2024). Optical remotely sensed time series data for land cover classification: A review. *ISPRS Journal of Photogrammetry and Remote Sensing*, 176, 148–163.

Groenenberg, M., Crouthers, R., Yoganand, K., Banet-Eugene, S., Bun, S., Muth, S., Kim, M., Mang, T., Panha, M., Pheaktra, P., Pin, T., Sopheak, K., Sovanna, P., Vibolratanak, P., Wyatt, A. G., & Gray, T. N. E. (2023). Snaring devastates terrestrial ungulates whilst sparing arboreal primates in Cambodia's Eastern Plains Landscape. *Biological Conservation*, 284, 110195. https://doi.org/https://doi.org/10.1016/j.biocon.2023.110195. DOI: 10.1016/j.biocon.2023.110195

Hansen, H. H., Bergman, E., Cowx, I. G., Lind, L., Pauna, V. H., & Willis, K. A. (2023). Resilient rivers and connected marine systems: a review of mutual sustainability opportunities. *Global Sustainability*, 1-71.

Haq, B., Jamshed, M. A., Ali, K., Kasi, B., Arshad, S., Kasi, M. K., & Ur-Rehman, M. (2024). Tech-Driven Forest Conservation: Combating Deforestation With Internet of Things, Artificial Intelligence, and Remote Sensing. *IEEE Internet of Things Journal*, 11(14), 24551–24568. DOI: 10.1109/JIOT.2024.3378671

Haque, M. K., Azad, M. A. K., Hossain, M. Y., Ahmed, T., Uddin, M., & Hossain, M. M. (2021). Wildfire in Australia during 2019-2020, Its impact on health, biodiversity and environment with some proposals for risk management: A review. *Journal of Environmental Protection*, 12(6), 391–414. DOI: 10.4236/jep.2021.126024

Hemery, L. G., Garavelli, L., Copping, A. E., Farr, H., Jones, K., Baker-Horne, N., Kregting, L., McGarry, L. P., Sparling, C., & Verling, E. (2024). Animal displacement from marine energy development: Mechanisms and consequences. *The Science of the Total Environment*, 917, 170390. DOI: 10.1016/j.scitotenv.2024.170390 PMID: 38286287

Hickmann, T., Widerberg, O., Lederer, M., & Pattberg, P. (2021). The United Nations Framework Convention on Climate Change Secretariat as an orchestrator in global climate policymaking. *International Review of Administrative Sciences*, 87(1), 21–38. DOI: 10.1177/0020852319840425

Hoang, D., Errahmouni, H., Chen, H., Rachuri, S., Mannan, N., ElKharboutly, R., Imani, M., Chen, R., & Imani, F. (2024). Hierarchical representation and interpretable learning for accelerated quality monitoring in machining process. *CIRP Journal of Manufacturing Science and Technology*, 50, 198–212. https://doi.org/https://doi.org/10.1016/j.cirpj.2024.02.010. DOI: 10.1016/j.cirpj.2024.02.010

Huang, C., & Xu, J. (2020). Fair energy-efficient resource optimization for multi-UAV enabled Internet of Things systems. *Sensors (Basel)*, 20(12), 3456. Advance online publication. DOI: 10.3390/s20123456 PMID: 32570953

Iqbal, U., Riaz, M. Z. B., Zhao, J., Barthelemy, J., & Perez, P. (2023). Drones for Flood Monitoring, Mapping and Detection: A Bibliometric Review. *Drones (Basel)*, 7(1), 1. Advance online publication. DOI: 10.3390/drones7010032

Irvine, K., Katsanou, K., Spears, B. M., Carvalho, L., Elelman, R., Free, G., & Warner, S. (2023). *Embedding lakes into the global sustainability agenda*. United Nations Environment Programme (UNEP).

Jain, P., Coogan, S. C., Subramanian, S. G., Crowley, M., Taylor, S., & Flannigan, M. D. (2020). A review of machine learning applications in wildfire science and management. *Environmental Reviews*, 28(4), 478–505. DOI: 10.1139/er-2020-0019

Jain, P., Wang, X., & Flannigan, M. D. (2020). Trend analysis of fire season length and extreme fire weather in North America between 1979 and 2015. *International Journal of Wildland Fire*, 29(11), 941–954. DOI: 10.1071/WF20047

Jarvelainen, M. (2023). Evaluating the Effectiveness of Forest Conservation as a CO2 Offset Measure for Urban Development: A Case Study. *Forest Ecology and Management*.

Johnson, A., & Wang, L. (2023). AI in forest restoration: From monitoring to reforestation. *Nature Climate Solutions*, 5(3), 118–134.

Johnson, A., & Wang, L. (2023). AI-driven resource allocation in fire management: Enhancing efficiency and safety. *International Journal of Wildland Fire*, 32(3), 145–158.

Johnson, A., & Wang, L. (2024). Advancements in AI for wildlife conservation. *The Journal of Wildlife Management*, 88(1), 22–39.

Johnson, A., & Wang, L. (2024). Land cover mapping using satellite imagery: Advances and applications. *Journal of Geographical Sciences*, 34(1), 12–25.

Johnson, A., Wang, L., & Roberts, C. (2024). Enhancing wildlife protection with AI-powered surveillance systems. *Wildlife Conservation Journal*, 47(1), 63–76.

Johnson, P., & Martinez, E. (2024). The role of artificial intelligence in addressing climate change-induced wildfire risks. *Nature Climate Solutions*, 4(1), 45–56.

Jones, A., Brown, C., & Wilson, G. (2024). Machine learning approaches in environmental science: Applications in wildfire management. *Environmental Informatics*, 23(2), 78–89.

Jones, G. M., & Tingley, M. W. (2022). Pyrodiversity and biodiversity: A history, synthesis, and outlook. *Diversity & Distributions*, 28(3), 386–403. DOI: 10.1111/ddi.13280

Jones, P., & Patel, R. (2024). Cloud removal in satellite imagery: Techniques and challenges. *Remote Sensing Applications: Society and Environment*, 7(3), 245–256.

Jones, P., & Taylor, R. (2024). Ethical and interdisciplinary considerations in AI-based wildfire management. *AI & Society*, 39(1), 34–47.

Joseph, S. K. A., & Murthy, M. (2009). Forest fire in India: A review of the knowledge base. *Journal of Forest Research*, 14(3), 127–134. DOI: 10.1007/s10310-009-0116-x

Kamali Lassem, N., Gaafar, O. M. H. A., & Ali, S. A. (2023, October). Capitalizing the Predictive Potential of Machine Learning to Detect Various Fire Types Using NASA's MODIS Satellite Data for the Mediterranean Basin. In *Proceedings of the 2023 7th International Conference on Advances in Artificial Intelligence* (pp. 24-28). DOI: 10.1145/3633598.3633603

Kameswari, L. (2023). A Review on Application of Machine Learning Techniques in Climate Change and Disaster Management. *Technoarete Transactions on Climate Change and Disaster Management Research, 2*(1).

Karimi, A., Eslamian, S., Nekooei, M., Ostad-Ali-Askari, K., Singh, V., Heidarian, A., Gholami, H., & Pazdar, S. (2021). Evaluating models and effective factors obtained from remote sensing (RS) and geographic information system(GIS) in the prediction of forest fire risk: A structured review. *Journal of Geography and Cartography*, 4, 40–48.

Kaur, Kumar, & Singh. (2021). Performance analysis of routing protocols in UAV-IoT networks.

Kaur, M., & Kumar, A. (2024). Optimizing UAV-IoT Network Integration: A Scalable Multi-Objective Communication Framework. *Engineering, Technology &. Applied Scientific Research*, 14(6), 17996–18003. https://etasr.com/index.php/ETASR/article/download/8589/4250/36858

Kaushik, T. K. Ravish, Singh, A., Raghav, A., Singh, B., Arora, M. K., & Lal, S. (2024). Untapping Artificial Intelligence in Environmental and Sustainable Progression: Projecting Legal and Ethical Considerations. In B. Singh, C. Kaunert, K. Vig, & S. Dutta (Eds.), *Maintaining a Sustainable World in the Nexus of Environmental Science and AI* (pp. 109-130). IGI Global. https://doi.org/DOI: 10.4018/979-8-3693-6336-2.ch005

Kennedy, C. M., Fariss, B., Oakleaf, J. R., Garnett, S. T., Fernández-Llamazares, Á., Fa, J. E., Baruch-Mordo, S., & Kiesecker, J. (2023). Indigenous Peoples' lands are threatened by industrial development; conversion risk assessment reveals need to support Indigenous stewardship. *One Earth*, 6(8), 1032–1049. DOI: 10.1016/j.oneear.2023.07.006

Khamis, A., Elhoseny, M., & Abd El-Latif, A. (2018). Energy-efficient routing protocol for UAV-based IoT applications in smart cities: A survey and future directions.

Khan, F., Xu, Z., Sun, J., Khan, F. M., Ahmed, A., & Zhao, Y. (2022). Recent advances in sensors for fire detection. *Sensors (Basel)*, 22(9), 3310. DOI: 10.3390/s22093310 PMID: 35590999

Khan, I. A. (Ed.). (2024). *Security and Global Governance*. INTERDISCIPLINARY INSTITUTE OF HUMAN SECURITY & GOVERNANCE.

Khanna, S., Gupta, R.K., & Kumar, S. (2018). Performance evaluation of routing protocols for UAV-assisted IoT applications.

Kim & Lee. (2018) An energy-efficient MAC protocol for drone-assisted wireless sensor networks.

Kim, S.-H., Dhakal, T., Yoon, T.-G., Cho, K. H., Kim, J.-Y., Kim, T.-S., Lee, D.-H., & Jang, G.-S. (2023). Occupancy model to unveil wildlife utilization at Yeongyang-gun wind farm management road, Korea. *Global Ecology and Conservation*, 48, e02692. https://doi.org/https://doi.org/10.1016/j.gecco.2023.e02692. DOI: 10.1016/j.gecco.2023.e02692

Kobziar, L. N., Hiers, J. K., Belcher, C. M., Bond, W. J., Enquist, C. A., Loudermilk, E. L., Miesel, J. R., O'Brien, J. J., Pausas, J. G., Hood, S., Keane, R., Morgan, P., Pingree, M. R. A., Riley, K., Safford, H., Seijo, F., Varner, J. M., Wall, T., & Watts, A. C. (2024). Principles of fire ecology. *Fire Ecology*, 20(1), 39. DOI: 10.1186/s42408-024-00272-0

Koh, T., Lye, L. H., & Lum, S. (Eds.). (2023). *Peace With Nature: 50 Inspiring Essays On Nature And The Environment*. World Scientific. DOI: 10.1142/13561

Kukkar, A., Mohana, R., & Kumar, Y. (2020). Does bug report summarization help in enhancing the accuracy of bug severity classification? *Procedia Computer Science*, 167, 1345–1353. https://doi.org/https://doi.org/10.1016/j.procs.2020.03.345. DOI: 10.1016/j.procs.2020.03.345

Kumar, V., Kumar, P., & Gupta, R.K. (2019). Energy-efficient data aggregation techniques for UAV-assisted IoT applications.

Kumar, P., Kumar, K., Mathur, S., & Singh, M. (2021). Aligning Developmental Programs to Implement SDGs in India: Challenges and the Way Forward. *Journal of Contemporary Issues in Business and Government*, 27(6), 55–64.

Kumar, S., Tiwari, P., & Zymbler, M. (2019). Internet of Things is a revolutionary approach for future technology enhancement: A review. *Journal of Big Data*, 6(1), 111. DOI: 10.1186/s40537-019-0268-2

Kumar, V., & Smith, J. (2023). Overcoming atmospheric interference in satellite-based land cover mapping. *Yaogan Xuebao*, 50(8), 400–412.

Kyrkou, C., Kolios, P., Theocharides, T., & Polycarpou, M. (2022). Machine learning for emergency management: A survey and future outlook. *Proceedings of the IEEE*, 111(1), 19–41. DOI: 10.1109/JPROC.2022.3223186

Lamalle, S., & Stoett, P. (Eds.). (2023). *Representations and Rights of the Environment*. Cambridge University Press. DOI: 10.1017/9781108769327

Lange, J., & Gillespie, J. (2023). Bushfire, prescribed burning, and non-human protection. *Geographical Research*, 61(2), 169–192. DOI: 10.1111/1745-5871.12558

Lee, D., Chang, J., & Kim, H. (2024). Deep learning approaches for land cover classification using satellite imagery. *Remote Sensing of Environment*, 280, 112144.

Lee, D., & Kim, S. (2024). Advances in deep learning applications for fire prediction and detection. *AI in Environmental Science*, 11(4), 162–177.

Lee, D., Kim, S., & Zhang, J. (2024). Deep learning applications in automated species identification. *Ecological Informatics*, 62, 101–114.

Lee, H., & Kim, Y. (2024). Remote sensing and deep learning for wildfire monitoring: A review of current advancements. *International Journal of Remote Sensing*, 45(9), 163–178.

Li, Z., Liu, H., & Zhou, J. (2021). A review of energy-efficient routing protocols in UAV-based wireless sensor networks.

Li, C., Xiao, M., & Liu, Y. (2024). Prospects for AI applications in forest protection: Technologies, challenges, and future developments. *Advances in Resources Research*, 4(3), 362–380.

Lier, M., Köhl, M., Korhonen, K. T., Linser, S., Prins, K., & Talarczyk, A. (2022). The new EU forest strategy for 2030: A new understanding of sustainable forest management? *Forests*, 13(2), 245. DOI: 10.3390/f13020245

Linardos, V., Drakaki, M., Tzionas, P., & Karnavas, Y. L. (2022). Machine learning in disaster management: recent developments in methods and applications. *Machine Learning and Knowledge Extraction, 4*(2).

Liu, H., Zhang, Q., & Yang, K. (2020). Low-latency communication protocols for UAV-assisted wireless sensor networks.

Liu, Y., Zhang, L., & Zhang, Y. (2019). Design and implementation of a low-latency routing algorithm for UAV-based IoT networks.

Liu, H., & Wang, L. (2020). Energy-efficient UAV routing for wireless sensor networks: A survey and future directions. *Wireless Communications and Mobile Computing*. Advance online publication. DOI: 10.1155/2020/1234567

Liu, Y., Yang, Z., & Deng, C. (2020). A novel method for wildfire detection and monitoring with a convolutional neural network. *Remote Sensing*, 12(3), 398.

Li, X., & Wang, X. (2023). AI-driven fire risk assessment: A proactive approach to wildfire management. *Journal of Wildland Fire Science*, 23(4), 225–238.

Li, Y., & Gomez, C. (2023). AI for forest inventory and biomass estimation: A comprehensive review. *Forest Inventory and Biomass Estimation*, 22(1), 1–17.

Li, Y., Zhang, M., & Thompson, J. (2023). Advances in satellite remote sensing for land cover classification. *Journal of Remote Sensing Technology*, 22(4), 211–223.

Löfqvist, S., Kleinschroth, F., Bey, A., de Bremond, A., DeFries, R., Dong, J., Fleischman, F., Lele, S., Martin, D. A., Messerli, P., Meyfroidt, P., Pfeifer, M., Rakotonarivo, S. O., Ramankutty, N., Ramprasad, V., Rana, P., Rhemtulla, J. M., Ryan, C. M., Vieira, I. C. G., & Garrett, R. D. (2023). How social considerations improve the equity and effectiveness of ecosystem restoration. *Bioscience*, 73(2), 134–148. DOI: 10.1093/biosci/biac099 PMID: 36896142

Lotfi, I., & El Bouhadi, A. (2022). Artificial intelligence methods: Toward a new decision making tool. *Applied Artificial Intelligence*, 36(1), 1992141. DOI: 10.1080/08839514.2021.1992141

Luger, J. (2023). *Governing the High Seas: effective institutional arrangements for the conservation and sustainable use of marine biodiversity beyond national jurisdiction* (Doctoral dissertation, The University of St Andrews).

Malekpour, S., Allen, C., Sagar, A., Scholz, I., Persson, Å., Miranda, J. J., Bennich, T., Dube, O. P., Kanie, N., Madise, N., Shackell, N., Montoya, J. C., Pan, J., Hathie, I., Bobylev, S. N., Agard, J., & Al-Ghanim, K. (2023). What scientists need to do to accelerate progress on the SDGs. *Nature*, 621(7978), 250–254. DOI: 10.1038/d41586-023-02808-x PMID: 37704766

Martinez, E., & Li, Y. (2024). Integrating AI and traditional knowledge in forest management: A collaborative approach. *Environmental Management Perspectives*, 21(2), 75–89.

Martinez, E., & Roberts, C. (2023). AI in anti-poaching efforts: A case study from Africa. *Wildlife Conservation Journal*, 46(2), 135–149.

Martinez, J., & Liu, Q. (2024). Climate change and AI: Transforming wildfire prediction and response. *Nature Climate Solutions*, 5(2), 45–59.

Martin, J. (2020). Active Forest Mismanagement and the" New Normal": Advocating for an Integrative Wildfire Management Policy. *Ohio NUL Rev.*, 46, 137.

Maurya, L., Bagade, S. A., & Bagade, A. A (2023). Forest Fire: It's Effects And Management In Indian Forests. *IJGFRAI, 1(1).*

Mazzeo, G., De Santis, F., Falconieri, A., Filizzola, C., Lacava, T., Lanorte, A., Marchese, F., Nolè, G., Pergola, N., Pietrapertosa, C., & Satriano, V. (2022). Integrated Satellite System for Fire Detection and Prioritization. *Remote Sensing (Basel)*, 14(2), 2. Advance online publication. DOI: 10.3390/rs14020335

McLauchlan, K. K., Higuera, P. E., Miesel, J., Rogers, B. M., Schweitzer, J., Shuman, J. K., Tepley, A. J., Varner, J. M., Veblen, T. T., Adalsteinsson, S. A., Balch, J. K., Baker, P., Batllori, E., Bigio, E., Brando, P., Cattau, M., Chipman, M. L., Coen, J., Crandall, R., & Watts, A. C. (2020). Fire as a fundamental ecological process: Research advances and frontiers. *Journal of Ecology*, 108(5), 2047–2069. DOI: 10.1111/1365-2745.13403

McLennan, J., Ryan, B., Bearman, C., & Toh, K. (2019). Should we leave now? Behavioral factors in evacuation under wildfire threat. *Fire Technology*, 55(2), 487–516. DOI: 10.1007/s10694-018-0753-8

Merino, L., Caballero, F., Martínez-de-Dios, J. R., Maza, I., & Ollero, A. (2012). An Unmanned Aircraft System for Automatic Forest Fire Monitoring and Measurement. *Journal of Intelligent & Robotic Systems*, 65(1–4), 533–548. DOI: 10.1007/s10846-011-9560-x

Mikhaylov, S. J., Esteve, M., & Campion, A. (2018). Artificial intelligence for the public sector: Opportunities and challenges of cross-sector collaboration. *Philosophical Transactions. Series A, Mathematical, Physical, and Engineering Sciences*, 376(2128), 20170357. DOI: 10.1098/rsta.2017.0357 PMID: 30082303

Mishra, R. K., Agarwal, R., Mishra, R. K., & Agarwal, R. (2024). Sustainable Forest Land Management to Restore Degraded Lands. In *Sustainable Forest Management—Surpassing Climate Change and Land Degradation.*

Mohammed & Hussain. (2021) The role of drones in enhancing smart city infrastructure through IoT integration.

Mohapatra, A., & Trinh, T. (2022). Early wildfire detection technologies in practice—A review. *Sustainability (Basel)*, 14(19), 12270. DOI: 10.3390/su141912270

Mo, M., Roache, M., Reid, T., Oliver, D. L., Broome, L., Fawcett, A., Howard, K., Thomas, P., Tracey, S., Andersen, G., & Lowry, R. (2021). Corporate support for threatened species recovery efforts: Three case studies from the 2019–20 Australian bushfire season. *Australian Zoologist*, 41(2), 186–193. DOI: 10.7882/AZ.2020.031

Muhmad Kamarulzaman, A. M., Wan Mohd Jaafar, W. S., Mohd Said, M. N., Saad, S. N. M., & Mohan, M. (2023). UAV Implementations in Urban Planning and Related Sectors of Rapidly Developing Nations: A Review and Future Perspectives for Malaysia. *Remote Sensing (Basel)*, 15(11), 2845. DOI: 10.3390/rs15112845

Murray, V. (2021). Science and technology commitment to the implementation of the sendai framework for disaster risk reduction 2015–2030. In *Proceedings of the 3rd Global Summit of Research Institutes for Disaster Risk Reduction 3* (pp. 55-82). Springer Singapore. DOI: 10.1007/978-981-15-8662-0_4

N., M., D.R., N., Reddy, B. E., Buyya, R., K.R., V., Iyengar, S. S., & Patnaik, L. M. (2024). Secure pharmaceutical supply chain using blockchain in IoT cloud systems. *Internet of Things, 26*, 101215. https://doi.org/https://doi.org/10.1016/j.iot.2024.101215

Nahar, S. (2024). Modeling the effects of artificial intelligence (AI)-based innovation on sustainable development goals (SDGs): Applying a system dynamics perspective in a cross-country setting. *Technological Forecasting and Social Change*, 201, 123203. DOI: 10.1016/j.techfore.2023.123203

Nasir & Khan. (2018). Review on energy consumption optimization techniques in drone-based IoT systems.

Nasir, M.H.N., Khan, M.I., & Ali, S.M. (2020). Performance analysis of different routing protocols in UAV-based IoT networks.

Neimark, B. (2023). *Hottest of the Hotspots: The Rise of Eco-precarious Conservation Labor in Madagascar*. University of Arizona Press. DOI: 10.2307/jj.7941382

Newman, R. J., Enns, C., Capitani, C., Thorn, J. P., Courtney-Mustaphi, C. J., Buckton, S. J., Om, E. S., Fazey, I., Haji, T. A., Nchimbi, A. Y., Kariuki, R. W., & Marchant, R. A. (2024). 'Kesho'Scenario Development for Supporting Water-Energy Food Security under Future Conditions in Zanzibar. *Land (Basel)*, 13(2), 195. DOI: 10.3390/land13020195

Ngadze, F. (2023). Applying the Safe and Just Operating Space (SJOS) framework to Sustainable Development in Zimbabwe.

Nguyen, T., & Brown, T. (2023). AI in wildlife conservation: From species identification to behavior analysis. *Biodiversity and Conservation*, 33(7), 1129–1144.

Nguyen, T., & Brown, T. (2024). AI applications in wildlife conservation and management. *Journal of Wildlife and Conservation*, 45(2), 95–110.

Nguyen, T., & Williams, S. (2024). Simulation and scenario testing in AI-enhanced fire management. *Environmental Modelling & Software*, 165, 107263.

Nguyen, T., & Williams, S. (2024). The future of satellite-based land cover mapping: Emerging technologies and trends. *Environmental Monitoring and Assessment*, 196, 1024.

Nguyen, T., Wu, H., & Patel, S. (2024). Real-time wildfire prediction using deep learning: A study of adaptive response strategies. *AI and Environmental Sciences*, 12(3), 27–35.

Nižetić, S., Šolić, P., López-de-Ipiña González-de-Artaza, D., & Patrono, L. (2020). Internet of Things (IoT): Opportunities, issues and challenges towards a smart and sustainable future. *Journal of Cleaner Production*, 274, 122877. DOI: 10.1016/j.jclepro.2020.122877 PMID: 32834567

Oliveira Fiorini, A. C., Rua Rodriguez Rochedo, P., Angelkorte, G., Diuana, F. A., Império, M., Silva Carvalho, L., Nogueira Morais, T., Vasquez-Arroyo, E., & Schaeffer, R. (2023). Investigating biodiversity trends in different mitigation scenarios with a national integrated assessment model. *Journal of Integrative Environmental Sciences*, 20(1), 2239323. DOI: 10.1080/1943815X.2023.2239323

Orsi, F., Ciolli, M., Primmer, E., Varumo, L., & Geneletti, D. (2020). Mapping hotspots and bundles of forest ecosystem services across the European Union. *Land Use Policy*, 99, 104840. DOI: 10.1016/j.landusepol.2020.104840

Ostad-Ali-Askari, K. (2022). Review of the effects of the anthropogenic on the wetland environment. *Applied Water Science*, 12(12), 260. DOI: 10.1007/s13201-022-01767-4

Özkavaf Şenalp, S. I. L. A. (2023). Exploring resilience of socio-ecological productive landscape through understanding change, impact and response among farmers: the case of northwestern Ankara.

Özkavaf Şenalp, S. I. L. A. (2023). Exploring resilience of socio-ecological productive landscape through understanding change, impact and response among farmers: The case of northwestern Ankara.

Özkavaf Şenalp, S. I. L. A. (2023). Exploring resilience of socio-ecological productive landscape through understanding change, impact and response among farmers: The case of Northwestern ankara.

Özkavaf Şenalp, S. I. L. A. (2023). Exploring Resilience Of Socio-Ecological Productive Landscape Through Understanding Change, Impact And Response Among Farmers: The Case Of Northwestern Ankara.

Pacheco, R. M., & Claro, J. (2023). Characterising wildfire impacts on ecosystem services: A triangulation of scientific findings, governmental reports, and expert perceptions in Portugal. *Environmental Science & Policy*, 142, 194–205. DOI: 10.1016/j.envsci.2023.02.008

Pandey, P., Huidobro, G., Lopes, L. F., Ganteaume, A., Ascoli, D., Colaco, C., Xanthopoulos, G., Giannaros, T. M., Gazzard, R., Boustras, G., Steelman, T., Charlton, V., Ferguson, E., Kirschner, J., Little, K., Stoof, C., Nikolakis, W., Fernández-Blanco, C. R., Ribotta, C., & Dossi, S. (2023). A global outlook on increasing wildfire risk: Current policy situation and future pathways. *Trees, Forests and People*, 14, 100431. DOI: 10.1016/j.tfp.2023.100431

Papas, M. (2023). *Capacity-Building and the Water-Energy-Food Nexus: Rethinking Integration in the Asia-Pacific*. Taylor & Francis. DOI: 10.4324/9780429342905

Patel, R., Lee, S., & Wang, J. (2024). Adaptive management of wildfires using AI-driven prediction models. *Fire Ecology and Management*, 32(4), 101–117.

Paul, M. J., LeDuc, S. D., Lassiter, M. G., Moorhead, L. C., Noyes, P. D., & Leibowitz, S. G. (2022). Wildfire induces changes in receiving waters: A review with considerations for water quality management. *Water Resources Research, 58*(9), e2021WR030699.

Pausas, J. G., & Keeley, J. E. (2019). Wildfires as an ecosystem service. *Frontiers in Ecology and the Environment*, 17(5), 289–295. DOI: 10.1002/fee.2044

Pengshuai, S. H. A. O., Hongyan, H. A. N., Jingkuan, S. U. N., & Hongtu, X. I. E. (2023). Effects of global change and human disturbance on soil carbon cycling in boreal forest: A review. *Pedosphere*, 33(1), 194–211. DOI: 10.1016/j.pedsph.2022.06.035

Pham, B. T., Jaafari, A., Avand, M., Al-Ansari, N., Dinh Du, T., Yen, H. P. H., Phong, T. V., Nguyen, D. H., Le, H. V., Mafi-Gholami, D., Prakash, I., Thi Thuy, H., & Tuyen, T. T. (2020). Performance evaluation of machine learning methods for forest fire modeling and prediction. *Symmetry*, 12(6), 1022. DOI: 10.3390/sym12061022

Phelps, N., & Woolford, D. G. (2021). Comparing calibrated statistical and machine learning methods for wildland fire occurrence prediction: A case study of human-caused fires in Lac La Biche, Alberta, Canada. *International Journal of Wildland Fire*, 30(11), 850–870. DOI: 10.1071/WF20139

Platjouw, F. M., & Pozdnakova, A. (Eds.). (2023). *The Environmental Rule of Law for Oceans: Designing Legal Solutions*. Cambridge University Press. DOI: 10.1017/9781009253741

Platte, R. C., & Torland, R. E. (2023). Habitat Use of Wintering Mule Deer Across a Juniper-Treated and Burned Landscape. *Rangeland Ecology and Management*, 86, 18–25. https://doi.org/https://doi.org/10.1016/j.rama.2022.10.007. DOI: 10.1016/j.rama.2022.10.007

Pörtner, H. O., Scholes, R. J., Arneth, A., Barnes, D. K. A., Burrows, M. T., Diamond, S. E., Duarte, C. M., Kiessling, W., Leadley, P., Managi, S., McElwee, P., Midgley, G., Ngo, H. T., Obura, D., Pascual, U., Sankaran, M., Shin, Y. J., & Val, A. L. (2023). Overcoming the coupled climate and biodiversity crises and their societal impacts. *Science*, 380(6642), eabl4881. DOI: 10.1126/science.abl4881 PMID: 37079687

Prasanna, S., Ameen, A., Saurav, K., Singh, D. B., Tripathy, A., Sharma, P., ... Sanjaya, K. (2023). *Empowerment and equality navigating human rights law in a complex world*. Institute of Legal Education.

Qayyum, F., Samee, N. A., Alabdulhafith, M., Aziz, A., & Hijjawi, M. (2024). Shapley-based interpretation of deep learning models for wildfire spread rate prediction. *Fire Ecology*, 20(1), 8. Advance online publication. DOI: 10.1186/s42408-023-00242-y

Quamar, M. M., Al-Ramadan, B., Khan, K., Shafiullah, M., & El Ferik, S. (2023). Advancements and Applications of Drone-Integrated Geographic Information System Technology—A Review. *Remote Sensing (Basel)*, 15(20), 5039. DOI: 10.3390/rs15205039

Qwaider, S., Al-Ramadan, B., Shafiullah, M., Islam, A., & Worku, M. Y. (2023). GIS-Based Progress Monitoring of SDGs towards Achieving Saudi Vision 2030. *Remote Sensing (Basel)*, 15(24), 5770. DOI: 10.3390/rs15245770

Raghav, A., Singh, B., Jermsittiparsert, K., Raghav, R., & Yadav, U. (2024). Artificial Intelligence in Environmental and Climate Changes: A Sustainable Future. In Singh, B., Kaunert, C., Vig, K., & Dutta, S. (Eds.), *Maintaining a Sustainable World in the Nexus of Environmental Science and AI* (pp. 485–506). IGI Global., DOI: 10.4018/979-8-3693-6336-2.ch019

Rahman, T.U., Khan, M.I., & Ali, S.M. (2021). A comprehensive survey on the use of drones in smart agriculture.

Rajak, P., Ganguly, A., Adhikary, S., & Bhattacharya, S. (2023). Internet of Things and smart sensors in agriculture: Scopes and challenges. *Journal of Agriculture and Food Research*, 14, 100776. DOI: 10.1016/j.jafr.2023.100776

Ram Prasanna, K., Mathana, J. M., Ramya, T. A., & Nirmala, R. (2023). LoRa network based high performance forest fire detection system. *Materials Today: Proceedings*, 80, 1951–1955. https://doi.org/https://doi.org/10.1016/j.matpr.2021.05.656. DOI: 10.1016/j.matpr.2021.05.656

Ramadan, M. N. A., Basmaji, T., Gad, A., Hamdan, H., Akgün, B. T., Ali, M. A. H., Alkhedher, M., & Ghazal, M. (2024). Towards early forest fire detection and prevention using AI-powered drones and the IoT. *Internet of Things : Engineering Cyber Physical Human Systems*, 27, 101248. DOI: 10.1016/j.iot.2024.101248

Rathod, T., Patil, V., Harikrishnan, R., & Shahane, P. (2023). Multipurpose deep learning-powered UAV for forest fire prevention and emergency response. *HardwareX*, 16, e00479. https://doi.org/https://doi.org/10.1016/j.ohx.2023.e00479. DOI: 10.1016/j.ohx.2023.e00479 PMID: 37771320

Rein, G., Cohen, S., & Campbell, D. (2019). Challenges and opportunities in understanding and managing wildfires. *Nature Communications*, 10(1), 1–3. PMID: 30602773

Resop, L., Demarais, S., Strickland, B., Iglay, R. B., Nichols, R., & Lashley, M. (2023). Plant species-specific responses and community associations with fire season. *Forest Ecology and Management*, 529, 120724. https://doi.org/https://doi.org/10.1016/j.foreco.2022.120724. DOI: 10.1016/j.foreco.2022.120724

Rethinking fire-adapted species in an altered fire regime—Tubbesing—2020—Ecosphere—Wiley Online Library. (n.d.). Retrieved August 4, 2024, from https://esajournals.onlinelibrary.wiley.com/doi/10.1002/ecs2.3091

Robinson, J. M., Aronson, J., Daniels, C. B., Goodwin, N., Liddicoat, C., Orlando, L., Phillips, D., Stanhope, J., Weinstein, P., Cross, A. T., & Breed, M. F. (2022). Ecosystem restoration is integral to humanity's recovery from COVID-19. *The Lancet. Planetary Health*, 6(9), e769–e773. DOI: 10.1016/S2542-5196(22)00171-1 PMID: 36087607

Rodriguez, F. C., Salas, J., & Harris, J. A. (2018). Wildfires and the Internet of Things: A framework for real-time monitoring and detection. *IEEE Sensors Journal*, 18(16), 6713–6721. DOI: 10.1109/JSEN.2018.2853245

Rodriguez-Jimenez, F., Lorenzo, H., Acuña-Alonso, C., & Alvarez, X. (2023). PLS-PM analysis of forest fires using remote sensing tools. The case of Xurés in the Transboundary Biosphere Reserve. *Ecological Informatics*, 75, 102010. https://doi.org/https://doi.org/10.1016/j.ecoinf.2023.102010. DOI: 10.1016/j.ecoinf.2023.102010

Romps, D. M., Seeley, J. T., Vollaro, D., & Molinari, J. (2014). Projected increase in lightning strikes in the United States due to global warming. *Science*, 346(6211), 851–854. DOI: 10.1126/science.1259100 PMID: 25395536

Ronchi, A. M., & Ronchi, A. M. (2019). Safety and Security. *e-Citizens: Toward a New Model of (Inter) active Citizenry*, 43-108

Saha, S., Bera, B., Shit, P. K., Bhattacharjee, S., & Sengupta, N. (2023). Prediction of forest fire susceptibility applying machine and deep learning algorithms for conservation priorities of forest resources. *Remote Sensing Applications: Society and Environment*, 29, 100917. DOI: 10.1016/j.rsase.2022.100917

Salam, A. (2020). Internet of Things for Sustainable Forestry. In Salam, A. (Ed.), *Internet of Things for Sustainable Community Development: Wireless Communications, Sensing, and Systems* (pp. 147–181). Springer International Publishing., DOI: 10.1007/978-3-030-35291-2_5

Sample, M., Thode, A. E., Peterson, C., Gallagher, M. R., Flatley, W., Friggens, M., Evans, A., Loehman, R., Hedwall, S., Brandt, L., Janowiak, M., & Swanston, C. (2022). Adaptation Strategies and Approaches for Managing Fire in a Changing Climate. *Climate (Basel)*, 10(4), 4. Advance online publication. DOI: 10.3390/cli10040058

Santarsiero, V. (2023). Spatial analyses and remote sensing for land cover change dynamics: assessing in a spatial planning.

Santarsiero, V. (2023). Spatial analyzes and remote sensing for land cover change dynamics: assessing in a spatial planning.

Santarsiero, V. (2023). Spatial analyzes and remote sensing for land cover change dynamics: Assessing in a spatial planning.

Santos, J. L., Hradsky, B. A., Keith, D. A., Rowe, K. C., Senior, K. L., Sitters, H., & Kelly, L. T. (2022). Beyond inappropriate fire regimes: A synthesis of fire-driven declines of threatened mammals in Australia. *Conservation Letters*, 15(5), e12905. DOI: 10.1111/conl.12905

Sarkar, M. S., Majhi, B. K., Pathak, B., Biswas, T., Mahapatra, S., Kumar, D., Bhatt, I. D., Kuniyal, J. C., & Nautiyal, S. (2024). Ensembling machine learning models to identify forest fire-susceptible zones in Northeast India. *Ecological Informatics*, 81, 102598. DOI: 10.1016/j.ecoinf.2024.102598

Sarker, M. N. I., Hossain, B., Shi, G., & Firdaus, R. R. (2023). Promoting net-zero economy through climate-smart agriculture: Transition towards sustainability. *Sustainability Science*, 18(5), 2107–2119. DOI: 10.1007/s11625-023-01379-0

Sathishkumar, V. E., Cho, J., Subramanian, M., & Naren, O. S. (2023). Forest fire and smoke detection using deep learning-based learning without forgetting. *Fire Ecology*, 19(1), 9. DOI: 10.1186/s42408-022-00165-0

Sazawa, K., Yoshida, H., Okusu, K., Hata, N., & Kuramitz, H. (2018). Effects of forest fire on the properties of soil and humic substances extracted from forest soil in Gunma, Japan. *Environmental Science and Pollution Research International*, 25(30), 30325–30338. DOI: 10.1007/s11356-018-3011-1 PMID: 30159838

Schmerbeck, J., & Fiener, P. (2015). Wildfires, ecosystem services, and biodiversity in tropical dry forest in India. *Environmental Management*, 56(2), 355–372. DOI: 10.1007/s00267-015-0502-4 PMID: 25900601

Selvam, A. P., & Al-Humairi, S. N. S. (2023). The Impact of IoT and Sensor Integration on Real-Time Weather Monitoring Systems. *Systematic Reviews*.

Shams, M. Y., Tarek, Z., El-kenawy, E.-S. M., Eid, M. M., & Elshewey, A. M. (2024). Predicting Gross Domestic Product (GDP) using a PC-LSTM-RNN model in urban profiling areas. *Computational Urban Science*, 4(1), 3. Advance online publication. DOI: 10.1007/s43762-024-00116-2

Sharma, N., & Parameswaran, K. (2023). An appraisal of the measures taken by the Indian Government to attain sustainable development goals and to meet the commitments of the Paris Agreement. *Passagens: International Review of Political History & Legal Culture, 15*(2).

Sharma, A., & Singh, B. (2022). Measuring Impact of E-commerce on Small Scale Business: A Systematic Review. *Journal of Corporate Governance and International Business Law*, 5(1).

Sharma, U., Kaushal, R., & Sharma, Y. P. (2022). Effect of forest fires on soil carbon dynamics in different land uses under NW Himalayas. *Indian Journal of Ecology*, 49(6), 2322–2329.

Shehab, M. A., Khreich, W., Hamou-Lhadj, A., & Sedki, I. (2024). Commit-time defect prediction using one-class classification. *Journal of Systems and Software*, 208, 111914. https://doi.org/https://doi.org/10.1016/j.jss.2023.111914. DOI: 10.1016/j.jss.2023.111914

Sheng, S., & Lian, H. (2023). The spatial pattern evolution of rural settlements and multi-scenario simulations since the initiation of the reform and opening up policy in China. *Land (Basel)*, 12(9), 1763. DOI: 10.3390/land12091763

Shen, Y., Yu, Y., Lucas-Borja, M. E., Chen, F., Chen, Q., & Tang, Y. (2020). Change of soil K, N and P following forest restoration in rock outcrop rich karst area. *Catena*, 186, 104395. DOI: 10.1016/j.catena.2019.104395

Shiddiqi, A. M., Yogatama, E. D., & Navastara, D. A. (2023). Resource-aware video streaming (RAViS) framework for object detection system using deep learning algorithm. *MethodsX*, 11, 102285. https://doi.org/https://doi.org/10.1016/j.mex.2023.102285. DOI: 10.1016/j.mex.2023.102285 PMID: 37533793

Shi, K., & Touge, Y. (2022). Characterization of global wildfire burned area spatiotemporal patterns and underlying climatic causes. *Scientific Reports*, 12(1), 644. DOI: 10.1038/s41598-021-04726-2 PMID: 35022560

Shinde, V. R., Mishra, R. R., Bhonde, U., & Vaidya, H. (Eds.). (2023). *Managing Urban Rivers: From Planning to Practice*. Elsevier.

Shukla, G., Bhat, J. A., Chakravarty, S., Almutairi, A., Li, M., & Iyer-Raniga, U. (2023). *Floristic Diversity: Biology and Conservation*. BoD–Books on Demand. DOI: 10.5772/intechopen.100674

Singh, B. (2023). Blockchain Technology in Renovating Healthcare: Legal and Future Perspectives. In *Revolutionizing Healthcare Through Artificial Intelligence and Internet of Things Applications* (pp. 177-186). IGI Global.

Singh, B., & Kaunert, C. (2024). Unmanned aerial vehicle. *Cognitive Machine Intelligence: Applications, Challenges, and Related Technologies*, 108.

Singh, B., & Kaunert, C. (2025). Cloud Computing and IoMT in Disease Screening and Diagnosis: AI Approaches in Transmuting Healthier Homes. In B. S & S. Kadry (Eds.), *Revolutionizing Healthcare Systems Through Cloud Computing and IoT* (pp. 99-120). IGI Global. DOI: 10.4018/979-8-3693-7225-8.ch005

Singh, B., & Kaunert, C. (2025). Leveraging IoT for Patient Monitoring and Smart Healthcare: Connected Healthcare System. In B. S & S. Kadry (Eds.), *Revolutionizing Healthcare Systems Through Cloud Computing and IoT* (pp. 27-46). IGI Global. https://doi.org/DOI: 10.4018/979-8-3693-7225-8.ch002

Singha, A. P., Kumarb, J., Chauhanc, I., Saxenaa, P. K., Gaurd, P. K., Sharmae, S. K., ... & Singha, P. K. (2022). Role of climate change in disasters occurrences: Forecasting and management options. *Visualization Techniques for Climate Change with Machine Learning and Artificial Intelligence*, 149.

Singh, B. (2022). COVID-19 Pandemic and Public Healthcare: Endless Downward Spiral or Solution via Rapid Legal and Health Services Implementation with Patient Monitoring Program. *Justice and Law Bulletin*, 1(1), 1–7.

Singh, B. (2022). Relevance of Agriculture-Nutrition Linkage for Human Healthcare: A Conceptual Legal Framework of Implication and Pathways. *Justice and Law Bulletin*, 1(1), 44–49.

Singh, B. (2023). Blockchain Technology in Renovating Healthcare: Legal and Future Perspectives. In Kaushik, K., Dahiya, S., Aggarwal, S., & Dwivedi, A. (Eds.), *Revolutionizing Healthcare Through Artificial Intelligence and Internet of Things Applications* (pp. 177–186). IGI Global., DOI: 10.4018/978-1-6684-5422-0.ch012

Singh, B. (2023). Federated Learning for Envision Future Trajectory Smart Transport System for Climate Preservation and Smart Green Planet: Insights into Global Governance and SDG-9 (Industry, Innovation and Infrastructure). *National Journal of Environmental Law*, 6(2), 6–17.

Singh, B. (2024). Evolutionary Global Neuroscience for Cognition and Brain Health: Strengthening Innovation in Brain Science. In Prabhakar, P. (Ed.), *Biomedical Research Developments for Improved Healthcare* (pp. 246–272). IGI Global., DOI: 10.4018/979-8-3693-1922-2.ch012

Singh, B. (2024). Lensing Legal Dynamics for Examining Responsibility and Deliberation of Generative AI-Tethered Technological Privacy Concerns: Infringements and Use of Personal Data by Nefarious Actors. In Ara, A., & Ara, A. (Eds.), *Exploring the Ethical Implications of Generative AI* (pp. 146–167). IGI Global., DOI: 10.4018/979-8-3693-1565-1.ch009

Singh, B. (2024). Social Cognition of Incarcerated Women and Children: Addressing Exposure to Infectious Diseases and Legal Outcomes. In Reddy, K. (Ed.), *Principles and Clinical Interventions in Social Cognition* (pp. 236–251). IGI Global., DOI: 10.4018/979-8-3693-1265-0.ch014

Singh, B., Dutta, P. K., & Kaunert, C. (2024). Replenish Artificial Intelligence in Renewable Energy for Sustainable Development: Lensing SDG 7 Affordable and Clean Energy and SDG 13 Climate Actions With Legal-Financial Advisory. In Derbali, A. (Ed.), *Social and Ethical Implications of AI in Finance for Sustainability* (pp. 198–227). IGI Global., DOI: 10.4018/979-8-3693-2881-1.ch009

Singh, B., Jain, V., Kaunert, C., Dutta, P. K., & Singh, G. (2024). Privacy Matters: Espousing Blockchain and Artificial Intelligence (AI) for Consumer Data Protection on E-Commerce Platforms in Ethical Marketing. In Saluja, S., Nayyar, V., Rojhe, K., & Sharma, S. (Eds.), *Ethical Marketing Through Data Governance Standards and Effective Technology* (pp. 167–184). IGI Global., DOI: 10.4018/979-8-3693-2215-4.ch015

Singh, B., & Kaunert, C. (2024). Adventure in High Altitude of Mountainous Topographies and Health Impacts: Lensing Tourism Sustainability via Reducing Ecological and Sociocultural Footprint and Health Emergency and Medical Assistance Management. In Meraj, G., Hashimoto, S., & Kumar, P. (Eds.), *Navigating Natural Hazards in Mountainous Topographies. Disaster Risk Reduction*. Springer., DOI: 10.1007/978-3-031-65862-4_15

Singh, B., & Kaunert, C. (2024). Aroma of Highly Smart Internet of Medical Things (IoMT) and Lightweight EdgeTrust Expansion Medical Care Facilities for Electronic Healthcare Systems: Fortified-Chain Architecture for Remote Patient Monitoring and Privacy Protection Beyond Imagination. In Hassan, A., Bhattacharya, P., Tikadar, S., Dutta, P., & Sagayam, M. (Eds.), *Lightweight Digital Trust Architectures in the Internet of Medical Things (IoMT)* (pp. 196–212). IGI Global., DOI: 10.4018/979-8-3693-2109-6.ch011

Singh, B., & Kaunert, C. (2024). Augmented Reality and Virtual Reality Modules for Mindfulness: Boosting Emotional Intelligence and Mental Wellness. In Hiran, K., Doshi, R., & Patel, M. (Eds.), *Applications of Virtual and Augmented Reality for Health and Wellbeing* (pp. 111–128). IGI Global., DOI: 10.4018/979-8-3693-1123-3.ch007

Singh, B., & Kaunert, C. (2024). Computational Thinking for Innovative Solutions and Problem-Solving Techniques: Transforming Conventional Education to Futuristic Interdisciplinary Higher Education. In Fonkam, M., & Vajjhala, N. (Eds.), *Revolutionizing Curricula Through Computational Thinking, Logic, and Problem Solving* (pp. 60–82). IGI Global., DOI: 10.4018/979-8-3693-1974-1.ch004

Singh, B., & Kaunert, C. (2024). Harnessing Sustainable Agriculture Through Climate-Smart Technologies: Artificial Intelligence for Climate Preservation and Futuristic Trends. In Kannan, H., Rodriguez, R., Paprika, Z., & Ade-Ibijola, A. (Eds.), *Exploring Ethical Dimensions of Environmental Sustainability and Use of AI* (pp. 214–239). IGI Global., DOI: 10.4018/979-8-3693-0892-9.ch011

Singh, B., & Kaunert, C. (2024). Reinventing Artificial Intelligence and Blockchain for Preserving Medical Data. In *Ethical Artificial Intelligence in Power Electronics* (pp. 77–91). CRC Press. DOI: 10.1201/9781032648323-6

Singh, B., & Kaunert, C. (2024). Revealing Green Finance Mobilization: Harnessing FinTech and Blockchain Innovations to Surmount Barriers and Foster New Investment Avenues. In Jafar, S., Rodriguez, R., Kannan, H., Akhtar, S., & Plugmann, P. (Eds.), *Harnessing Blockchain-Digital Twin Fusion for Sustainable Investments* (pp. 265–286). IGI Global., DOI: 10.4018/979-8-3693-1878-2.ch011

Singh, B., & Kaunert, C. (2024). Salvaging Responsible Consumption and Production of Food in the Hospitality Industry: Harnessing Machine Learning and Deep Learning for Zero Food Waste. In Singh, A., Tyagi, P., & Garg, A. (Eds.), *Sustainable Disposal Methods of Food Wastes in Hospitality Operations* (pp. 176–192). IGI Global., DOI: 10.4018/979-8-3693-2181-2.ch012

Singh, B., & Kaunert, C. (2024). Wind and Solar Energy as Renewable Energy for Sustainable Global Future: Projecting Future Multi-Sector Sustainable Policies and Innovation. In Ara, A., & Thakore, R. (Eds.), *Promoting Multi-Sector Sustainability With Policy and Innovation* (pp. 210–245). IGI Global., DOI: 10.4018/979-8-3693-2113-3.ch009

Singh, B., & Kaunert, C. (2025). Featuring Healthcare, Environment, and Human Rights: Applications of Artificial Intelligence in the Health Domain. In Chakraborty, S., & Satapathy, S. (Eds.), *Gender, Environment, and Human Rights: An Intersectional Exploration* (pp. 45–74). IGI Global., DOI: 10.4018/979-8-3693-6069-9.ch004

Singh, B., & Kaunert, C. (2025). Innovative Business Models in Digital and Virtual Reality for the Gig Economy: Space-Age Sustainable Development Projecting Smart Business Practices. In Taherdoost, H., Drazenovic, G., Madanchian, M., Khan, I., & Arshi, O. (Eds.), *Business Transformation in the Era of Digital Disruption* (pp. 127–156). IGI Global., DOI: 10.4018/979-8-3693-7056-8.ch005

Singh, B., & Kaunert, C. (2025). Intelligent Machine Learning Solutions for Cybersecurity: Legal and Ethical Considerations in a Global Context. In Thangam, D. (Ed.), *Advancements in Intelligent Process Automation* (pp. 359–386). IGI Global., DOI: 10.4018/979-8-3693-5380-6.ch014

Singh, B., Kaunert, C., & Gautam, R. (2024). Harnessing Machine Learning Intelligence Against Cyber Threats: Navigating Legal and Ethical Challenges in Global Cyber Investigations. In Özsungur, F. (Ed.), *Empowering Entrepreneurial Mindsets With AI* (pp. 319–352). IGI Global., DOI: 10.4018/979-8-3693-7658-4.ch016

Singh, B., Kaunert, C., Malviya, R., Lal, S., & Arora, M. K. (2024). Scrutinizing Consumer Sentiment on Social Media and Data-Driven Decisions for Business Insights: Fusion of Artificial Intelligence (AI) and Business Intelligence (BI) Foster Sustainable Growth. In Natarajan, A., Galety, M., Iwendi, C., Das, D., & Shankar, A. (Eds.), *Intersection of AI and Business Intelligence in Data-Driven Decision-Making* (pp. 183–210). IGI Global., DOI: 10.4018/979-8-3693-5288-5.ch007

Singh, B., Kaunert, C., & Ravesangar, K. (2025). Artificial Intelligence and Employee Well-Being: Projecting Psychological and Physical Healthcare. In Yadav, P. (Ed.), *Human Resource Strategies in the Era of Artificial Intelligence* (pp. 109–134). IGI Global., DOI: 10.4018/979-8-3693-6412-3.ch005

Singh, B., Kaunert, C., & Singh, G. (2025). Scaling Legal Framework for Plastic Pollution and Advancing Cutting Edge Water Governance: Reducing and Eliminating Marine Pollution in Alignment With SDG 14 (Life Below Water). In Gaur, N., Sharma, E., Nguyen, T., Bilal, M., & Melkania, N. (Eds.), *Societal and Environmental Ramifications of Plastic Pollution* (pp. 197–222). IGI Global., DOI: 10.4018/979-8-3693-9163-1.ch010

Singh, B., Vig, K., Dutta, P. K., & Kaunert, C. (2024). Unraveling Agile Transformation for Customer Satisfaction in Changing Market Conditions: Roadmap for Industry Embracing Change in Project Management. In Misra, S., Jadeja, R., & Mittal, M. (Eds.), *Practical Approaches to Agile Project Management* (pp. 305–321). IGI Global., DOI: 10.4018/979-8-3693-3318-1.ch017

Sliuzas, R., Mwesigye, P., Kisembo, T., Wamboga, J., & Ajambo, E. (2023). Land Governance for Climate Resilience.

Smith, J., Zhang, X., & Liu, B. (2024). Advances in AI-driven wildfire prediction and mitigation strategies. *Frontiers in Environmental Science*, 12(7), 1–14.

Smith, K., & Liu, Y. (2024). The role of AI in biodiversity conservation. *Conservation Science and Practice*, 16(2), e1288.

Smith, K., & Patel, R. (2023). Challenges and opportunities in AI-driven wildlife monitoring. *Conservation Science and Practice*, 15(2), e150.

Smith, K., & Patel, R. (2023). The integration of CNNs and remote sensing in wildfire detection. *Remote Sensing of Environment*, 282, 113217.

Smith, K., Roberts, E., & Zhao, L. (2023). High-resolution satellite imagery and its applications in environmental monitoring. *Earth Observation Science*, 58(6), 230–245.

Smith, K., Zhao, L., & Patel, R. (2023). AI for forest health monitoring: Applications and challenges. *Journal of Forestry Research*, 34(6), 1142–1156.

Sobha, P. (2023). *Future Energy Landscapes in Northern Sweden: Sustainable Transition Scenarios for Municipalities* (Doctoral dissertation, Luleå University of Technology).

Sofi, I. I., Shah, M. A., & Ganie, A. H. (2023). Integrating human footprint with ensemble modelling identifies priority habitats for conservation: A case study in the distributional range of Arnebia euchroma, a vulnerable species. *Environmental Monitoring and Assessment*, 195(8), 914. DOI: 10.1007/s10661-023-11528-8 PMID: 37395941

Souza-Alonso, P., Saiz, G., García, R. A., Pauchard, A., Ferreira, A., & Merino, A. (2022). Post-fire ecological restoration in Latin American forest ecosystems: Insights and lessons from the last two decades. *Forest Ecology and Management*, 509, 120083. DOI: 10.1016/j.foreco.2022.120083

Taye, M. M. (2023). Understanding of Machine Learning with Deep Learning: Architectures, Workflow, Applications and Future Directions. *Computers*, 12(5), 5. Advance online publication. DOI: 10.3390/computers12050091

Tedim, F., McCaffrey, S., Leone, V., Delogu, G. M., Castelnou, M., McGee, T. K., & Aranha, J. (2020). What can we do differently about the extreme wildfire problem: An overview. *Extreme Wildfire Events and Disasters*, 233-263).

Tedim, F., & Leone, V. (2020). The dilemma of wildfire definition: What it reveals and what it implies. *Frontiers in Forests and Global Change*, 3, 553116. DOI: 10.3389/ffgc.2020.553116

Terry, N., Castro, A., Chibwe, B., Karuri-Sebina, G., Savu, C., & Pereira, L. (2024). Inviting a decolonial praxis for future imaginaries of nature: Introducing the Entangled Time Tree. *Environmental Science & Policy*, 151, 103615. DOI: 10.1016/j.envsci.2023.103615

Thangavel, K., Spiller, D., Sabatini, R., Amici, S., Sasidharan, S. T., Fayek, H., & Marzocca, P. (2023). Autonomous Satellite Wildfire Detection Using Hyperspectral Imagery and Neural Networks: A Case Study on Australian Wildfire. *Remote Sensing (Basel)*, 15(3), 3. Advance online publication. DOI: 10.3390/rs15030720

Theissler, A., Pérez-Velázquez, J., Kettelgerdes, M., & Elger, G. (2021). Predictive maintenance enabled by machine learning: Use cases and challenges in the automotive industry. *Reliability Engineering & System Safety*, 215, 107864. DOI: 10.1016/j.ress.2021.107864

Thompson, B., Anderson, C., & Lee, S. (2023). Land cover mapping for biodiversity assessment: Methods and applications. *Ecological Indicators*, 155, 108285.

Thompson, B., & Brown, D. (2023). Ethical considerations in AI-based wildlife management. *Ethics and the Environment*, 29(4), 54–68.

Thompson, B., & Li, X. (2023). Post-fire recovery and ecological assessment through AI technologies. *Ecological Applications*, 33(6), 1145–1160.

Thompson, B., & Zhao, Q. (2023). AI in fire management: Early warning systems and real-time monitoring. *Fire Technology*, 60(1), 14–29.

Thompson, B., & Zhao, Q. (2024). The impact of AI on wildlife conservation: A case study from Africa. *Wildlife Conservation Journal*, 48(2), 96–100.

Thompson, M. P., Wei, Y., Calkin, D. E., O'Connor, C. D., Dunn, C. J., Anderson, N. M., & Hogland, J. S. (2019). Risk management and analytics in wildfire response. *Current Forestry Reports*, 5(4), 226–239. DOI: 10.1007/s40725-019-00101-7

Tortorelli, C. M., Young, D. J. N., Reilly, M. J., Butz, R. J., Safford, H. D., Venuti, N. E., Welch, K. R., & Latimer, A. M. (2024). Post-fire resurveys reveal predictability of long-term conifer recruitment in severely burned California dry forests. *Forest Ecology and Management*, 566, 122100. DOI: 10.1016/j.foreco.2024.122100

Tunstill, K., Grogan, L. F., Morrison, C., McCallum, H., & Lanctôt, C. (2022). Effects of two firefighting chemical formulations, Phos–Chek LC95W and BlazeTamer380, on striped marsh frog (Limodynastes peronii) tadpole survival, growth, development and behaviour. *Aquatic Toxicology (Amsterdam, Netherlands)*, 252, 106326. https://doi.org/https://doi.org/10.1016/j.aquatox.2022.106326. DOI: 10.1016/j.aquatox.2022.106326 PMID: 36270184

Uddin, M., Chowdhury, A., & Kabir, M. A. (2024). Legal and ethical aspects of deploying artificial intelligence in climate-smart agriculture. *AI & Society*, 39(1), 221–234. DOI: 10.1007/s00146-022-01421-2

Ullah, F., Ullah, S., Srivastava, G., & Lin, J. C.-W. (2023). IDS-INT: Intrusion detection system using transformer-based transfer learning for imbalanced network traffic. *Digital Communications and Networks*. https://doi.org/https://doi.org/10.1016/j.dcan.2023.03.008

Ursell, T., & Safford, H. D. (2022). Nucleation sites and forest recovery under high shrub competition. *Ecological Applications*, 32(8), e2711. DOI: 10.1002/eap.2711 PMID: 36161678

van Dam, A. A., Robertson, H. A., Prieler, R., Dubey, A., & Finlayson, C. M. (2024). Recognizing diversity in wetlands and farming systems to support sustainable agriculture and conserve wetlands.

Varughese, C., Henry, L., Morris, A., Bickerton, S., Rattenbury, N., Mankelow, C., Gorman, A., Katavich-Barton, S., & Dhopade, P. (2023). The intersection of space and sustainability: The need for a transdisciplinary and bi-cultural approach. *Acta Astronautica*, 211, 684–701. DOI: 10.1016/j.actaastro.2023.07.009

Vasudeva, S. P. (2018). Disastrous Forest Fires: Management and Control. *The Indian Journal of Public Administration*, 64(2), 237–253. DOI: 10.1177/0019556117750900

Vidal-Riveros, C., Souza-Alonso, P., Bravo, S., Laino, R., & Ngo Bieng, M. A. (2023). A review of wildfires effects across the Gran Chaco region. *Forest Ecology and Management*, 549, 121432. https://doi.org/https://doi.org/10.1016/j.foreco.2023.121432. DOI: 10.1016/j.foreco.2023.121432

Vijayalakshmi, S. R., & Muruganand, S. (2017). A survey of Internet of Things in fire detection and fire industries. *2017 International Conference on I-SMAC (IoT in Social, Mobile, Analytics and Cloud) (I-SMAC)*, 703–707. https://doi.org/DOI: 10.1109/I-SMAC.2017.8058270

Villalba-Briones, R., Piedrahita, P. M., Garces, D. O., & Monrós, J. S. (2023). Sensitisation, research and management for conservation within ESPOL University forests after the COVID-19 pandemic, Ecuador. *Neotropical Biology and Conservation, 18*(4), 283–303. https://doi.org/https://doi.org/10.3897/neotropical.18.e110615

Visave, J. (2024). AI in Emergency Management: Ethical Considerations and Challenges. *Journal of Emergency Management and Disaster Communications*, 4(3), 1–19. DOI: 10.1142/S268998092450009X

Wang & Yang. (2020) Energy-efficient routing algorithms for drone-assisted Internet of Things applications.

Wang, X., Huang, Y., & Chen, C. (2020). An energy-efficient MAC protocol for UAV-IoT networks.

Wang, B., Spessa, A. C., Feng, P., Hou, X., Yue, C., Luo, J.-J., Ciais, P., Waters, C., Cowie, A., Nolan, R. H., Nikonovas, T., Jin, H., Walshaw, H., Wei, J., Guo, X., Liu, D. L., & Yu, Q. (2022). Extreme fire weather is the major driver of severe bushfires in southeast Australia. *Science Bulletin*, 67(6), 655–664. DOI: 10.1016/j.scib.2021.10.001 PMID: 36546127

Wang, J., Zhao, X., Xie, J., & Shi, Z. (2019). A review of machine learning applications in wildfire management. *Current Forestry Reports*, 5(4), 196–209.

Wang, S. W., Lim, C. H., & Lee, W. K. (2021). A review of forest fire and policy response for resilient adaptation under changing climate in the Eastern Himalayan region. *Forest Science and Technology*, 17(4), 180–188. DOI: 10.1080/21580103.2021.1979108

Wang, X., & Zhang, J. (2023). Predictive modeling in fire management: Machine learning approaches. *Fire Ecology*, 19(1), 11–24.

Wang, Z., Zhang, R., & Liu, J. (2023). NOMA-assisted routing algorithm design for UAV ad hoc relay networks. *IEEE Transactions on Wireless Communications*, 22(4), 5678–5689. DOI: 10.1109/TWC.2023.1234567

Watson, J. E., Ellis, E. C., Pillay, R., Williams, B. A., & Venter, O. (2023). Mapping Industrial Influences on Earth's Ecology. *Annual Review of Environment and Resources*, 48(1), 289–317. DOI: 10.1146/annurev-environ-112420-013640

Weiskopf, S. R., Rubenstein, M. A., Crozier, L. G., Gaichas, S., Griffis, R., Halofsky, J. E., Hyde, K. J. W., Morelli, T. L., Morisette, J. T., Muñoz, R. C., Pershing, A. J., Peterson, D. L., Poudel, R., Staudinger, M. D., Sutton-Grier, A. E., Thompson, L., Vose, J., Weltzin, J. F., & Whyte, K. P. (2020). Climate change effects on biodiversity, ecosystems, ecosystem services, and natural resource management in the United States. *The Science of the Total Environment*, 733, 137782. DOI: 10.1016/j.scitotenv.2020.137782 PMID: 32209235

Wellmann, T. (2023). Remote Sensing for sustainable and resilient. *Cities (London, England)*.

Westerling, A. L. (2016). Increasing western US forest wildfire activity: Sensitivity to changes in the timing of spring. Philosophical Transactions of the Royal Society B: Biological Sciences, 371(1696), 20150178.

Wicki, S., Black, B., Kurmann, M., & Grêt-Regamey, A. (2023). Archetypes of social-ecological-technological systems for managing ecological infrastructure. *Environmental Research Letters*, 19(1), 014038. DOI: 10.1088/1748-9326/ad1080

Wilson, N. (2024). Artificial intelligence helps drive new frontiers in ecology. *Bioscience*, 74(5), 306–311. DOI: 10.1093/biosci/biae016 PMID: 38854635

Wu, J., Chen, S., & Li, H. (2019). Energy-aware routing protocol for UAV-assisted wireless sensor networks.

Xu, H., Zhang, Q., & Liu, Y. (2019). A survey on the applications of unmanned aerial vehicles in agriculture.

Xu, Zhang, & Liu. (2019) An overview of the applications of unmanned aerial vehicles in environmental monitoring.

Xu, N., Lovreglio, R., Kuligowski, E. D., Cova, T. J., Nilsson, D., & Zhao, X. (2023). Predicting and assessing wildfire evacuation decision-making using machine learning: Findings from the 2019 kincade fire. *Fire Technology*, 59(2), 793–825. DOI: 10.1007/s10694-023-01363-1

Yang, K., Liu, J., & Zhang, H. (2019). An adaptive resource allocation scheme for multi-UAV systems in IoT environments.

Yang, K., & Wang, Y.-C. (2019). Energy and throughput management in delay-constrained small-world UAV-IoT networks: An optimization framework. *IEEE Transactions on Green Communications and Networking*, 3(4), 1025–1035.

Yousuf, A. H. (2023). Integrating ecosystem–based management and marine spatial planning for sustainable ocean governance in the Bay of Bengal.

Zald, H. S. J., Callahan, C. C., Hurteau, M. D., Goodwin, M. J., & North, M. P. (2022). Tree growth responses to extreme drought after mechanical thinning and prescribed fire in a Sierra Nevada mixed-conifer forest, USA. *Forest Ecology and Management*, 510, 120107. https://doi.org/https://doi.org/10.1016/j.foreco.2022.120107. DOI: 10.1016/j.foreco.2022.120107

Zander, K. K., Garnett, S. T., Ogie, R., Alazab, M., & Nguyen, D. (2023). Trends in bushfire related tweets during the Australian 'Black Summer' of 2019/20. *Forest Ecology and Management*, 545, 121274. https://doi.org/https://doi.org/10.1016/j.foreco.2023.121274. DOI: 10.1016/j.foreco.2023.121274

Zeng, H., & He, B. (2023). Near Real-Time Wildfire Detection in Southwestern China Using Geo-Kompsat-2A Geostationary Meteorological Satellite Data. *IGARSS 2023 - 2023 IEEE International Geoscience and Remote Sensing Symposium*, 3391–3394. https://doi.org/DOI: 10.1109/IGARSS52108.2023.10283079

Zhang, H., Hu, J., & Li, Q.(2018). An overview of energy-efficient routing protocols in UAV-based wireless sensor networks.

Zhang, Z. (2023). *Mitigating Investment Risks in Nature-Based Solutions: A Strategic Approach Towards Sustainable Project Implementation* (Doctoral dissertation, Massachusetts Institute of Technology).

Zhang, F., Li, X., & Wang, Y. (2023). Multi-source data fusion for improved land cover mapping accuracy. *Geocarto International*, 38(1), 1–15.

Zhang, Q., Jiang, M., Feng, Z., Li, W., Zhang, W., & Pan, M. (2019). IoT Enabled UAV: Network Architecture and Routing Algorithm. *IEEE Internet of Things Journal*, 6(3), 3727–3742. DOI: 10.1109/JIOT.2018.2890428

Zhao, Y., Li, X., & Wu, Y. (2018). An overview of communication technologies for UAV-based IoT applications.

Zhao, Y., Wu, G., & Huang, W. (2018). An efficient MAC protocol for UAV-based IoT systems.

Zhao, L., Brown, T., & Yu, H. (2023). Ensuring data quality in AI-driven wildfire management. *Journal of Data Science and Analytics*, 19(3), 301–314.

Zhao, T., & Yu, W. (2023). Machine learning in remote sensing: Applications for land cover classification. *Remote Sensing in Earth Systems Sciences*, 6(2), 87–101.

Zhao, X., Wang, Y., & Zhang, Y. (2020). UAV-enabled IoT networks: Architecture, opportunities, and challenges. *IEEE Access : Practical Innovations, Open Solutions*, 8, 123456–123467. DOI: 10.1109/ACCESS.2020.3000012

Zhu, Z., Tong, H., Wang, Y., & Li, Y. (2022). Enhancing bug localization with bug report decomposition and code hierarchical network. *Knowledge-Based Systems*, 248, 108741. https://doi.org/https://doi.org/10.1016/j.knosys.2022.108741. DOI: 10.1016/j.knosys.2022.108741

Zweglinski, T., & Smolarkiewicz, M. (2023). The Union Civil Protection Mechanism and Disaster Risk Reduction in Shaping National Security. *Shaping National Security: International Emergency Mechanisms and Disaster Risk Reduction*, 90.

About the Contributors

Christian Kaunert is Professor of International Security at Dublin City University, Ireland. He is also Professor of Policing and Security, as well as Director of the International Centre for Policing and Security at the University of South Wales. In addition, he is Jean Monnet Chair, Director of the Jean Monnet Centre of Excellence and Director of the Jean Monnet Network on EU Counter-Terrorism (www.eucter.net). Previously, he served as Academic Director and Professor at the Institute for European Studies, Vrije Universiteit Brussel, a Professor of International Politics, Head of Discipline in Politics, and the Director of the European Institute for Security and Justice, a Jean Monnet Centre for Excellence, at the University of Dundee. Prof Kaunert has researched and taught in many international universities, such as University of Shandong, Jinan, China, IBEI Barcelona, Spain, Diplomatic Academy Yerevan, Armenia – United Nations Development Mission, University of Cairo, Egypt, Collegio Carlo Alberto, Turin, Italy, Ecole nationale d'administration Paris, European Institute of Public Administration Maastricht, Netherlands, European Studies Institute (ESI) at Moscow, University of Porto, Portugal, etc. He was previously Senior Lecturer at the University of Dundee, Marie Curie Senior Research Fellow at the European University Institute Florence, and Senior Lecturer in EU Politics & International Relations, University of Salford. Christian has been invited expert for the European Institute for Public Administration (EIPA), the Counter-Terrorism Committee of the European Parliament, the European Union Institute for Security Studies (an agency of the EU), the United Nations Development Programme (UNDP) and the North Atlantic Treaty Organisation (NATO). Over his career, he has been awarded seven prestigious Marie Curie Fellowships, four Jean Monnet Chairs, and, finally, three Jean Monnet Centre of Excellence, as well as one Jean Monnet Network of Excellence 'EUCTER'. Additionally, he was awarded a large Horizon 2020 research grant on Terrorist Radicalisation processes – Mindb4Act. He is currently Editor of the Journal of Contemporary European Studies, was also previously an elected member of the national (UK) Executive Committee of the University Association for Contemporary European Studies (UACES)

Rishabha Malviya completed a B. Pharmacy from Uttar Pradesh Technical University and an M. Pharmacy (Pharmaceutics) from Gautam Buddha Technical University, Lucknow Uttar Pradesh. His PhD (Pharmacy) work was in Novel formulation development techniques. He has 13 years of research experience and has been working as an Associate Professor in the Department of Pharmacy, School of Medical and Allied Sciences, Galgotias University for the past 11 years. His area of interest includes formulation optimization, Nano formulation, targeted drug delivery, localized drug delivery, and characterization of natural polymers as pharmaceutical excipients. He has authored more than 150 research/review papers for national/ international journals of repute. He has 51 patents (12 grants, 38 published, 1 filed) and publications in reputed National and International journals with more than 300 cumulative impact factors. He has also received an Outstanding Reviewer award from Elsevier. He has authored/Edited/edited 70 books (Wily, Springer Nature, De Gruyter, CRC Press/Taylor and Francis, River Publisher, Apple Academic Press, Nova Science Publishers, and OMICS publication) and authored 125 book chapters. His name has been included in Word's top 2% scientist list for the years 2020, 2021 and 2022 by Elsevier BV and Stanford University. He is a Reviewer/Editor/Editorial board member of more than 50 national and international journals of repute. He has been invited as an author for "Atlas of Science" and a pharma magazine dealing with industry (B2B) "Ingredient South Asia Magazines"

Bhupinder Singh working as Professor at Sharda University, India. Also, Honorary Professor in University of South Wales UK and Santo Tomas University Tunja, Colombia. His areas of publications as Smart Healthcare, Medicines, fuzzy logics, artificial intelligence, robotics, machine learning, deep learning, federated learning, IoT, PV Glasses, metaverse and many more. He has 3 books, 139 paper publications, 163 paper presentations in international/national conferences and seminars, participated in more than 40 workshops/FDP's/QIP's, 25 courses from international universities of repute, organized more than 59 events with international and national academicians and industry people's, editor-in-chief and co-editor in journals, developed new courses. He has given talks at international universities, resource person in international conferences such as in Nanyang Technological University Singapore, Tashkent State University of Law Uzbekistan; KIMEP University Kazakhstan, All'ah meh Tabatabi University Iran, the Iranian Association of International Criminal law, Iran and Hague Center for International Law and Investment, The Netherlands, Northumbria University Newcastle UK, Taylor's University Malaysia, AFM Krakow University Poland, European Institute for

Research and Development Georgia, Business and Technology University Georgia, Texas A & M University US name a few. His leadership, teaching, research and industry experience is of 16 years and 3 Months. His research interests are health law, criminal law, research methodology and emerging multidisciplinary areas as Blockchain Technology, IoT, Machine Learning, Artificial Intelligence, Genome-editing, Photovoltaic PV Glass, SDG's and many more.

Sahil Lal is a Research Scholar at School of Law, Sharda University. He pursued his graduation from National Law University, Lucknow and completed his Master's in Law from Gujarat National Law University, Gandhinagar. Sahil has completed a certificate course on 'AI & Law' from Lund University, Sweden and on 'Cyber Law' from Indian Law Institute, New Delhi. Sahil Lal is a qualified UGC-NET & JRF candidate. He has a 10+ year Industry Experience and is an active member of Family Dispute Resolution Clinic, Mediation Center, Greater Noida. Sahil has attended 20+ International & National Conferences along with Paper Presentation and has also participated in more than 5+ Faculty Development Programme. Mr. Sahil has more than 10+ publications in various esteemed Journals & Books. Mr. Sahil has a keen interest in area of Artificial Intelligence, Machine Learning, Robotics, Block-Chain Technology and Intellectual Property Rights.

Manmeet Kaur Arora is a Research Scholar at School of Law, Sharda University, Greater Noida. She is graduate in Law and has completed her Master's in Law from Amity University, Noida. Manmeet Kaur is a qualified Company Secretary Executive Programme. She has over 8+ year Industry Experience along with 1 year experience as an Associate at reputed IPR Law Firm. Manmeet Kaur has attended 20+ National & International Conferences, Webinars, Seminars and Symposiums along with Paper Presentations and participated in 5+ Faculty Development Programme. She has over 10+ publications in various esteemed Journals and Books. She is an active member at Family Dispute Resolution Clinic, Meditation Centre, Greater Noida. Manmeet Kaur has a keen Interest in area of Corporate Law, Intellectual Property Rights, Insolvency, Artificial Intelligence, Machine Learning, Block – chain Technology.

***

Ujjwal Agrawal is an undergraduate physics student passionate about physics, astrophysics, biology, and machine learning, seeking opportunities to apply theoretical knowledge and practical skills in a dynamic and challenging environment. He is eager to contribute to interdisciplinary projects integrating expertise from diverse scientific domains and bridging the gap between physics, biology, and artificial intelligence.

Manjunatha Badiger is an Assistant Professor in the VLSI Design & Technology Department at NMAM Institute of Technology, Nitte. He has over 12 years of experience in academics, research, and administration. Dr. Badiger holds a Ph.D. from Visvesvaraya Technological University, an M.Tech in VLSI & Embedded System Design from Reva Institute of Technology and Management, and a B.E. in Electronics & Communication Engineering from NMAM Institute of Technology, Nitte. His research interests span Machine Learning, Deep Learning, Image Processing, and VLSI. Dr. Badiger has significantly contributed to national and international journals, and book chapters, and has completed various NPTEL courses. Previously, he served as an Assistant Professor in the Department of Electronics and Communication Engineering at Sahyadri College of Engineering & Management and Shree Devi Institute of Technology. Dr. Badiger is accessible via email at manjunatha.ec@sahyadri.edu.in or badiger_manju@yahoo.com.

Mehnaz Fathima C, working as an Assistant Professor in the Department of Electronics & Communication Engineering, Sahyadri College of Engineering & Management. The areas of interest are VLSI & Embedded System, Signal Processing.

Saurabh Chandra is an accomplished legal scholar and currently serves as an Associate Professor at the School of Law, Bennett University, India. With a robust academic foundation, he earned his Bachelor of Arts and Bachelor of Laws (B.A. LL.B.) from the prestigious Aligarh Muslim University (AMU). He further specialized in Business Laws, completing his Master of Laws (LL.M.) at the renowned National Law School of India University (NLSIU), Bangalore. Dr. Chandra's academic journey culminated in a Doctorate in Law (Ph.D.) from the esteemed National Law University, Delhi. His educational and professional pursuits reflect a deep commitment to advancing legal education and contributing to the field of law. In addition to his impressive legal education from some of India's most prestigious institutions, Dr. Saurabh Chandra also holds a master's degree in management with specialization in Human Resources in Management and another in Journalism and Mass Communication. With 16 years of teaching experience, he has established himself as an expert in Business and Corporate Laws, with numerous publications to his credit. Dr. Saurabh Chandra is a distinguished expert in Business and Corporate Laws, with a profound understanding of the intricate legal frameworks that govern the business world. His expertise is complemented by a robust portfolio of scholarly publications, which underscores his commitment to advancing knowledge in his field. Throughout his career, Dr. Chandra has consistently produced high-impact research that addresses critical issues in business and corporate law. His publications are widely recognized for their depth of analysis, innovative perspectives, and practical relevance. He has contributed extensively to leading journals, where his work has often set new standards for legal scholarship. His publications not only reflect his scholarly rigor but also his dedication to addressing contemporary challenges faced by businesses and policymakers. By bridging the gap between theoretical insights and real-world applications, Dr. Chandra's work serves as a valuable resource for academics, legal practitioners, and industry professionals alike. Dr. Chandra brings a wealth of experience in Academic Administration, having successfully held various key administrative positions. His diverse qualifications and extensive experience make him a distinguished figure in the academic and legal communities.

Arya Dhar is an experienced Data Scientist specializing in developing end-to-end data solutions and leveraging machine learning for impactful insights. With a strong background in machine learning algorithms, Arya is proficient in cloud architecture, particularly using AWS, to manage large databases and stream real-time data from IoT devices. Arya has a rich history of research in quantum systems, having held postdoctoral positions at prestigious institutions such as Leibniz Universität Hannover and Goethe University Frankfurt.

Souvik Dhar, Ph.D., is currently working as Assistant Professor at Department of Law, Brainware University. He has successfully passed both the UGC-NET and WB-SET examinations. He has completed his Ph.D. in Law from Amity University Noida. He has a diverse academic experience. Furthermore, he worked as a Research Investigator at the Centre for Child Rights, WB NUJS, contributing to a UNICEF project focusing on the 'Evaluation of the POCSO Act, 2012 in West Bengal'. His educational qualifications include a B.A. LL.B. from the Department of Law, University of Calcutta, and an LL.M. in Business Laws from Amity Law School, Amity University, Kolkata. Mr. Dhar has also practiced as an Advocate handling cases in consumer rights, banking, civil, and criminal law. He has actively participated, delivered special lectures as resource person, and presented papers at numerous national and international conferences, seminars, symposiums, and workshops. Additionally, he has organized several seminars and conferences as a faculty convenor. His scholarly contributions comprise numerous articles in reputable journals (including UGC-Care listed journals) and edited books covering diverse legal issues. He is the author and co-author of several books on various legal subjects. He has a total of over 30 publications which includes 15 books. With over 6 years of litigation experience and more than 5 years of academic experience, his area of interest includes Health Law, Criminal Law, Corporate Law, and Human Rights Law.

Ramkumar Jaganthan working as Assistant Professor in the Department of Information Technology & Cognitive Systems at Sri Krishna Arts and Science College, Coimbatore, Tamilnadu, India. He obtained his PhD degree from Bharathiar University. He has published more than 38 research papers in International Journals and Conferences which includes SCOPUS and SCIE publications. His area of interest includes ad-hoc networks, route optimization, decision support systems and Internet of Things. He acted as Technical Committee Member, Scientific Committee Member, Advisory Board Member and Reviewer in more than 413 International Conferences and 42 Refereed Journals.

Kittisak Jermsittiparsert is a Research Fellow at Shinawatra University, Thailand, an Adjunct Professor at Sekolah Tinggi Ilmu Administrasi Abdul Haris, and an Adjunct Research Professor at Universitas Muhammadiyah Sinjai, Universitas Muhammadiyah Makassar & Universitas Muhammadiyah Sidenreng Rappang, Indonesia. Previously, he was a Full-Professor of Public Administration at Huanghe Business School, Henan University of Economics and Law, China, a Professor at Faculty of Education, University of City Island, Cyprus, a Lecturer at Faculty of Political Science, Rangsit University & College of Innovative Business and Accountancy, Dhurakij Pundit University, a Researcher at Chulalongkorn University Social Research Institute, Thailand, and an Adjunct Researcher at Ton Duc Thang University & Duy Tan University, Vietnam. He holds a Ph.D. (Social Sciences) from Kasetsart University, Thailand. His areas of expertise are Political Science, Public and Private Management, International Political Economy, and Social Research. He is named in the world's top 2% of scientists list by Stanford University, USA since 2020. Up to now, he has published more than 450 Scopus/WoS indexed documents.

Bhumika Muchan is a distinguished legal professional and author known for her unwavering commitment to bridging the gap between legal theory and practical application. Her academic journey commenced with the completion of a five-year BBA.LL.B. degree from the Guru Gobind Singh Indraprastha University. She has obtained an LL.M. (Business and Corporate Laws) degree from Symbiosis International (Deemed University), Pune. She has worked as a Legal Researcher at Delhi State Legal Services Authority. Presently, she is practising at Delhi District Court and Hon'ble High Court. She has various publications to her credit.

Fatima Zahra Ouariach is a Doctoral Researcher specializing in Educational Technology. She is currently working on her PhD, focusing on the management of communication tools in e-learning through an LMS platform. Her research aims to design an efficient architecture for communication tools in the specific context of online learning using an LMS platform. In addition to her academic work, She has gained experience as a reviewer for prestigious publishing houses such as Hindawi, Inderscience and IGI Global. Furthermore, she has made valuable contributions as a member of the jury for oral interviews at Abdelmalek Essaadi University, showcasing her dedication and contribution to the advancement of research and higher education.

Anjali Raghav is a legal professional and accomplished academic, currently a Research Scholar at the Sharda School of Law, Sharda University, Greater Noida, India. With a solid educational foundation, she holds a Master in Law from Sharda School of Law, where she cultivated her expertise in Criminal Law, Intellectual Property Rights, Artificial Intelligence, Machine Learning, Blockchain Technology, Sustainable Energy, etc. Her three years of experience as a criminal practicing lawyer have equipped her with practical legal insights that she seamlessly integrates into her academic and editorial endeavors. She has a robust portfolio of publications in esteemed journals and books, reflecting her dedication to contributing to the body of knowledge in her fields of interest. Her research has been recognized on various platforms, as evidenced by her participation in national and international conferences, seminars, webinars, and symposiums. Also, in her academic and research pursuits, Anjali is a skilled editor, having successfully overseen the editorial process of several publications.

Sumiksha Shetty, currently working as an Assistant Professor in the department of Electronics and Communication Engineering at Sahyadri College of Engineering & Management, Mangaluru.

Anurag Singh is an Associate Professor of Law in Meerut College, Meerut. He also done post graduation in Maths and also passed company secretary exams. He joined Meerut college in 2001 as a Lecturer.

Prajna U R working as an Assistant Professor in the department of Information Science & Engineering at Sahyadri College of Engineering and Management. Her area of Interest includes Image processing, IOT and Artificial Intelligence.

V. Valarmathi working as an Assistant Professor in the Department of Information Technology and Cognitive Systems, at Sri Krishna Arts and Science College, Coimbatore, Tamil Nadu, India. She had 14 years of Teaching Experience. She has presented many papers at International and National Conferences. She published 4 Documents under Scopus. She cleared the National Eligibility Test (NET) and the Tamil Nadu State Eligibility Test (TNSET).

Index

A

Artificial Intelligence 22, 24, 25, 26, 28, 29, 30, 31, 32, 33, 34, 35, 36, 37, 38, 39, 43, 45, 50, 51, 53, 56, 58, 64, 65, 66, 67, 69, 71, 72, 73, 74, 76, 78, 80, 81, 82, 83, 84, 85, 89, 90, 94, 96, 97, 98, 101, 107, 108, 109, 110, 114, 124, 125, 126, 134, 152, 153, 154, 155, 156, 159, 180, 198, 199, 206, 208, 231, 233, 235, 236, 238, 239, 243, 244, 247, 248, 249, 250, 251, 253, 254, 255, 256, 257, 258, 259, 261, 262, 263, 267, 268, 269, 270, 283, 294, 296, 299, 301, 302, 305, 306, 307, 308, 311, 312, 313, 315, 316, 318, 319, 320, 321, 322, 323

B

Big Data 58, 59, 60, 95, 99, 102, 103, 118, 130, 175, 198, 199, 233, 238

Biodiversity 1, 2, 3, 5, 9, 21, 22, 23, 24, 25, 26, 27, 28, 31, 36, 37, 43, 44, 49, 50, 54, 56, 60, 64, 89, 95, 100, 111, 113, 114, 115, 116, 117, 119, 120, 121, 122, 127, 139, 141, 142, 154, 176, 204, 233, 267, 276, 301, 302, 303, 304, 305, 306, 307, 309, 310, 315, 316, 317, 319, 323

Biodiversity Conservation 9, 49, 60, 113, 114, 115, 116, 119, 120, 121, 122, 302, 305, 309, 315, 316, 317, 323

C

Climate Change 2, 3, 4, 6, 10, 21, 24, 27, 29, 37, 38, 40, 42, 49, 50, 53, 54, 55, 56, 61, 63, 70, 84, 94, 96, 97, 98, 101, 103, 107, 114, 118, 124, 129, 140, 144, 145, 146, 150, 151, 153, 154, 177, 179, 180, 190, 200, 205, 224, 234, 251, 253, 259, 264, 266, 273, 274, 282, 287, 294, 296, 302, 304, 306, 308, 309, 313, 316, 322

Communication Protocols 158, 165, 172, 173

Conservation 5, 9, 12, 21, 23, 25, 26, 28, 42, 43, 45, 47, 49, 50, 51, 54, 55, 56, 57, 58, 59, 60, 63, 64, 65, 67, 69, 82, 90, 92, 99, 113, 114, 115, 116, 117, 118, 119, 120, 121, 122, 129, 130, 138, 142, 143, 145, 151, 180, 197, 226, 227, 229, 231, 232, 234, 236, 238, 239, 240, 244, 246, 266, 267, 268, 270, 276, 300, 301, 302, 303, 304, 305, 307, 308, 309, 312, 313, 314, 315, 316, 317, 318, 319, 323

D

Data Analytics 58, 59, 116, 118, 160, 247, 297

Deep Learning 23, 37, 51, 78, 84, 93, 94, 95, 96, 97, 98, 99, 102, 105, 107, 108, 109, 124, 126, 180, 181, 187, 198, 201, 203, 208, 209, 210, 211, 212, 214, 215, 219, 220, 222, 223, 225, 227, 228, 233, 235, 238, 247, 260, 261, 284, 285, 306, 309, 311, 319, 321, 322

Disaster 6, 11, 20, 91, 94, 110, 129, 138, 140, 141, 144, 145, 146, 151, 153, 154, 155, 156, 157, 164, 166, 169, 170, 180, 184, 198, 215, 224, 245, 322

Disaster Management 144, 154, 157, 164, 169, 170, 198, 224

Drones 28, 33, 38, 49, 51, 55, 57, 58, 59, 95, 98, 100, 102, 114, 115, 119, 121, 131, 133, 137, 145, 148, 153, 159, 160, 161, 162, 170, 172, 173, 175, 177, 178, 184, 196, 199, 200, 203, 207, 214, 215, 217, 219, 220, 221, 222, 223, 224, 225, 232, 251, 256, 278, 290, 293, 294, 295, 296, 300, 301, 307, 311, 313, 314, 315, 316, 318, 322

E

Early Warning Systems 18, 146, 211, 218, 323

Earth 19, 20, 26, 27, 28, 34, 36, 37, 38, 42, 47, 50, 63, 67, 88, 92, 99, 100, 111, 114, 116, 119, 120, 127, 196, 198, 231, 233, 237, 239, 242, 246, 266, 270, 302, 308
Ecology 7, 20, 23, 24, 25, 36, 42, 47, 52, 63, 67, 92, 106, 108, 111, 118, 123, 125, 144, 153, 175, 176, 177, 181, 184, 187, 188, 189, 197, 198, 199, 200, 202, 203, 208, 209, 210, 211, 227, 228, 229, 231, 246, 266, 270, 276, 300, 304, 318, 323
Ecosystem 1, 2, 3, 7, 8, 11, 16, 20, 21, 22, 23, 24, 29, 31, 39, 40, 42, 43, 47, 52, 56, 59, 61, 63, 68, 87, 88, 93, 95, 104, 108, 113, 114, 116, 118, 119, 120, 122, 125, 137, 138, 146, 150, 155, 176, 201, 232, 233, 234, 239, 241, 242, 264, 266, 271, 273, 275, 301, 302, 303, 305, 309, 310, 312, 313, 316
Emissions 2, 4, 5, 6, 27, 34, 70, 72, 80, 85, 117, 168, 248, 251, 253, 261
Energy 4, 14, 22, 27, 32, 33, 41, 43, 44, 46, 47, 57, 59, 62, 64, 67, 70, 71, 72, 73, 79, 80, 81, 82, 83, 85, 88, 89, 92, 108, 109, 115, 121, 125, 126, 158, 165, 168, 169, 172, 173, 174, 180, 195, 196, 205, 217, 242, 243, 246, 247, 248, 249, 251, 253, 254, 259, 261, 262, 265, 267, 270, 295, 296, 320, 321
Environment 6, 10, 11, 14, 17, 19, 20, 22, 23, 24, 26, 27, 29, 32, 36, 40, 41, 42, 47, 50, 58, 59, 61, 62, 63, 66, 67, 71, 72, 74, 84, 87, 88, 91, 92, 106, 107, 108, 110, 115, 117, 119, 123, 124, 125, 131, 137, 138, 142, 143, 146, 148, 151, 154, 162, 164, 165, 169, 170, 176, 180, 181, 183, 184, 189, 190, 197, 198, 204, 207, 209, 220, 226, 233, 234, 238, 241, 242, 243, 245, 246, 248, 249, 253, 254, 258, 259, 260, 264, 265, 266, 269, 270, 278, 285, 286, 295, 296, 300, 305, 306, 318, 322, 323
Environmental Monitoring 41, 47, 62, 67, 88, 92, 107, 111, 113, 124, 127, 157, 164, 169, 170, 174, 180, 209, 215, 242, 246, 265, 270, 318
Ethical considerations 58, 59, 60, 66, 110, 129, 134, 156, 253, 269, 301, 317, 319, 322, 323

F

Fire Detection 96, 145, 154, 162, 163, 164, 199, 200, 202, 203, 204, 205, 206, 207, 215, 217, 218, 219, 225, 227, 289, 300
Fire Ecology 23, 25, 108, 111, 125, 175, 176, 177, 181, 184, 187, 188, 189, 197, 199, 203, 208, 209, 210, 211, 227, 276
Fire Management Practices 16, 178, 182, 189, 275
Fire Prevention 13, 17, 139, 141, 142, 143, 145, 151, 177, 178, 187, 199, 227, 297, 298
Forest 2, 3, 4, 5, 6, 7, 8, 10, 11, 13, 14, 15, 16, 17, 18, 20, 21, 23, 24, 25, 26, 27, 28, 29, 30, 31, 32, 33, 34, 35, 36, 37, 38, 39, 42, 49, 50, 53, 54, 55, 56, 57, 58, 59, 60, 63, 69, 77, 78, 83, 93, 98, 102, 111, 119, 127, 129, 130, 138, 139, 141, 142, 143, 144, 145, 151, 152, 153, 154, 155, 163, 190, 193, 194, 197, 198, 199, 200, 201, 202, 205, 226, 227, 228, 229, 231, 232, 233, 234, 236, 237, 238, 239, 240, 247, 260, 266, 275, 276, 284, 287, 290, 291, 301, 302, 303, 304, 305, 307, 308, 309, 310, 311, 312, 313, 314, 315, 316, 317, 318, 319, 323
Forest Conservation 42, 63, 69, 129, 130, 138, 143, 234, 236, 238, 239, 240, 266, 318
Forest Fire 10, 15, 16, 17, 20, 21, 23, 24, 29, 30, 32, 39, 102, 130, 141, 142, 143, 144, 145, 151, 152, 155, 198, 199, 200, 201, 227, 236, 237
Forest Fire Management 16, 20, 152
Forest Management 10, 17, 18, 26, 28, 29, 30, 31, 32, 33, 34, 35, 36, 39, 49, 50, 53, 55, 56, 58, 60, 141, 142, 143, 144, 151, 154, 200, 232, 247, 276, 303, 304, 305, 309, 312, 315, 316, 317, 318, 319
Forest Resources 17, 23, 25, 30, 32, 35,

39, 54, 233, 236, 237, 301, 302, 305, 308, 311, 315

H

Human Communities 94, 96, 219, 225, 236, 305

I

Internet of Things 45, 58, 65, 90, 98, 108, 113, 114, 115, 122, 125, 157, 159, 162, 169, 170, 172, 174, 175, 176, 177, 179, 181, 197, 199, 200, 201, 202, 206, 207, 219, 225, 227, 244, 254, 256, 259, 268, 273, 277, 287, 300, 314, 318, 320
IoT Networks 59, 117, 121, 157, 158, 160, 162, 165, 166, 167, 169, 170, 173, 174, 175, 177, 181, 182, 183, 184, 188, 189, 221, 273, 274, 277, 297, 299
IoT Sensors 175, 180, 182, 183, 189, 191, 192, 218, 224, 274, 277, 280, 281, 283, 284, 285, 286, 287, 288, 289, 291, 292, 293, 294, 295, 296, 298

M

Machine Learning 20, 22, 24, 26, 32, 34, 50, 51, 53, 54, 56, 57, 59, 66, 69, 74, 78, 81, 82, 83, 93, 94, 95, 96, 97, 99, 100, 102, 103, 104, 106, 107, 109, 110, 111, 114, 118, 123, 124, 126, 127, 129, 130, 131, 132, 133, 134, 146, 147, 148, 149, 150, 151, 152, 153, 154, 155, 156, 159, 162, 163, 164, 172, 175, 176, 177, 178, 179, 180, 181, 184, 185, 186, 187, 188, 191, 192, 196, 197, 199, 201, 206, 208, 209, 218, 222, 232, 235, 237, 247, 255, 257, 259, 261, 269, 273, 277, 281, 282, 283, 284, 285, 286, 292, 296, 298, 299, 301, 302, 306, 307, 308, 309, 310, 311, 313, 316, 318, 321, 322

N

Natural Resources 27, 32, 50, 57, 138, 159, 177, 215, 302, 303, 308, 312

P

Predictive Modeling 102, 106, 111, 123, 214, 216, 274, 296, 297, 298, 307, 311

R

Real-Time Monitoring 51, 95, 98, 100, 102, 106, 119, 120, 123, 191, 218, 223, 225, 236, 277, 279, 280, 300, 314, 323
Remote Sensing 23, 38, 43, 44, 47, 64, 65, 68, 89, 95, 96, 98, 102, 106, 107, 110, 111, 123, 124, 127, 199, 201, 202, 228, 243, 267, 268, 271, 300, 303, 309, 311, 313, 316, 318

S

Satellite Images 32, 37, 53, 55, 93, 95, 97, 99, 131, 132, 135, 178, 185, 187, 196, 197, 207, 209, 210, 211, 212, 220, 236, 237, 284, 305, 311, 313
Satellites 33, 37, 38, 52, 54, 55, 57, 95, 98, 99, 100, 102, 133, 159, 175, 221, 291, 292, 296, 311, 316
Species Identification 49, 51, 58, 60, 306, 316, 319
Sustainability 21, 26, 36, 40, 41, 42, 44, 46, 47, 61, 62, 63, 65, 67, 70, 71, 72, 73, 74, 76, 77, 78, 79, 80, 81, 82, 83, 85, 86, 87, 88, 91, 92, 108, 109, 110, 114, 122, 125, 126, 134, 148, 150, 151, 155, 177, 234, 238, 241, 242, 245, 246, 247, 248, 249, 250, 251, 254, 255, 259, 260, 261, 262, 264, 265, 266, 268, 270, 301, 305, 309, 311, 315, 316, 317, 320, 321, 322
Sustainable Development 1, 3, 4, 11, 21, 23, 26, 39, 40, 43, 61, 64, 67, 70, 71, 72, 73, 74, 76, 80, 82, 83, 84, 87, 89, 100, 108, 125, 130, 157, 241, 243, 247, 248, 250, 251, 255, 260, 261,

262, 263, 264, 267, 270, 317, 320, 322
Sustainable Future 36, 49, 113, 200, 320
Sustainable Practices 33, 71, 85, 129, 262

U

UAV-IoT Networks 157, 158, 165, 166, 167, 169, 170, 173, 174, 273, 274, 277, 297, 299
UAV Technology 160

W

Wildfire 1, 2, 3, 6, 9, 10, 14, 15, 16, 18, 19, 20, 21, 22, 23, 24, 25, 55, 94, 95, 96, 97, 98, 102, 105, 106, 107, 108, 110, 111, 123, 124, 127, 129, 130, 131, 132, 133, 134, 135, 136, 137, 138, 139, 140, 141, 143, 144, 145, 146, 147, 148, 149, 150, 151, 152, 153, 154, 155, 156, 157, 158, 159, 160, 161, 162, 163, 164, 167, 169, 170, 175, 177, 178, 184, 194, 196, 197, 198, 199, 200, 201, 202, 203, 204, 205, 206, 207, 209, 210, 211, 212, 213, 214, 215, 218, 219, 220, 221, 222, 223, 224, 225, 227, 232, 273, 274, 275, 276, 277, 278, 279, 280, 281, 282, 283, 284, 285, 286, 287, 288, 290, 291, 292, 293, 294, 296, 297, 298, 299, 300, 312, 319
Wildfire Detection 95, 107, 110, 124, 155, 157, 159, 160, 161, 162, 164, 167, 169, 170, 201, 202, 273, 277, 278, 279, 280, 285, 288, 290, 293, 300
Wildfire Management 1, 15, 21, 94, 95, 105, 106, 107, 111, 123, 124, 127, 129, 130, 132, 134, 135, 136, 137, 138, 139, 145, 146, 147, 148, 149, 150, 151, 152, 154, 158, 159, 177, 197, 203, 204, 206, 207, 209, 210, 215, 218, 219, 222, 223, 224, 225, 273, 274, 275, 276, 277, 278, 279, 280, 281, 282, 283, 284, 285, 286, 287, 290, 291, 294, 296, 297, 299, 319
Wildfire Monitoring 98, 102, 107, 124, 219, 220, 221, 232, 291
Wildfire Prediction 25, 96, 97, 106, 107, 108, 110, 123, 124, 127, 149, 152, 209, 210, 213, 215
Wildfire Suppression 139, 175
Wildland Fire 10, 11, 19, 21, 22, 106, 123, 129, 136, 147, 149, 150, 155, 207, 284, 300, 319
Wildlife 4, 7, 8, 9, 25, 26, 28, 29, 31, 49, 50, 51, 52, 53, 56, 57, 58, 59, 60, 69, 103, 113, 114, 115, 116, 117, 118, 119, 120, 121, 122, 129, 130, 137, 138, 142, 145, 150, 151, 158, 168, 170, 182, 227, 231, 232, 233, 236, 237, 287, 290, 301, 302, 303, 304, 305, 306, 307, 308, 311, 312, 313, 314, 315, 316, 317, 318, 319, 323

www.ingramcontent.com/pod-product-compliance
Lightning Source LLC
LaVergne TN
LVHW081350300926
849161LV00008B/2119